Implosion

Implosion
Lessons from National Security, High Reliability Spacecraft, Electronics, and the Forces Which Changed Them

L. Parker Temple III

IEEE Press

A John Wiley & Sons, Inc., Publication

Cover Design: Michael Rutkowski
Cover Illustration: Crystal Ball © David Vernon/iStockphoto, Spaceships © Stefan Schulze/
iStockphoto, Spaceship © Iurii Kovalenko/iStockphoto, American Flag © Bryndon Smith/iStockphoto,
Explosion © Soubrette/iStockphoto

Published by John Wiley & Sons, Inc., Hoboken, New Jersey.
Published simultaneously in Canada.

For general information on our other products and services or for technical support, please contact our
Customer Care Department within the United States at (800) 762-2974, outside the United States at
(317) 572-3993 or fax (317) 572-4002.

Wiley also publishes its books in a variety of electronic formats. Some content that appears in print
may not be available in electronic formats. For more information about Wiley products, visit our web
site at www.wiley.com.

Library of Congress Cataloging-in-Publication Data is available.
ISBN 978-1-118-46242-3

10 9 8 7 6 5 4 3 2 1

Contents

List of Figures   vii

List of Tables   ix

Preface   xi

Acknowledgments   xv

Acronyms, Abbreviations, and Program Names   xvii

Part I   Activation Energy (1931–1968)

1. Washington . . . We Have a Problem . . .   3

2. The Quantum Leap   6

3. Preparation   21

4. The Final Frontiers   29

5. Minuteman Means Reliability   58

6. Skinning Cats   68

Part II   Startup Transient (1969–1980)

7. Changing the Sea State   87

8. Space Parts: From A to S   93

9. There's S, and Then There's S 122

10. A Little Revolution Now and Then Is Good   140

11. Quality on the Horizon   144

Part III   Switching Transient (1980–1989)

12. Crossing the Operational Divide 153

13. Stocking the Shelves 168

14. Hammered 184

15. Battlegrounds: Reorganization and Reform 187

16. Implementing Change in a Changing World 207

Part IV   Shorting To Ground (1989–2002)

17. Leap First, Look Later 231

18. Hardly Standing PAT 248

Part V   Resetting the Circuit Breakers

19. Brewing the Perfect Storm 277

20. Summing the Parts 301

Epilogue: Can One Ever Truly Go Home Again? 309

Index 322

List of Figures

Figure 4.1   Density of Early Solid State Electronics   43
Figure 5.1   MINUTEMAN Program Reliability Improvement
              Effectiveness   66
Figure 6.1   The 1962 View of Sources of Failures   71
Figure 8.1   The Crossover Concept   96
Figure 8.2   Waterfall Approach   98
Figure 8.3   The Classic Failure Rate "Bathtub" Curve   108
Figure 17.1   The Space Systems Engineering V-Diagram   233
Figure 18.1   An Example of Satellite Cost Growth   271
Figure 19.1   Changing Skills in Acquisition Programs   291

List of Tables

Table 2.1	Breakdown of Early Semiconductor Patents by Firms	16
Table 5.1	Distribution of U.S. Semiconductor Sales by End Use	63
Table 5.2	Government Purchases of Integrated Circuits Through the End of MINUTEMAN II Procurements	64
Table 8.1	Experience of Failure Rates versus Calculated Rates for Some Selected Part Types	109
Table 8.2	Reliability Predictions and Failure Experience	110
Table 8.3	Cost of Higher Reliability	111
Table 8.4	Class A versus Class S Requirements	117
Table 13.1	Responses to SD Survey of Class S Need Projections	182
Table 16.1	SD Program Parts Usage, May 1988	213

Preface

The first expectation of any historian is that the story of the events he or she is relating is worth reading. In that light, the expectation that a history of high reliability electronics will enflame the passions of a reader (at least, one who is not in the high reliability electronics business) is probably a stretch. The sad part of this truth, though, is this is not the time for one's passions to be enflamed. The time is long gone for us to get so outraged at what has happened that we run to the window, fling it open, and yell that we are damned mad about what occurred and we will not put up with it any longer. That time is gone, never to be recovered. The time for outrage has passed.

When, in the early part of the 21st century, extremely important satellites began to fail, reviews of the failures pointed to elimination of some military standards about a decade earlier. The failed satellites not only cost a great deal of money, the problems created by their loss affected our national security and national defense. Whether the reader is aware of the specific satellite failures or not, he or she must understand just how dependent the United States is on these machines and how serious was the jeopardy in which the nation was placed when they failed. The essential thing for the reader is to see that the failures were not simply due to the elimination of the standards.

How the failures occurred is, of course, the central story told here. Just as important for everyone who reads this is the understanding of what was lost. High reliability electronic parts were and are fundamental elements of U.S. national security and defense, the economy, and long-lived spacecraft. A convergence of many different factors, forces if you will, combined to bring a highly successful enterprise to an end. The task to recover some of what was lost depends in great measure on understanding what was lost. Some of the loss can be recovered; some cannot. The difference informs us about how to avoid similar mistakes stemming from the best of intentions.

This history mainly follows the military portion of high reliability electronics which, at first, went into missiles and shortly thereafter into satellites. Clearly, not all high reliability electronics went into the military, as we consider the civil space program's human spaceflight efforts such as the Apollo program of sending people to the moon. Such civil space programs of the National Aeronautics and Space Agency are generally outside the scope of this history, because so much depends on the evolution of the military standards. Also, the emphasis on high reliability electronic parts is mainly about active components such as microcircuits and transistors, with less emphasis on passive devices such as resistors. The reason is passive component technology did not experience the same rapid acceleration of capability, technology, and shrinking feature size that the active components did. A massive convergence of a number of forces hit active components hard but left only a little sea sickness for passives. To tell a coherent story, passives are not considered extensively, except as they were included in hybrid devices.

The heavy focus of the story is how electronic component technology developed high reliability as a response to the criticality of the uses (especially for national security and national defense), evolved processes and the necessary controls to improve over time, and managed increasing complexity. Then, in nearly the blink of an eye, well-meant changes caused the system to collapse. The full set of consequences cannot be described adequately for a variety of reasons, so the story pauses at times to benchmark the importance of the end uses of the parts—that is, for the national security and defense of the United States.

Numerous studies have examined the history of the semiconductor industry. At first, most paid little or only passing attention to the role of the military.[1] Later studies began to recognize a simplistic relationship. Eventually, as the scope of the involvement and influence became clear, historians saw that the scope, focus, and direction of the semiconductor industry was due to the early and continued interest by the military.[2] Even that level of insight was insufficient, because little, if any, distinction was made in terms of classes of parts. Most discussion has been on commercial parts, and the involvement of the military was in terms of reliability.[3] That is correct, as far as it goes.

This history goes the next important step, and has benefited from the gradually increasing awareness and understanding of the role of national security and national defense with industry. The awareness has increased with the extent to which information about the military programs was declassified and released.

At first, the Cold War precluded release of extensive details on parts, processes, and program relationships. Then, as more information became available about Minuteman and related missile developments, the understanding of the military influence

[1] Typical of this literature are Douglas W. Webbink, *The Semiconductor Industry: A Survey of Structure, Conduct and Performance* (Washington, D.C.: Federal Trade Commission, 1977); Ernest Braun and Stuart Macdonald, *Revolution in Miniature: The History and Impact of Semiconductor Electronics* (New York: Cambridge University Press, 1982); Robert W. Wilson, P.K. Ashton, and T.P. Egan, *Innovation, Competition, and Government Policy in the Semiconductor Industry* (Lexington, Massachusetts: Lexington Books, 1980); David C. Mowery and Nathan Rosenberg, *Technology and the Pursuit of Economic Growth* (New York: Cambridge University Press, 1989).

[2] This literature is well represented by Norman J. Asher and Leland D. Strom, *The Role of the Department of Defense in the Development of Integrated Circuits* Paper P-1271 (Arlington, Virginia: Institute for Defense Analysis, May 1977), and James M. Utterback and A. Murray, *Influence of Defense Procurement and Sponsorship of Research and Development on the Civilian Electronics Industry* (Cambridge, Massachusetts: MIT Center for Policy Alternatives, 1977).

[3] For an outstanding history, though focused only on the transistor, see Thomas J. Misa, "Military Needs, Commercial Radios, and the Development of the Transistor, 1948–1958", in *Military Enterprise and Technological Change, Perspectives on the American Experience*, ed. Merritt Roe Smith (Cambridge: Massachusetts: MIT Press, 1987). See also R.E. Anderson and R.M. Ryder, "Development History of the Transistor in the Bell Laboratories and Western Electric (1947–1975)" (unpublished manuscript, AT&T Archives, 45 11 01 03), 186, cited in Daniel Holbrook, "Government Support of the Semiconductor Industry: Diverse Approaches and Information Flows," *Business and Economic History* 24, no. 2 (Winter 1995): 138–139.

on the semiconductor industry matured. For instance, we will see and appreciate the importance of "Minuteman reliability" as a descriptor for discrete parts well into the 1970s, and see that program's reliability emphasis did not end after the fielding of the Minuteman III. In fact, Minuteman processes were generalized and fed directly into the creation of space quality parts.

The extreme compartmentalization of the national security space program, whose existence was not even acknowledged until the early 1990s, precluded a real appreciation of the importance of high reliability electronics, from the standpoint of how they were used. With the declassification of the National Reconnaissance Office and its early satellite reconnaissance programs, a new appreciation of the role of space programs became possible. Not only did these programs help keep the nation secure, they helped shape the direction of the semiconductor industry through the processes and materials developed to meet the needs of the highest reliability parts.

This present history places into context the story of reliability engineering, the highest reliability parts, how they came about, their influence on the larger commercial industry, and their importance in the context of national security and national defense.

To keep the story manageable, the early scope was limited to missiles and eventually space, but mostly as part of the Air Force's efforts in these areas. Although aircraft also require high reliability parts, they are maintainable, and do not require the same extremes of high reliability needed by satellites. Therefore, though much of what is said about spacecraft is true of aircraft, the latter deal with the consequences of the breakdown in electronics reliability in terms of higher maintenance, lower in-service rates, and longer logistics tails. None of these consequences have corollaries in satellites, which once in orbit are essentially on their own from a mechanical replacement standpoint. Also, the strategic missile systems of the Navy (Polaris, Poseidon, and Trident) are not extensively covered because the consequences to them are essentially the same as those of the Air Force covered in this part of the story. The story is complicated enough using the exemplar of the Air Force's efforts, but there is no denying that the Navy's efforts also made important contributions, particularly as these programs were in the design and buying phases. Where necessary, the Navy troops across the stage, but not as a main theme.

Keeping in mind the complexity of the subject and the myriad of military standards, choices had to be made about which standards to discuss. Inevitably, some important standards were not included. The standards that are discussed should adequately illustrate the evolution of standards and their importance.

Implementation of policy takes center stage at times, and properly so. The history demonstrates several important principles. Among these are that an acquisition system optimized to attain the highest level of reliability cannot also be the least expensive. Further, no matter how well intentioned, top-level acquisition reforms implemented as political expedients without understanding the implications at the lowest levels of the acquisition processes cannot succeed. Few enterprises evidence these truths more than high reliability space programs did. The best of intentions generating well-meant policies followed by flawed implementation rarely lead to positive results. Public policy analysts will find this an extensive discourse on implementation. Implementation of the policies and procedures leading to the high degree of optimization creating highly reliable space systems at the lowest level contrasts with the imperfect changes to acquisition at the highest. Everything starts with parts, but in the end, it is the implementation of policies that makes the biggest difference.

Tracing the evolution of the highest reliability parts provides the framework that eventually ties together such disparate topics as overseas production, extremely small feature sizes, the Cold War (or lack thereof), marketplace economics, failure modes and effects, and much more. Along the course of these events the baby was thrown out with the bathwater, and a very successful enterprise got off track due to a number of coinciding factors to yield the current state of affairs.

Thus, occasionally interspersed in sections dealing with the evolution of the high reliability parts in spacecraft, the reader will find references to events on the world stage. Our hope is these references, though sometimes intentionally jarringly out of place with the immediate story, will help remind the reader these events took place against, and contributed to, some of the most dramatic events of the Cold War and its aftermath. Additionally, as the story unfolds, coming to the nexus of the 1990s, these seemingly irrelevant strands of events will come to make sense in a larger picture of many forces at work affecting high reliability electronics and spacecraft.

Important distinctions must be held in mind. The manufacture of high reliability electronics continues, of course. Such manufacturing is significantly different from that of the late 1980s and early 1990s. Thus, an impression that highly reliable parts are no longer available is only partially true. More to this history's overall point is that highly reliable spacecraft depend on the best possible parts as the warp and woof of their creation. But that creation has been affected by much more than changes to the reliability of parts.

Finally, this history distinguishes between national security and national defense space programs. This is a convention based in history, when the Air Force developed most military satellites, and the National Reconnaissance Office developed those for the Intelligence Community and the National Command Authority. Recent practice has merged references to the two different space programs under the rubric of national security space. However, the usage here mirrors and complements that of an earlier work, *Shades of Gray: National Security and the Evolution of Space Reconnaissance*, because the distinctions remain meaningful. This history and its historian are ever mindful of things that were and remain classified because the purpose here is to improve national security, not to exploit or damage it in any way. Thus, our mention of things needing continued security protection is at best deliberately vague. Inquiring minds might find some place where lack of detailed specifics is annoying, for which the author begs tolerance. Such specifics are unnecessary for understanding the larger history and what it has to teach us.

L. PARKER TEMPLE III

Acknowledgments

When it became clear to me that this topic had never been approached in anything remotely like the way I had intended, there were two immediate choices for research, and I must express my gratitude to Dr. George "Skip" Bradley and his Space Command History Office staff for their help, and especially Dr. Harry Waldron of the Space and Missile Command with his staff. Too much of the history of the United States in space has been devoted to the civil space program, while the more vast and critically important military and national security space programs' history resources remain virtually untapped, resulting in a skewed perspective.

My subject grew in complexity rapidly as the extent of the disaster known as Acquisition Reform became clear. It became equally clear that not all of the problems were created by Acquisition Reform, though the most damaging ones surely were. To help clarify the issues and separate them, and then to make them available to a audience wider than the microelectronics parts community, I could not have accomplished this without the help of my reviewers and contributors: Lawrence I. Harzstark, Melvin H. Cohen, John Ingram-Cotton, James F. Bockman, and S. Lynn Sanchez.

As with any work of this scope, there are a few key individuals without whom things simply could not have proceeded. First, I owe a debt of gratitude to Michael J. Sampson, Manager NASA EEE Parts Program at Goddard Space Flight Center. His constant involvement in a broad range of topics culminates in a weekly telephone conference as enlightening about current problems as it is about how things got to be the way they did. His management of the NASA program, though not directly on point for this view of the military involvement in high reliability spacecraft, inspired the idea that perhaps this topic was worthwhile and achievable. Second, the fact that the current state of the world did not reflect its history in detail, but that those details could be recovered (with some pain), was an idea that I owe to my colleague David M. Peters, who inspired parts of this work and also was gracious enough to review and improve it. Third, my words are inadequate to express all that I owe to two career-long friends and colleagues whose intellectual inspiration for this and other works of mine simply cannot be explained in a few words. Without Russell C. Cykoski and Julius F. Sanks, so much in my professional career would not have happened that I am in awe of their unselfish sharing, encouragement, and constancy. Don't worry, it's okay—I have mine. What about the Navy?

And if anyone thinks that he or she can do this kind of thing regularly without the support, or at least acquiescence, of his or her spouse, they are, frankly, nuts. I would be nowhere were it not for the love of my life, Betty.

L.P.T.

Acronyms, Abbreviations, and Program Names

ADVENT	Early communication satellite
AEDC	Arnold Engineering Development Center
AEHF	Advanced Extremely High Frequency (communications satellites)
AFB	Air Force Base
AFBMD	Air Force Ballistic Missile Division
AFCMD	Air Force Contract Management Division
AFPRO	Air Force Plant Representative Office
AFSAB	Air Force Scientific Advisory Board
AFSC	Air Force Systems Command
AFSPACE	Air Force Space Command
AGREE	Advisory Group on the Reliability of Electronic Equipment
AJAX	Army Nike anti-aircraft missile
ALERT	NASA parts problem information sharing program
ALMV	Air Launched Miniature Vehicle antisatellite
Apollo	NASA human lunar spaceflight program
ARDC	Air Research and Development Command
ARPA	Advanced Research Projects Agency
ASD	Aeronautical Systems Division or Assistant Secretary of Defense
ASIC	Application specific integrated circuits
ATLAS	Air Force Intercontinental Ballistic Missile
BAMBI	Boost-phase Antiballistic Missile Interceptor program
BAR	Broad Area Review
BRAC	Base Realignment and Closure Commission
BrigGen	Air Force brigadier general
BSTS	Boost Surveillance and Tracking System
C4I	Command, Control, Communications, Computers, and Intelligence
Centaur	Launch vehicle upper stage
Challenger	NASA space shuttle
CMOS	Complementary MOS
Corona	First U.S. photoreconnaissance satellite
COTS	Commercial-Off-The-Shelf
CPSU	Communist Party of the Soviet Union
CQAP	Component Quality Assurance Program
CRHS	Navy Component Reliability History Survey Program

DCAS	Defense Contract Administration Services
DDOU	DLA Depot Ogden
DDR&E	Director of Defense Research and Engineering
DESC	Defense Electronic Supply Center
Discoverer	Unclassified name of the Corona photoreconnaissance satellite
DLA	Defense Logistics Agency
DMSP	Defense Meteorological Satellite Program
DoD	Department of Defense
DODAC	DoD Activity Code number
DODI	DoD Instruction
DSA	Defense Supply Agency
DSB	Defense Science Board
DSCC	Defense Supply Center Columbus
DSCS	Defense Satellite Communication System
DSP	Defense Support Program
Dyna-Soar	Air Force spaceplane, also called Program 624A and X-20
EIA	Electronic Industries Association
ELV	Expendable Launch Vehicle
ERD	Electronics Reliability Division of Rome Laboratory
ERDA	Reliability and Diagnostics Branch of Rome Laboratory
ERDB	Design and Diagnostics Branch of Rome Laboratory
ERDR	Reliability Physics Branch of Rome Laboratory
ERDS	Design Analysis Branch of Rome Laboratory
ESD	Electronic Systems Division
EW	Early Warning
Explorer	First U.S. satellite
FAR	Federal Acquisition Regulation
FARADA	Failure Rate Data program to collect and analyze reliability
FASA	Federal Acquisition Streamlining Act
FCRC	Federal Contract Research Center
FFRDC	Federally Funded Research and Development Center
FIA	Future Imagery Architecture
FLTSATCOM	Fleet Satellite Communication program
Hercules	Army Nike anti-aircraft missile
Hubble	NASA space observatory program
GAO	General Accounting Office
GEIA	Government-Electronic Industries Association
Gen	Air Force general
GIDEP	Government–Industry Data Exchange Program
GMDEP	Navy Guided Missile Data Exchange Program
GPS	Global Positioning System
HQ	Headquarters
IC	Integrated Circuit
ICBM	Intercontinental Ballistic Missile
IDCSP	Initial Defense Communication Satellite Program

IDEP	Interservice Data Exchange Program
IEEE	Institute of Electrical and Electronics Engineers
IG	Inspector General
IRBM	Intermediate Range Ballistic Missile
IRE	Institute of Radio Engineering
ITAR	International Traffic in Arms Regulations
IUS	Inertial Upper Stage
IW	Indications and Warning
JAN	Joint Army Navy
JEDEC	Joint Electron Device Engineering Councils (originally, now no longer an acronym)
JHU/APL	Johns Hopkins University, Applied Physics Laboratory
kHz	Kilohertz, or thousand cycles per second
LtGen	Air Force lieutenant general
LSI	Large Scale Integration
MajGen	Air Force major general
Mercury	First NASA human spaceflight program
MFG	Manufacturing
MHV	Miniature Homing Vehicle antisatellite
MHz	Megahertz, or million cycles per second
MIDAS	Infra-red early warning satellite
Milstar	Air Force communications satellite program
MILSTRIP	Military Standard Requisitioning and Issue Procedures
Minuteman	Air Force solid fuel Intercontinental Ballistic Missile
MIPS	Minimum In-plant Surveillance program
MIT	Massachusetts Institute of Technology
MMIC	Monolithic Microwave Integrated Circuits
MOL	Manned Orbiting Laboratory
MOS	Metal oxide semiconductor
MOSFET	Metal oxide semiconductor field effect transistors
MOSIS	Metal Oxide Silicon Implementation Service
MOU	Memorandum of understanding
MQPL	Military Qualified Products List
MSI	Medium Scale Integration
NASA	National Aeronautics and Space Agency
NAD	Naval Ammunition Depot
NESRC	Nuclear and Space Radiation Effects Conference
NOL-Corona	Naval Ordnance Laboratory at Corona, California
NPOESS	Naval Polar Orbiting Environmental Satellite System
NRL	Naval Research Laboratory
NRO	National Reconnaissance Office
NSC	National Security Council
NSN	National Stock Number
NSR	National Security Review (for President George H.W. Bush)
OASD	Office of the Assistant Secretary of Defense

OASD (S&L)	Office of the Assistant Secretary of Defense, Supply and Logistics
ODM/SAC	Office of Defense Mobilization's Scientific Advisory Committee
OEM	Original equipment manufacturer
PARC	Palo Alto Research Corporation
PAT	Process Action Team
PCP	Parts Control Plan
Peacekeeper	Air Force Intercontinental Ballistic Missile
PEO	Program Executive Officers
PM&P	Parts, Materials, and Processes
Polaris	Navy ballistic missile program
Poseidon	Navy ballistic missile program
PRB	Program Review Board
PSAC	Presidential Science Advisory Committee
QA	Quality assurance
QML	Qualified Manufacturers List
QPL	Qualified Parts List *or* Qualified Products List
RADC	Rome Air Development Center
RAND	Project R and D; *later* RAND Corporation
RCA	Radio Corporation of America
RDB	Research and Development Board
SAC	Strategic Air Command
SAFSP	Secretary of the Air Force Special Projects
SAGE	Semi-Automatic Ground Environment
SAINT (SATIN)	Satellite Interceptor program
SAMSO	Space and Missile Systems Organization
SBIRS	Space-Based Infra-Red Satellite
SCD	Source Control Drawing
SD	Space Division
SDI	Strategic Defense Initiative ("Star Wars")
SGLS	Space-Ground Link System
SMD	Standardized Military Drawing (occasionally, Standard Microcircuit Drawing)
SOLAR MAX	NASA space observatory program
SPO	System Program Offices
Sputnik	World's first artificial satellite
SPWG	Space Parts Working Group
SSI	Small Scale Integration
SSTS	Space Surveillance and Tracking System
STL	Space Technologies Laboratory
STS	Space Transportation System
TENCAP	Tactical Exploitation of National Capabilities
Thor	Air Force Intermediate Range Ballistic Missile
TI	Texas Instruments

Tinkertoy	Early Navy program on microelectronics
Titan	Air Force ICBM and space launch vehicle
TOPEX	NASA Topological Experiment Program
TPPC	Total Package Procurement Concept
TQM	Total Quality Management
TRADIC	Transistorized Digital Computer
TRANSIT	Early communications satellite
TSPR	Total System Performance Responsibility
TWT	Travelling wave Tube
USSPACECOM	U.S. Space Command
Vanguard	U.S. satellite program intended to be first U.S. artificial satellite
Vela	Nuclear detonation detection system
VHSIC	Very High Speed Integrated Circuit
VIKING	A NASA space exploration program
VLSI	Very Large Scale Integration
WDD	Western Development Division
WGEPP	Working Group for Electronic Piece Parts
ZEUS	Army Nike anti-ballistic missile and anti-satellite missile

Part I

Activation Energy (1931–1968)

Discoveries in quantum physics lay fallow until breakthroughs in electronics technology made solid state electronics real. As that first began to happen, the Cold War began. The United States needed information about the intentions and preparations of the Soviet Union to determine the severity of the threat of a nuclear, Pearl Harbor–style attack against the heartland of the United States. Overflights of the Soviet Union by U.S. bombers and reconnaissance aircraft were dangerous and provocative. They would eventually have to stop, but hopefully not before some more feasible alternative was found. Based on the technologies developed by the Germans in World War II, feasibility studies of long range missiles turned into strategic realities. The military at first provided the critical funding for the nascent solid state electronics industry, which held the keys necessary for practical long range missiles. As those missiles became practical, their capabilities also enabled other studies to turn concepts of orbital space vehicles into realities to relieve the manned aircraft overflights of their most dangerous missions. The demands of strategic missiles and critically important spacecraft drove the electronics industry to make advances in reliability. Proliferation of solid state devices led to their incorporation into the military standards program, which had been created to solve similar problems with vacuum tubes during World War II.

The period through 1968 brought together these many threads of what would be the national security, national defense, and civil space programs. All of these programs relied on available technology to get started, but

Implosion: Lessons from National Security, High Reliability Spacecraft, Electronics, and the Forces Which Changed Them, First Edition. L. Parker Temple III.
© 2013 the Institute of Electrical and Electronics Engineers. Published 2013 by John Wiley & Sons, Inc.

moved rapidly toward solid state electronics for advantages of weight, volume, and power consumption. Initially, national security and national defense space vehicles and strategic missiles had a clear effect on the development of commercial solid state devices. The growth of commercial solid state electronics quickly outgrew that influence except in the highest reliability uses. There, a symbiotic relationship began and secondarily nurtured the commercial electronics industry. By the end of the period, the limits of high reliability were approached within the constraints of current technology, understanding of system complexity, and the mechanisms to control these.

Chapter 1

Washington . . . We Have a Problem . . .

Toward the end of the 20th century, failure rates in space systems began to climb. Within the first years of the new century, the failures had not abated. Initial analysis showed the space systems affected had all begun to deteriorate around 1994 or thereafter. Spates of failures were not something new to the national security and defense space programs. Several episodes had occurred since the U.S. space programs began with the launch of Explorer I in 1958. The nation's first photoreconnaissance satellite, Corona, failed more than a dozen times in a row before a picture was successfully returned. Corona's time was one of inventing space programs. Each failure improved understanding, moving us closer to success. Extensive processes evolved over more than four decades since that time to identify the causes of failures, learn from failures, and then, as in a closed-loop control process, feed lessons back into the processes to be used in subsequent space systems. This learning and feedback process had always worked, and gotten even better over time. By the last decade of the 20th century, U.S. space program reliability was the envy of all other spacefaring nations. The United States had shown the way to get the most out of every ounce of every spacecraft or expendable launch vehicle, and had been emulated by most emerging spacefaring countries.

Starting as robotic spacecraft whose value had to be proven, the national security and national defense space systems evolved to a position of great importance in the affairs of state, national policy, national security, national defense, and the

Implosion: Lessons from National Security, High Reliability Spacecraft, Electronics, and the Forces Which Changed Them, First Edition. L. Parker Temple III.

commercial marketplace. The run-up to and performance of Operation Desert Storm demonstrated that the American way of war critically depended on space systems. They had also stood the test of time, serving to maintain the strategic way of peace. Space systems were an American way of life.

Failures occur, whether they are an option or not. Whenever so much power is crammed into one place and expended over so short a time as in a launch vehicle, even the smallest problem can cascade into a devastating loss. No failures are acceptable, some are worse than others, and every failure damages some important user's needs and expectations. Failures draw significant attention.

This is even more true when failures have an upward trend over a short time. Then, each failure does not exist in its own isolated microcosm of each U.S. space program. Each begins to be seen in light of the overall space programs. When multiple agencies' space systems are affected, the trend gathers the greatest attention of the brightest minds. Reversal of the trend must be had in the very shortest time. The nation's defense and security depend on it.

Near the end of the 1990s, a Titan IV, the nation's largest expendable launch vehicle and workhorse of the national security and defense space programs, failed with loss of its critical payload. The loss of the mission affected organizations from the White House throughout the Executive Branch of government, the Intelligence Community, and the military establishment. Then another Titan IV failed, with loss of another satellite's mission. These occurred amidst the losses of other, smaller, expendable launch vehicle payloads.

The cause had to be found and corrected immediately, as a national security and defense crisis loomed because of the satellites' loss. The satellites were to replace older versions, and were to provide enhanced new capabilities with wide-ranging impacts. Replacements could not be ready for years to come, as some represented the most complex machines ever put into orbit. A team of the best minds had to be called together and put to correcting the problem before another launch failure.

Then, orbiting satellites began failing early. Not the old ones near the end of their missions, but ones near the beginning of their lifetimes. The ones with the best, most advanced technologies. More satellites on whose data national decision makers from the President down depended on a daily basis. The problem commanded the immediate attention of the President, and thus everyone working with and for the national security and defense space programs. The root of the problems had to be found and fixed without any delay.

A high powered team began to examine the problem. That is how it had worked for more than forty years. Things had always been set right by finding each event's process flaw, immature technology, design error, or act of God, as well as could be done. This time was different.

The timing of the start of most of the failed systems suggested some cause in the early 1990s. Some sort of cause that took years to express itself. Almost immediately, the focus narrowed to the changes wrought by Acquisition Reform. Promoting a streamlined approach to acquisition, the Acquisition Reformers claimed better systems could be acquired more quickly and at less cost. Acquisition Reform had been well supported and well intentioned, driven by the end of the Cold War and the need to economize. Acquisition Reform did away with the military standards. Perhaps some military standards needed to be reinstated.[1]

This time, things were not so simple. This time was very different. As the analysis continued, the realization dawned that much more had contributed to the failures than anyone thought. This time, a fix might not even be possible. How had this happened?

---

[1] Liam P. Sarsfield, *The Application of Best Practices to Spacecraft Development: An Exploration of Success and Failure in Recent Missions* (Arlington, Virginia: RAND Corporation, 2002); William F. Tosney, *Faster, Better, Cheaper: An Idea Without a Plan* (El Segundo, California: Aerospace Corporation, 2000); Steve Pavlicka and William Tosney, "An Assessment of NRO Satellite Development Practices" (El Segundo, California: Aerospace Corporation, 2003), cited in A. Thomas Young, *Report of the Defense Science Board/Air Force Scientific Advisory Board Joint Task Force on Acquisition of National Security Space Programs* (Washington, D.C.: Office of the Under Secretary of Defense for Acquisition, Technology, and Logistics, May 2003), 49.

Chapter 2

The Quantum Leap

Space systems in service to national defense and national security have long been held as the epitome of high reliability. Achievement of high reliability equates to long operational lifetimes in space. Needless to say, that was an evolutionary process. First came the evolution of a set of feedback mechanisms to learn lessons and improve future generations based on problems, failures, and solutions. Second, evolution was Darwinian, guided by an unseen hand. No one was truly in charge of all parts of space systems from starting with electronic parts through development, to launch, operations, and then disposal at the end of the mission. The unseen hand created an optimal mix of products, processes, and applications responding to market forces and, at times, political pressures.

Since the launch of Explorer I in 1958, the government, contractors, and suppliers supporting high reliability spacecraft became partners in learning lessons from failures and anomalies as technologies emerged and as inspection and detection technologies improved. At every step, demands of mission success moved the enterprise closer to optimality.

In the modern age, the complex mixture of people, technology, and processes has created the assumption that high reliability spacecraft are the norm. What was accomplished with so much hard work and diligence in the past seems to have become routine—little more than what Henry Ford started with the modern assembly line for manufacturing cars. The fact is, complex satellites are marvels of science and technology and have never been the norm. Each is a sophisticated mixture of several kinds of constituent pieces combined using what some would claim rivals a modern "black art." Contrary to that claim, though, stands the long-term process that created ever more reliable spacecraft. Only through a disciplined and concerted effort to understand every individual part from manufacturing at the vendors through the final assembly and launch did this happen.

This story reflects the times during which high reliability spacecraft became possible, achieved extraordinary successes, when suddenly the recipe for their reliability seemed to have been lost. Understanding this necessitates describing the roots of electronic piece parts and the evolution of high reliability for spacecraft.

Implosion: Lessons from National Security, High Reliability Spacecraft, Electronics, and the Forces Which Changed Them, First Edition. L. Parker Temple III.
© 2013 the Institute of Electrical and Electronics Engineers. Published 2013 by John Wiley & Sons, Inc.

Explicitly, space qualified parts are a specialized subset of high reliability parts, the latter a term associated generally with military avionics, and also referred to as military grade parts. Space qualified parts are but one (very essential) means of achieving high reliability at the system level—after all, as the quip goes, everything starts with parts.

Every history has to choose a point at which some seminal event occurred as a starting point for what follows. Because so many historical processes are really little more than continuations of things that went before but with some novel twist, this is often very hard to do. Describing the U.S. space program as starting with its first satellites can be done, but it ignores the extensive studies, technology preparation, and development of the missiles that would become the workhorse launch vehicles. That would take the story back only to the late 1940s and early 1950s, and that is too late. High reliability spacecraft reach even farther back, since they were made possible by the development of solid state electronics. Solid state electronics emerged from work in companies, laboratories, and universities exploring how different materials transmit, alter, or resist the flow of electrons. Understanding the behavior of electrons was essential to be able to do these things, and understanding required a quantum leap.

Several advances in physics had to come together to make practical use of the quantum theory of electrons. Following the work in quantum physics done by Rudolf Peierls and Felix Bloch, the British theorist Alan Wilson published two papers in 1931 on the behavior of electrons in semiconductor materials. This provided the key insights necessary to make a practical application based on understanding the peculiar natures of quantum physics and electrons.[1]

Transitioning solid state theory into real devices took some time—about a decade and a half. In the interim, vacuum tubes carried us through World War II. Important ideas about standardization's usefulness began in that War. As the electronics industry and the Services scrambled to meet the overwhelming supply demands of a burgeoning military, the 1930s way of doing business had to change. Standardization was critical for the military use of radio electron tubes. Until 1942, the Army and Navy each had their own individual unique tube nomenclatures. These worked reasonably well for the peacetime military. But war production demands were very high for radio applications, and separate systems for essentially the same devices were an unsustainable vanity. Combining systems would allow for commonality, improving and enabling interchangeability, common stockpiles, joint inspection, and efficiency. Thus, in 1942, the Services began cooperative development of JAN-1, whose title, in the arcane language of standards, was "Specification for Tubes, Radio Electron." "JAN" stood for "Joint Army-Navy," indicating the parts specification was controlled by both Services. It was published in 1943, and the Army Signal Corps and Navy Bureau of Ships became its first users in March of

---

[1] Michael Riordan and Lillian Hoddeson, *Crystal Fire: The Invention of the Transistor and the Birth of the Information Age* (New York: W.W. Norton, 1997), 66. Wilson published two papers with the same title in 1931, "The Theory of Electronic Semi-Conductors."

that year. A rapid update came by the end of 1943, issued as the "Joint Army-Navy Specification JAN-1A for Radio Electron Tubes." The first revision held an important additional feature: a "Preferred List of Vacuum Tubes." In addition to reducing the number of tube types, the ultimate objectives included "savings to the Services in space and manpower for handling stock; also tube manufacturers can, by concentrating on the production of those types, achieve greater production efficiency, higher quality, and lower shrinkage." Any given manufacturer could, by choice, concentrate on specialization in a few tube types, thereby gaining greater expertise in them, further enhancing these benefits of standardization.[2]

The positive benefits of this approach lasted long after the end of the War. The momentum carried forward, and would eventually adapt easily to solid state electronics.

Emerging from the war, Dr. Vannevar Bush, Director of the Office of Scientific Research and Development, transmitted his organization's response to several questions asked by President Franklin D. Roosevelt. The study's title, *Science, the Endless Frontier*, was dated 5 July 1945, by which time Harry S. Truman was president. The title, though, came from Roosevelt's November 1944 remarks, and captured the prevailing view of the future role of science and technology. Roosevelt said:

> *New frontiers of the mind are before us, and if they are pioneered with the same vision, boldness, and drive with which we have waged this war we can create a fuller and more fruitful employment and a fuller and more fruitful life.*[3]

Bush painted a picture of a world in which life was more fulfilling, with greater productivity in everything from farming through manufacturing, resulting in workers with more leisure time, greater security, and peace. His hopeful message depended on continuing the strong emphasis on science and technology. World War II had been won because of this, making the United States the new world leader in scientific research.

He went on:

> *In 1939 millions of people were employed in industries which did not even exist at the close of the last war – radio, air conditioning, rayon and other synthetic fibers, and plastics are examples of the products of these industries. But these do not mark the end of progress – they are but the beginning if we make full use of our scientific resources. New manufacturing industries can be started and many older industries*

---

[2] C.W. Martel and J.W. Greer, "Joint Army-Navy Tube Standardization Program," *Proceedings of the I.R.E.* (July 1944), 430, 434. Martel was a captain in the Army Signal Corps and Greer was assigned to the Navy's Bureau of Ships. The use of JAN markings became a U.S. military registered mark of certification, under Patent Registration number 504,860, registered 14 Dec. 1948, and renewed the third time in 2008.

[3] Vannevar Bush, *Science, the Endless Frontier: A Report to the President by Vannevar Bush, director of the Office of Scientific Research and Development, July 1945* (Washington, D.C.: Office of Scientific Research and Development, 1945), reprinted by the University of Michigan Library and Hewlett-Packard, ix.

greatly strengthened and expanded if we continue to study nature's laws and apply new knowledge to practical purposes.... Advances in science will also bring higher standards of living, will lead to the prevention or cure of diseases, will promote conservation of our limited natural resources, and will assure means of defense against aggression.[4]

All wonderful new technologies, but he overlooked one advance that pointed to an unexpected new technology.

The important and greatly unappreciated development (except by those who used them in battle) was the proximity fuze. Previously, artillery and bombs relied on impact or timing. Proximity fuzes changed that, making near misses nearly as good as hits and, in some cases, even better. The difficulty of creating the proximity fuze rivaled that of creating radar and the Manhattan Project's atomic bomb. These devices started as tiny, self-powered radio transmitters, triggered by energy reflected from a target. In restricted use at first, lest any duds fall into enemy hands and be reverse engineered, proximity fuzes were widely used to blunt kamikaze attacks in the Pacific. They were credited with a major success in anti-aircraft artillery against the V-1 missiles over England in 1944. General George S. Patton called them the decisive weapons against massed infantry and armor at the Battle of the Bulge, saving Liege and causing the need for a full revision of land warfare tactics.[5]

General Electric developed specialized computation circuits dependent on three small vacuum tubes individually about the size of the metal and eraser of a pencil. Production in the millions made them inexpensive enough to be expended in artillery shells or bombs. In 1945, the National Bureau of Standards and Globe-Union, Inc.'s Centralab collaborated in techniques enabling easily manufactured and affordable sensors and controls for munitions. They replaced some of the tiny vacuum tubes using a technology called a thick film. Silk-screening conductive inks onto a ceramic substrate created either resistive or capacitive shapes. This was the first use of a mask to lay down various shapes in a conductive layer on a semiconductor material, and foreshadowed the (not too) distant use of metal shapes on silicon as integrated circuits.[6]

While resistors and capacitors were necessary, they were insufficient. Building an industry on this gossamer technique required additional technological

[4] Bush, *Science*, 5.

[5] Ralph B. Baldwin, *The Deadly Fuze: Secret Weapon of World War II* (San Rafael, California: Presidio Press, 1980), xxxi; Vannevar Bush, *Pieces of the Action* (New York: William Morrow, 1970), 55–56, 112; Patrick E. Hagerty, *Management Philosophies and Practices of Texas Instruments* (Dallas, Texas: Texas Instruments, 1965), 162. Hagerty cites Prof. P.M.S. Blackett, "Tizard and the Science of War," *Nature* (5 March 1960). The origins of the proximity fuze were British, and are described in James W. Brenna, "The Proximity Fuze: Whose Brainchild?" *United States Naval Institute Proceedings* 94, no. 9 (Sept. 1958).

[6] Jack S. Kilby, "Invention of the Integrated Circuit," *IEEE Transactions on Electron Devices* ED-23, no. 7 (July 1976): 648; John A. Miller, *Men and Volts at War: The Story of General Electric in World War II* (New York: Bantam Books, 1948), 118; Hagerty, *Management Philosophies*, 123. Centralab later became a division of Johnson Controls, Inc.

breakthroughs with some of the smallest known bits of matter. A quantum leap was needed.

Successful application of the quantum physics breakthroughs of the early 1930s awaited a combination of materials breakthroughs with William Shockley, John Bardeen, and Walter Brattain at Bell Telephone Laboratories after World War II.

The materials breakthrough was the discovery by Jack H. Scaff and Henry C. Theuerer of a method to grow large single crystals of silicon or germanium by melting the material in a vacuum. The resulting material demonstrated a curious property: current rectification. "That is, they would act as a one-way valve for the passage of electricity."[7] Shortly, two variations on this rectification were realized. In some, the one-way valve behavior was observed when in a negative electrical field; in others, only when in a positive electrical field. To distinguish the behaviors, the first was called "n-type" and the latter, "p-type." The phenomenon was based on the presence of a small amount of other materials. Elements from the third and fifth columns of the periodic table of elements straddle the materials known as semiconductors (so called because they are conductors only under certain circumstances). By the presence of elements from the third column, p-type semiconductors resulted. The fifth column provided the excess electrons to make n-type semiconductors. When the doped material was diced into smaller pieces, electronic devices could be made.

The key event in the practical beginning of the solid state revolution can be traced accurately to the week before 23 December 1947. The breakthrough was a solid state device that could amplify electrical signals passing through it. Shockley, Bardeen, and Brattain first demonstrated the device to management on 23 December.[8] That demonstration was really an application of some key electronic principles in an impractical arrangement. Nevertheless, it proved the new device could be made to work, and was patentable.

Development continued and, on 23 June 1948, Bell Telephone Laboratory's president Oliver E. Buckley briefed the Armed Services on the technology breakthrough, in advance of a public announcement. Two representatives from each Service were sworn to top secrecy prior to an unclassified public announcement that would follow in a matter of days. Ralph Bown gave demonstrations of telephone amplifiers, radio receivers, and an oscillator. The Service representatives understood the power of what they had seen. Later that day, the Army Signal Corps representatives discussed a possible Army contract with Bell for additional work on the device.

---

[7] Bo Lojek, *History of Semiconductor Engineering* (New York: Springer-Verlag, 2007), 23, 65; Thomas J. Misa, "Military Needs, Commercial Radios, and the Development of the Transistor, 1948–1958," in *Military Enterprise and Technological Change, Perspectives on the American Experience*, ed. Merritt Roe Smith (Cambridge, Massachusetts: MIT Press, 1987), 268–269. This section owes a great deal to the excellent and very accessible essay by Misa.

[8] Riordan and Hoddeson, *Crystal Fire*, 1; Lojek, *History of Semiconductor Engineering*, 65. The story as told necessarily overlooks the contributions of several individuals working from theory to materials preparation and then fabrication of the device. On 27 Dec. 1949, W.S. Gorton, who worked for the Bell Labs director of research, listed 12 individuals responsible for the complete development. See Lojek for their names and specific contributions.

Donald A. Quarles, vice president of Bell Labs, said their research and development was not for sale.[9] Events show he probably came to regret that quickly, especially when, a few years later, he became the Assistant Secretary of Defense for Research and Development. However, at the moment, Quarles and others indicated they would work under contract for supplying applications information to the Services. For several weeks after the demonstration, the Services discussed classifying the work, along the lines of the Manhattan Project, which had produced the atomic bomb. Data sharing and competitive accomplishment won the day, however.[10]

In any case, a week later, on 30 June 1948, Bown announced the new technology at a press conference. "We have called it the Transistor, because it is a resistor or semiconductor device which can amplify electrical signals as they are transferred through it." The news splashed across a lone column on page 46 of the *New York Times* the following day. Hardly the herald of a revolution, the importance of the device was not obvious, appearing under "The News of Radio" on 1 July 1948. The *Times* understood enough, however, to explain that the device "has several applications in radio where a vacuum tube ordinarily is employed."[11]

Military interest in solid state technology had come early and remained high. The Army Signal Corps was hard at work on mobile and hand-held communication devices and other applications that would benefit from electronics miniaturization and solid state's lower power. Very early in their work, the Signal Corps' miniaturization efforts turned to circuit assembly. Initially, the approach taken on proximity fuzes was attractive, but the result was only simple resistor-capacitor circuits, and therefore too limited for general use.[12] The transistor, though, pointed the way to much more.

The Army was not alone in its interest. The first applied use of transistors may have been in the work Bell Labs did for the Navy Bureau of Ordnance. A Navy simulated warfare computer used 40 transistors, "nearly all that were available," at its development in late 1949.[13]

The Air Force wanted maintainable electronic systems because complex vacuum tube-based systems were a problem. For instance, the most complex military project to date had been the development of the airplane to deliver the atomic bomb, the Boeing B-29 Superfortress. Each of these bombers had about a thousand vacuum tubes and tens of thousands of capacitors and resistors. The number of connections alone was a problem, as were the tubes themselves.[14] In World War II, electron tubes

[9] Misa, "Military Needs," 264.

[10] Harold E. Zahl, "Birth of the Transistor," *Microwave Journal* (July 1966): 94–95; Misa, "Military Needs," 258. Zahl was at the Service demonstration, and was at the time the Chief of the Research Branch, Engineering Division, of the Army Signal Corps Engineering Laboratories at Fort Monmouth, N.J.

[11] Riordan and Hoddeson, *Crystal Fire*, 8.

[12] Misa, "Military Needs," 263.

[13] Ibid., 265.

[14] Kilby, "Invention of the Integrated Circuit," 648; William Denson, "The History of Reliability Prediction," *IEEE Transactions on Reliability* 47, no. 3-SP (Sept. 1998): SP-321.

were the most unreliable components.[15] These bombers, and their more advanced successors, depended on failure-prone tube-based analog computers. Functions from autopilots to bomb dropping failed every 70 hours on average. The Air Force hoped digitalization would solve a variety of these ills. Also, as the Air Force looked to the needs of its crucial long range ballistic missile developments, digital computers offered the potential for more accuracy and far greater speeds than mechanical and electromechanical machines could offer.[16]

Consequently, in 1949, all the Services accepted Bell's idea of a data sharing contract. They jointly established a contract with Bell Lab, and went on to renew it annually over the following decade. The Joint Services contract provisions had an early and important impact on sharing information about the new technology, as will be seen shortly.

* * *

In parallel with the development of solid state electronics, but neither its impelling force nor its reason for being, was the increasing tension of the Cold War.

The German use of tactical ballistic missiles in World War II ushered in a new era of military weaponry. Immediately following the war, work began in the Services both to exploit the new technology and defend against it. Both proved harder than expected.

By 1946, the Army was working on a defense against attack by V-2 type rockets, which it said constituted "a grave threat to national security."[17] Its defensive missile work also covered a series of air defense missiles to defend against bombers attacking the United States. These missiles became part of the Nike family comprising a series of different missiles. The Army Signal Corps was developing the electronics, and soon realized the importance of electronics miniaturization for the successful development of the missiles. Bell Labs was an important contributor to the Army's studies and development efforts.

In addition to the Army's interest in defensive missiles, all of the Services wanted long range ballistic missiles. By 1946, the cooperation among the Services during the recent war had largely evaporated. Inter-Service rivalry existed on many levels, but most notably in Washington, D.C., where budget battles were hottest. Each Service vied for its share of the budget in the race to develop long range missiles and capture a part of this important new mission area.

In 1947, as part of a military reorganization creating the Department of Defense and the Department of the Air Force, a number of important agreements had to be reached, not the least of which involved strategic warfare. The Services agreed on

---

[15] Denson, "The History of Reliability Prediction," SP-321.

[16] Chong-Moon Lee et al., *The Silicon Valley Edge: A Habitat for Innovation and Entrepreneurship* (Stanford, California: Stanford University Press, 2000), 167.

[17] Commanding General, Army Ground Forces, Letter to Commanding General, Army Service Forces, Pentagon, Washington, D.C., 4 Jan. 1946; L. Parker Temple, *Shades of Gray: National Security and the Evolution of Space Reconnaissance* (Reston, Virginia: American Institute of Aeronautics and Astronautics, 2005), 9.

a mutual definition by 1949, but the Air Force and the Navy were both contesting the other's exclusivity in strategic nuclear capabilities. By the spring of 1949, this rivalry reached extremes. The Navy's new aircraft carrier, the *United States*, was to have a new strategic bomber designed to deliver nuclear weapons. That threatened the Air Force's immense strategic nuclear bomber, the Convair B-36 Peacemaker. The B-36 was an essential element of strategic planning because it was the only means (other than ships or submarines) of delivering the 41,000-pound Mark 11 hydrogen bomb.[18]

When the first Secretary of Defense, James Forrestal, left office, Louis Johnson replaced him. Johnson favored the B-36. On 23 April 1949, having received consultation from the Army and the Air Force (and apparently not the Chief of Naval Operations), Johnson cancelled the *United States* and with it the Navy's atomic bomber. Led by Chief of Naval Operations Admiral Louis E. Denfield, the Pentagon admirals took their complaints to Congress in an episode known as the "Revolt of the Admirals."[19]

Competition can help incentivize the highest quality and levels of performance. However, over the following decade, as the technology for extremely long range ballistic missiles neared reality, such competition was to become excessive and counter-productive. This merely foreshadowed the inter-Service rivalry of the coming decade.

*   *   *

Just as the Services' long range missile designs differed, so did their procurement policies. Different procurement policies led to different specifications for essentially the same kind of parts, leading to the lack of interchangeability and higher costs that JAN-1 had sought to avoid. Inter-Service rivalry only exacerbated interchangeability problems by reducing the likelihood of agreement. Counterproductive rivalry temporarily turned back some of the advantages of standardization, requiring them to be relearned later. These differences festered, and would eventually draw the attention of economy-minded senior leadership.

*   *   *

As Bell Labs announced the transistor, the Soviet Union was developing its first atomic device. No one in the West knew the extent to which the Manhattan Project had been penetrated. Most experts thought the Soviets were many years away from detonating a fission device. Nevertheless, reconnaissance flights on the periphery of the Soviet Union began carrying radiation sampling sensors just in case.

On 3 September 1949, an aircraft off the Kamchatka Peninsula detected the radioactive debris from the first Soviet fission device's detonation. As Bell Labs advanced its transistor and others began work in solid state electronics, a sobering

---

[18] Temple, *Shades of Gray*, 16, 77; Richard Rhodes, *Dark Sun: The Making of the Hydrogen Bomb* (New York: Simon and Schuster, 1995), Fig. 86 and p. 541.

[19] Temple, *Shades of Gray*, 16; Neil Sheehan, *A Fiery Peace in a Cold War* (New York: Random House, 2009), 136–139.

1951 Central Intelligence Agency estimate announced, "We believe that the ultimate Soviet Objective is a Communist world dominated by the U.S.S.R. . . . and that an armed conflict [with the United States and its allies] is eventually inevitable."[20]

The Soviet Union became an increasingly hostile closed society. U.S. leaders recognized that a Pearl Harbor-like surprise attack by the Soviet Union would be far more devastating in the new nuclear age. The 1951 intelligence estimate underscored President Truman's decision in National Security Council document NSC 139, ordering the construction of a continental defense early warning system to be operational in 1955.

*   *   *

An organizational move that would hold incredibly important consequences for the history of high reliability electronics took place in 1951, when the Army Watson Laboratories became the Rome Air Development Center (RADC).[21] RADC was located on Griffiss AFB, near Rome, New York. RADC's initial responsibilities were focused on the increasingly important ground systems depending on electronics.

Close on the heels of the Air Force organizational change, Bell Labs' Quarles, on 27 August 1951, announced a symposium on the transistor and its military applications.

There was a great deal to talk about. The Services had been funding development of switching circuits to replace vacuum tubes for digital computers and other applications. For instance, by May 1951, Bell Labs had demonstrated the potential of "point-contact" transistors with sufficiently long lifetimes, with "a fourfold reduction in volume and an eightfold reduction in power requirements" over a 370-tube data transmission set.[22]

Additionally, Bell Labs' Gordon Teal and John B. Little had refined their crystal growing process to allow sandwiching together p-type and n-type semiconductors in alternating layers, making entirely new kinds of transistors. The "junction transistor" had shown the way, but with the combination of n-p-n materials, a "grown junction" transistor resulted. Then p-n-p layers were fabricated as "alloy junction transistors."[23] The theoretical and practical experience of the United States' transistor development resided mainly in Bell Labs. Theory and experience were about to be shared broadly at the military's insistence.

Invitations for the September 1951 event went to government laboratories, military commands, universities, 88 commercial firms, and selected individuals in Britain, Canada, and France. This was the seminal event in sharing technology information across the national defense community. When the event arrived, attendance included 139 industrial and 121 military representatives, with 41 university

---

[20] "Probable Soviet Courses of Action to Mid-1952," *National Intelligence Estimate* 24 (2 Aug. 1951): 1; Temple, *Shades of Gray*, 23–30.

[21] "Pioneering Reliability Laboratory Addresses Information Technology," *Journal of the Reliability Analysis Center* 6, no. 1 (1998): 1, 5.

[22] Misa, "Military Needs," 265.

[23] Ibid., 270.

participants. Twenty-five lectures and demonstrations covered a wide range of transistor topics. By November, the proceedings were available in a 792 page volume. In addition to one copy per participant, the Services distributed a further 5,500 copies, and an unknown number were disseminated under license from Bell Labs.[24]

Close on its heels was an April 1952 symposium "on transistor science, engineering and manufacturing" for licensees of AT&T, covering much more fundamental technical topics. The proceedings were a large two-volume set, known as "the Bible" within Bell Labs, which stood as the authoritative source for semiconductor technology into the late 1950s.[25]

Both of these were "crucial early events in the history of this technology and industry," primarily through the flow of information and the leveling of knowledge across a wide community.[26]

An example of the stimulus of these meetings was the Royal Radar Establishment's Geoffrey William Arnold Dummer's oft quoted prophecy about the future to the Institute of Radar Engineers' annual Electronic Components Conference in Washington, D.C., in May 1952:

> *With the advent of the transistor and the work in semiconductors generally, it seems now possible to envisage electronics equipment in a solid block with no connecting wires. The block may consist of layers of insulating, conducting, rectifying, and amplifying materials, the electrical functions being connected directly by cutting out areas of the various layers.*

This inspired Shockley to predict transistors being built for a nickel apiece.[27] Exactly how such layers of n-type and p-type material could be assembled, no one yet knew.

As important as these symposia were, sustained information sharing rapidly advanced solid state technology. The Joint Services agreement provided government rights to Bell Labs' data, and explicitly allowed for government disclosure of the technological information. As a result, other contractors could, should the government decide, benefit from the solid state electronics work at Bell Labs. Although the government had the responsibility, then as now, to protect proprietary information, the "contract system was . . . inherently 'leaky.'"[28]

The effectiveness of the symposia and the continued sharing of information can be seen in Table 2.1. This shows, by year, the growth of the number of semiconductor patents by Bell Labs and others. Of particular note is the third line, that

---

[24] Ibid., 267. Accounts of attendance vary, but all agree it was very widely attended. Holbrook cites 119 defense contractors, but does not expand on that much; Misa's numbers are used here.

[25] Ibid., 268.

[26] Daniel Holbrook, "Government Support of the Semiconductor Industry: Diverse Approaches and Information Flows," *Business and Economic History* 24, no. 2 (Winter 1995): 137–138; Misa, "Military Needs," 267.

[27] Kilby, "Invention of the Integrated Circuit," 648; Lojek, *History of Semiconductor Engineering* (New York: Springer-Verlag, 2007), 37. Lojek noted that Shockley was pessimistic in his prediction.

[28] R.E. Anderson and R.M. Ryder, "Development History of the Transistor in the Bell Laboratories and Western Electric (1947–1975)" (unpublished manuscript, AT&T Archives, 45 11 01 03), 186, cited in Holbrook, *Business and Economic History*, 138–139.

Table 2.1.   Breakdown of Early Semiconductor Patents by Firms

	1952	1953	1954	1955	1956
Bell Laboratories	56%	51%	46%	37%	26%
Tube Firms[a]	37%	40%	38%	42%	54%
New Firms[b]	7%	9%	16%	21%	20%
Total Number of Patents Granted	60	92	79	73	186

[a] Includes Radio Corporation of America, General Electric, Westinghouse, Sylvania, Philco-Ford, and Raytheon, all firms with previous experience manufacturing vacuum tubes.

[b] Includes International Business Machines, Motorola, Hughes, International Telephone and Telegraph, and Clevite, firms that had been set up explicitly to manufacture semiconductor components or that had no previous experience manufacturing vacuum tubes.

Source: After John Tilton, *International Diffusion of Technology* (Washington, D.C.: Brookings Institution, 1971), 57.

for the new entries into the industry, showing their rapid growth in technology contributions:

High military interest in transistors was expressed as financial support. Military support was not limited to fundamental and applied research, as the Army Signal Corps began underwriting development of manufacturing facilities. Western Electric, the manufacturing alter ego of Bell Labs, built a huge facility in Laureldale, Pennsylvania, in 1953 at Signal Corps expense. Overall, the Army invested about $13 million in facilities for pilot plants and production facilities for General Electric, Raytheon, Radio Corporation of America, and Sylvania.[29]

Proliferation of research across many companies meant an inevitable problem in lack of standardization, just as vacuum tubes had experienced ten years earlier. Already in 1953, some 60 different types of transistors existed. They were not different just in size, but in physical layout, underlying materials, and technology. Attempting to bring some order to impending chaos, in 1953 the Signal Corps sponsored discussions among the Services and the Radio Electronics Television Manufacturers Association. The attempt was well intended, and the correct thing to do, but too early. Lack of agreement on how to standardize, and lack of willingness by the industry to pay for retooling in anticipation of a standard, left the issue unsolved for the rest of the decade. So the number of different transistor types exploded to 275 by 1956, and over 900 two years after that.[30]

* * *

Intelligence estimates of the Soviet Union's intentions rapidly became even more gloomy and dire. The year of maximum danger of a surprise attack by the Soviet Union narrowed to 1954. A nuclear Pearl Harbor seemed ever closer when, years

[29] Misa, "Military Needs," 275.
[30] Ibid., 275–276.

ahead of predictions, the Soviet Union set off a thermonuclear fusion device on 12 August 1953. Apparently, the Soviet Union was not only rapidly catching up to the United States, but its rate of advance was faster and accelerating. In the very near future, so the estimate warned, the Soviet Union might surpass the capabilities of the United States and its allies. Once in that strategically superior position, a strategic nuclear conflict might become a reality.[31]

*   *   *

Meanwhile, a small electronics firm in Dallas, Texas, set up a research laboratory in 1953 to explore the potentials of transistor devices. Drawing on expertise from Bell Telephone Laboratories, Texas Instruments (TI) began developing a small-signal transistor. The company had already narrowed down the materials selection and made silicon the basis of its microelectronics work. By the spring of 1954, the company was able announce commercial availability of silicon transistors. The choice of silicon was important for the military.

> *The ambient temperature in jet aircraft and guided missiles . . . often exceeded 75°C, the maximum operating temperature for germanium transistors. Equally important, military applications required equipment to be of the small size so easily achieved by using solid-state devices. Consequently, silicon transistors, with a maximum operating temperature above 150°C, sold briskly to the military services. . . . The military's preference for silicon transistors allowed their chief manufacturer, Texas Instruments, to carve out a niche in the semiconductor market.*[32]

This was the first of many times when the company was ready at precisely the right time with an innovation.[33]

*   *   *

The German and Allied wartime technology advances set the tone for the next decades of military and intelligence developments. Sophisticated computing machines grew from the very basic efforts of John von Neumann in support of the Manhattan Project. As technology became more advanced, so did the need for computing. Vacuum tube computers were few in number, and each filled a large room. Solid state devices held a promise of reducing the size, heat, and power demands of such machines.

By 1954, the Air Force took a major step forward with a computer of very limited computational capability. The fully solid state electronics "TRansistorized DIgital Computer," or TRADIC, was first operated in January 1954.[34] Though not as computationally fast as vacuum tube computers, TRADIC approached their speed. At the time, TRADIC's computations were more important than its foreshadowing

---

[31] "Likelihood of the Deliberate Initiation of Full-Scale War by the Soviet Union," *National Intelligence Estimate* 48 (8 Jan 1952).

[32] Misa, "Military Needs," 280.

[33] Hagerty, *Management Philosophies*, 52.

[34] Riordan and Hoddeson, *Crystal Fire*, 204.

the move away from vacuum tubes. TRADIC and its successors worked on the complex computations necessary to make nuclear weapons more practical.

Practical nuclear weapons demanded a new delivery means more practical than the B-36: intercontinental ballistic missiles (ICBM).[35] Initially, nuclear weapons were immensely big and heavy, especially the thermonuclear fusion devices. By August 1952, the weight of the atomic fission warhead, which had started in the range of 11,000 pounds, had slowly dropped to 8,000 pounds. The size of a missile to launch such devices was gargantuan—over 670,000 pounds.[36] Although possible to build, these very large missiles were clearly militarily impractical. For instance, as Strategic Air Command (SAC) was formulating the requirements for military operational long range missiles, the time necessary to fuel and launch was critical. Based on the expected warning time of an attack by the Soviets, the time allocated was a few tens of minutes.[37] Fueling and checking immense missiles capable of delivering a fusion bomb would take much longer. Missiles highlighted a lesson from World War II's development of the Boeing B-29 Superfortress, designed to deliver the first atomic bombs. Weapons, delivery platforms, and ground processing were all interrelated, and part of a weapon *system*. So, as warhead weight began to reduce, the missile size could be shrunk, allowing faster fueling and launching. Computers performed difficult warhead design analyses, making them more efficient, reducing their total size and weight, and then iteratively optimizing the missile design.

Von Neumann chaired the Air Force Strategic Missiles Evaluation Group. By 1954, successive iterations finally produced a breakthrough in size which meant the ICBM development could proceed. As the weight of the warhead dropped, so did that of the missile, and by 1953, the total weight was down to 440,000 pounds. By the middle of 1953, with warhead weight nearing 1,500 pounds, the missile's weight projection approached a more manageable 240,000 pounds. The committee's report, released 10 February 1954, announced the feasible development of an ICBM by 1960.[38]

---

[35] The initial acronym was IBM, but that created confusion with International Business Machines, known by its initials of IBM. Hence the C was added.

[36] Neufeld, *The Development of Ballistic Missiles*, 73–74, 98. At that weight, such missiles would have been about 1/10 the size of the Saturn V moon rocket—so while feasible, they were not militarily practical. The lengthy time to fuel them, and the difficulty of protecting them above ground, contributed to such an assessment.

[37] James J. Ripley, *The Thor History Douglas Report SM-41860* (Santa Monica, California: Douglas Aircraft Company, May 1962), 1. The timelines were eventually refined to require both ICBM and the shorter range Intermediate Range Ballistic Missiles to be fueled and ready to launch in 15 minutes.

[38] Temple, *Shades of Gray*, 21; George B. Kistiakowsky, *A Scientist at the White House: The Private Diary of President Eisenhower's Special Assistant for Science and Technology* (Cambridge, Massachusetts: Harvard University Press, 1976), xxxiii; U.S. Congress, Office of Technology Assessment, *A History of the Department of Defense Federally Funded Research and Development Centers* (Washington, D.C.: U.S. Government Printing Office, June 1995), 25; Neufeld, *The Development of Ballistic Missiles*, 98.

Eight months after TRADIC began operating, in August 1954, Brigadier General (BrigGen) Bernard A. Schriever and a small group of military officers set up the Air Research and Development Command's (ARDC) Western Development Division (WDD) in an abandoned schoolhouse in the Los Angeles suburb of Inglewood. Secretly, they began the effort to build an ICBM.[39] Schriever's mission was to design, build, and deploy a fully operational ICBM within six years.

The complexity and urgency of the missile developments called for a new way of managing a development program. His approach was known as "concurrency." Essentially, the approach was to take every activity that could be done in parallel and do it as soon as possible, reducing as much as possible the serial dependencies of one project to another. The plan was for the warheads, missiles, equipment, crews, and launch sites to complete development at about the same time. Inside each of those paths, subprojects were also subject to concurrency. Schriever knew he needed a great deal of technically excellent advice and support to manage the complexity and keep things on track. He contracted with the Ramo Wooldridge Corporation for systems engineering. Ramo Wooldridge formed a separate Guided Missile Division just to handle the specialized additional effort.[40]

The first U.S. ICBM was Convair's Atlas. Impressively complex and expensive for the time, Atlas' 300,000+ parts were built by more than 20,000 companies employing more than 40,000 workers. ICBM development was on a fast track. Shortly after Atlas was started, the next generation of missiles began development. In 1955, the Martin company began to build the next liquid-fueled ICBM, Titan, in a completely new plant near Denver.[41]

This massive strategic buildup was under intense time pressure, since estimates of a similar race inside the Soviet Union fueled nuclear Pearl Harbor worries. The buildup had some important and beneficial side effects in the solid state electronics industry.

In October 1954, Texas Instruments, in collaboration with IDEA Corporation, produced the first commercially available pocket transistor radio, known as the Regency. This and the company's earlier research projects took credit for accelerating semiconductor device utilization worldwide by two years. The Regency showed that a relatively minor company, through innovative research and proper business strategy and tactics could affect a nascent world market.[42] For the public, small transistor radios first became a status symbol, and then a necessity. International competition began to affect the industry when, in 1954, Sony entered the fray.[43]

---

[39] History of the Minuteman Missile, National Park Service.

[40] Ivan A. Getting, *All in a Lifetime: Science in the Defense of Democracy* (Los Angeles, California: Vantage Press, 1989), 385; U.S. Congress, Office of Technology Assessment, *A History of the Department*, 25.

[41] Alan J. Levine, *The Missile and Space Race* (Westport, Connecticut: Praeger, 1994), 32–33.

[42] Hagerty, *Management Philosophies*, 53.

[43] Riordan and Hoddeson, *Crystal Fire*, 216.

Military applications were growing alongside this new commercial industry. Michael Riordan, an historian of early solid-state devices, said of the early military steps:

> *These military applications provided an immediate market for transistors where cost was not a concern. This was a crucial factor in the early 1950s, when point-contact transistors might cost nearly $20 each, compared to only a dollar or so for a vacuum tube. After 1951 the armed services began pouring millions of dollars into transistor development at the labs. From 1953 to 1955, for example, almost half the money spent there for this purpose came from the military. And in 1953 the Signal Corps even underwrote the construction costs of a Western Electric plan in Reading, Pennsylvania, devoted to mass producing transistors. To ensure reliable supplies, the Corps also spent millions on similar production lines at General Electric, RCA, Raytheon, and Sylvania. Suddenly, a new industry had been born.*[44]

[44] Ibid., 204.

Chapter 3

Preparation

President Dwight D. Eisenhower took office in January 1953, committing himself to preventing another Pearl Harbor strategic surprise attack. Eliminating strategic surprise meant that a wide array of problems needed solutions.

First, he had to end inter-Service rivalry and its destructive effects. This meant reordering military priorities. As Supreme Allied Commander in World War II, General of the Army Dwight Eisenhower deplored any lack of cooperation among the Services. He knew firsthand how such destructive rivalry affected accomplishment of the most important missions, so as President, he was committed to eliminating any continuing vestiges of this problem. Such rivalry would seriously diminish the chance information would flow through inter-Service channels fast enough to prepare once an attack was detected. Second, and closely related to fostering the rapid flow of information, he had to find ways to provide Indications and Warning (IW) and Early Warning (EW).

IW was intelligence information about a potential adversary's preparations for major military operations. Such preparations would include, for instance, flushing strategic bombers to forward operational bases, erecting and fueling ICBMs, or simply increasing military communications. IW intelligence would point to a attack potential, but would probably not indicate the exact time of the attack. The most likely scenario held that the next step that could be observed would be the attack's getting under way with the strategic bomber force coming over the North Pole. An observation of the bombers on the way constituted EW intelligence—an attack was actually under way, but there was some time before the bombers reached their targets. Eventually, once the Soviet Union fielded its own ICBM force, the only EW indication of the attack would be the launch of the missile forces with only minutes until they were at their targets.

Consequently, IW intelligence held a high priority but presented a difficult set of problems. The peacetime watching of the Soviet Union had to go on routinely and frequently, but the Soviet Union had closed borders. Furthermore, IW meant watching events across the broad expanse of the Soviet landmass. Nearly as daunting was the timely EW observation of the ICBM launches. How could these crucial needs be met?

Implosion: Lessons from National Security, High Reliability Spacecraft, Electronics, and the Forces Which Changed Them, First Edition. L. Parker Temple III.

Since the end of World War II, these needs had been fulfilled in part by radar and listening posts around the periphery of the Soviet Union, but nothing fills the need for certainty in the same way a picture does. Consequently, Strategic Air Command, with the approval of the President, occasionally sent flights of strategic bombers into the interior of the Soviet Union to detect Soviet radars and air defenses and take pictures of Soviet military installations. Needless to say, such flights were intrusive, provocative, and dangerous to a high degree. Other penetration flights were conducted by Navy patrol aircraft and some Air Force fighters, but to lesser degrees than their large bomber brethren. Although none of the bombers was ever lost in the interior, numerous U.S. aircraft were intercepted, and either damaged or ultimately disappeared. Before the end of the Cold War, more than 20 U.S. aircraft were shot down.[1]

No less important was the EW detection of an attack under way. The earlier the detection, the more likely U.S. forces would survive and retaliate.

Of course, all of this presupposed the third of Eisenhower's priorities, which was to build the U.S. strategic forces to hold the Soviet Union at risk in case of a Pearl Harbor-like attack. And everything had to be done in a fiscally responsible way, which was his fourth related priority. Eisenhower was very concerned about an arms race and the necessary size of strategic forces. When he entered office, there were simply no precise answers to the "how much is enough" sorts of questions, in large part because there was insufficient intelligence about the situation inside the Soviet Union.

These priorities were inextricably tied together, and each affected advancing technologies. World War II's legacy tied security and defense to the most advanced technologies available. Without knowing the Soviet Union's actual capabilities, but with the frightening realities of their nuclear developments at a rate which seemed accelerating past the United States' own, nothing but the best and most advanced capabilities were acceptable. National survival hung in the balance.

*   *   *

Fearing the worst, the United States and its closest allies committed to a major increase in strategic nuclear forces. In addition to new bombers and ICBMs, the United States began to pursue sea-based ballistic missiles. The U.S. intelligence estimates considered the ICBM a capability where surely the United States still led the Soviet Union by four to five years.

To sort through these issues, Eisenhower depended more heavily on science advice than any of his predecessors (and most of his successors). He was fortunate

---

[1] R. Cargill Hall and Clayton D. Laurie, eds., *Early Cold War Overflights, 1950-1956, Symposium Proceedings, Vol. II,* (Washington, D.C.: National Reconnaissance Office, 2003), 22-23; L. Parker Temple, *Shades of Gray: National Security and the Evolution of Space Reconnaissance* (Reston, Virginia: American Institute of Aeronautics and Astronautics, 2005), 27–33. The full details of U.S. and British overflights in fighters, naval reconnaissance aircraft, strategic bombers, cargo aircraft, and even specially-modified aircraft demonstrates the great need for such intelligence as aircraft and aircrews flew in harm's way.

to have available a number of the finest minds working for him in support of the U.S. Office of Defense Mobilization's Scientific Advisory Committee (ODM/SAC).

ODM/SAC included current and future luminaries such as Detlev W. Bronk (National Academy of Sciences); William Webster (Research and Development Board); Alan Waterman (director of the National Science Foundation); Hugh Dryden (Interdepartmental Committee on Scientific Research and Development); James B. Conant (Office of Scientific Research and Development); Lee A. DuBridge (future Presidential Science Advisor); J. Robert Oppenheimer (father of the atomic bomb); and James R. Killian, Jr. (Massachusetts Institute of Technology (MIT). Several were from the original "best and brightest" of the General Advisory Committee of the Atomic Energy Commission and the Manhattan Project.[2] Among the brightest was Edwin Land, Polaroid camera inventor, an energetic genius and innovator who would affect the next three decades of U.S. national security in crucially important ways. Eisenhower elevated the ODM/SAC to be his own advisory group, redesignating it the Presidential Science Advisory Committee (PSAC). On 26 July 1954, though, as ODM/SAC's Technological Capabilities Panel (TCP), they began studying surprise attack prevention, led by Killian. They emphasized gathering intelligence on Soviet actions and intentions (that is, IW), forming a strategic nuclear force second to no other nation, and a continental defense capability relying on EW to exact a high price from any airborne attack.[3]

With such people involved, not surprisingly, results came fast. Even before they completed their study, at Land's urging, they seized a technological opportunity. Land chaired Panel Three, the Intelligence Panel. In November 1954, he convinced the group of the feasibility of a high-altitude reconnaissance aircraft that could overfly the Soviet Union undetected by Soviet radars. The U-2, as it became known, was the first of the specially designed, developed, and operated IW systems. At its start in 1954, U.S. officials expected the U-2 to fly so high (above 60,000 feet) and be so small that Soviet radars could not track it. With such an ability to evade detection, the U-2 would eliminate the need for the provocative bomber overflights. Less than two years later, on 4 July 1956, the U-2's first operational flight flown by the Central Intelligence Agency's Harvey Stockman crossed Poland and the Byelorussian Soviet Socialist Republic all the way to Leningrad.[4] Stockman's flight, to the U.S. officials' surprise, was detected and tracked by Soviet air defenses almost from takeoff. The Soviets tried and failed to intercept Stockman, and subsequently filed

[2] David Z. Robinson, "The Presidency and Science Advising: Before and After Kennedy," in *The Presidency and Science Advising*, vol. 2, ed. Kenneth W. Thompson (New York: University Press of America, 1987), 26; Herbert F. York and G. Allen Greb, "Military Research and Development: A Postwar History," *Bulletin of the Atomic Scientists* (January 1977): 13–26; James R. Killian, Jr., *Sputnik, Scientists, and Eisenhower* (Cambridge, Massachusetts: MIT Press, 1987), 63.

[3] James R. Killian, Jr., *Meeting the Threat of Surprise Attack: Report of the Technological Capabilities Panel of the Scientific Advisory Committee*, President's Scientific Advisory Committee, 14 Feb. 1955; Temple, *Shades of Gray*, 70–71.

[4] Temple, *Shades of Gray*, 69.

a protest with the U.S. The U-2 could clearly be only an interim measure, as the clock began to count down until a successful Soviet intercept.[5]

The TCP also recognized that the maturity of large missiles enabled a revolutionary IW system—a film-return reconnaissance satellite—in the relatively near term. This U-2 successor was the Corona spy satellite, known to the world as the Discoverer scientific research satellites, facts that remained highly classified for twenty-five years.[6]

Following closely on the heels of the Corona were several other satellites. Among these was MIDAS, an infra-red sensing satellite for detection of the rocket exhaust of large ICBMs, meeting the initial needs for EW. Another satellite series, known as SAMOS, investigated electronic intelligence and different methods of photographic reconnaissance for both IW and EW. Research into the blast phenomena of nuclear weapons allowed development of the VELA satellites, from whose extremely high altitude orbits even underground nuclear blasts could be detected worldwide.

Many of these early Cold War revolutionary capabilities became highly classified, known to only a small circle of the highest officials in the United States. The classification obscured what the systems were actually able to do, worsening the Soviet Union's ability to understand them and thus develop the means to counter them. But importantly, the Soviet Union's leaders, in theory, would never be sure just how well aware the United States' leaders were about the Soviet Union's actions. This was a key element of deterrence.

In 1955, newly promoted Major General (MajGen) Bernard A. Schriever, still WDD commander, held a significant portion of the strategic offensive developments with his responsibilities for reconnaissance satellite development and the ICBM.[7]

Donald A. Quarles, former Assistant Secretary of Defense for Research and Development, had become the Secretary of the Air Force. Before taking over the reins of the Air Force, he had also been deeply involved in another, nearer-term missile project: the Thor Intermediate Range Ballistic Missile (IRBM).[8] The Thor traced its roots to a meeting of the ODM/SAC in January 1955, and was soon part of Schriever's portfolio as well, as the United States and the United Kingdom

---

[5] G.W. Pedlow and D.E. Welzenbach, eds., *The CIA and the U-2 Program 1951–1974* (Washington, D.C.: Central Intelligence Agency Center for the Study of Intelligence, 1998), 72–73; Temple, *Shades of Gray*, 173. The U-2 was understood as an interim measure at its outset, but intelligence officials had no idea that it would prove to be as readily detected. The U-2 had flown in the Los Angeles Air Traffic Control airspace, and U.S. radars had not picked it up, creating false hope that the Soviets would be similarly unable. Their radars, however, were older and operated in different frequencies, and the U-2's operational lifetime proved much shorter than expected as a result.

[6] Temple, *Shades of Gray*, 173; Kenneth E. Greer, "Corona," *Studies in Intelligence*, Supplement 17 (Washington, D.C.: Central Intelligence Agency, Spring 1973).

[7] Jacob Neufeld, *The Development of Ballistic Missiles in the United States Air Force, 1945–1960* (Washington, D.C.: Office of Air Force History, U.S. Air Force, 1990), 121.

[8] Neufeld, *The Development of Ballistic Missiles*, 126.

agreed to deploy the missiles in England.[9] As important and broad ranging as his responsibilities were, Schriever's systems were only part of a larger set of U.S. developments.

The Army's responsibilities included some tactical ballistic missiles, the Jupiter IRBM, as well as the Nike defensive missile programs, starting with the Nike AJAX anti-aircraft missile. Ajax was intended to defend key areas, known as point targets because of their limited size. Point targets included military installations, industry, and government centers. Outgrowths of this program included the Nike Hercules for defense against high altitude aircraft, which might also include defense against ballistic missiles. By 1955, a specialized missile based on the Hercules, the Nike Zeus, began development as an anti-ballistic missile. Such developments were possible only because of advances in radar and specialized computers.[10] All, in turn, relied on increasingly advanced electronics.

Electronics, however, could be as much of a curse as a blessing, depending on their reliability.

* * *

One thing strategic offensive, defensive, and intelligence capabilities began demanding was *reliability*. Once the weapons or intelligence systems' capabilities were real, it had to become something on which the nation's security could depend. Reliability took on different forms, but the solutions would hold many features in common.

Reliability had not been a foremost concern of the systems' developers. The force driving rapid technological advance excused many faults, as long as each failure moved towards the needed goal. For example, Thor's high risk revealed itself when the first four launches ended in failure.[11] The fifth Thor flight on 20 September 1957 was the first completely successful launch,[12] and the first full range test took place on 24 October 1957.[13] The Atlas ICBM was legendary for its repeated failures before its first successful launch. That should not be surprising in light of its need to have over 300,000 parts all function perfectly and in proper sequence. Corona needed 14 launches before an image was returned. Zeus, depending as it did on very high acceleration forces, stressed to the maximum its vacuum tube avionics, and often fell short of achieving its objectives. Once these capabilities were in place, each had to be made reliable.

[9] L. Parker Temple and Peter L. Portanova, "Project Emily and Thor IRBM Readiness in the United Kingdom, 1955–1960," *Air Power History* 56, no. 3 (Fall 2009): 41.

[10] Temple, *Shades of Gray*, 103.

[11] Robert Jackson, *Strike Force: The U.S.A.F. in Britain since 1948* (London: Robson Books, 1986), 93.

[12] James J. Ripley, *The Thor History*, Douglas Report SM-41860 (Santa Monica, California: Douglas Aircraft Company, May 1962), 4.

[13] W.M. Arms, *Thor: The Workhorse of Space: A Narrative History MD.C. G3770* (Huntington Beach, California: McDonnell Douglas Astronautics Company-West, 1972), 3–4; Temple and Portanova, "Project Emily," 30.

In the case of the ballistic missiles, for instance, reliability translated to readiness to launch when commanded. The ICBM and IRBM had to be ready to launch nearly instantaneously with high reliability, and then to hit their targets accurately.[14] For the Air Force's Thor IRBM, this translated to a requirement to launch fifteen minutes after the start of a countdown.[15] The short time to erect, checkout, and fuel a missile was driven by the simplicity of the launch facility—the sites could not withstand any direct attack, nuclear or otherwise.[16] Therefore, they had to be fired quickly.[17]

Meeting such reliability was demanding. Reliable first attempt launches determined survivability rates, targets destroyed, and a myriad of other factors that went into planning SAC's Emergency War Plan.[18] This in turn affected the numbers of weapons, and controlling those numbers was on the President's agenda.

Of no less importance was the accuracy of the long range missiles once they were launched. Once in flight, little more could be done from the ground to affect the performance of the missile. A successful launch but a missed target was of little good, so the electronics controlling the missile in flight also had to be reliable enough to survive launch and continue functioning during the short flight of the missile.

Conventionally, improving reliability meant providing backup subsystems in the case of the loss of a primary. But vacuum-tube technologies were heavy and launch weights restricted, so redundancy was limited to only a few key subsystems.

All of these demands for reliability meant that some new technology was needed, and that was, of course, one of the things solid state electronics brought to bear.

Nike missiles were among the first beneficiaries. Nike Ajax's targets were 400–500 mile-per-hour bombers, but Zeus' missile targets moved faster than 3 miles each second. At the time, computers' dominant cost and performance metric was the computing second. Short warning and flight times precluded waiting for vacuum tubes to warm up to operating temperature.[19] Although Zeus could get targeting information from ground radars after launch, its very short time of flight demanded high speed calculations from vacuum tube-based mainframe computers. Computation speed was critical when everything depended on the fastest possible determination of where the threat was, which missile to fire, and what the best targeting

---

[14] Neufeld, *The Development of Ballistic Missiles*, 121.

[15] Ripley, *The Thor History*, 1.

[16] Arms, *Thor*, 3–7.

[17] Neufeld, *The Development of Ballistic Missiles*, 161.

[18] Lt.Col. Kenneth F. Gantz, ed., *The United States Air Force Report on the Ballistic Missile: Its Technology, Logistics and Strategy* (New York: Doubleday, 1958), 181. The Emergency War Plan, in the words of General Thomas S. Power, "covers the target system assigned to SAC and in turn assigns accomplishment of specified strategic missions to various elements of SAC's strike forces." The EWP evolved into the Single Integrated Operations Plan (SIOP).

[19] To reduce warm up times, some vacuum tubes contained heaters, but that had consequences for tube lifetime and maintenance. Solid state components were still faster than the pre-warmed tubes.

solution would be. Solid state devices were almost instantaneously on when power was applied. Not surprisingly, though, the earliest solid state devices did not actually replace all the functions of vacuum tubes, partly because vacuum tubes worked at faster speeds.

Thus, initial solutions compromised between the two different electronics technologies. Critical functions of getting targeting information from widespread radar sites to command centers depended on vacuum tube and solid state hybrid electronics. The Army's Nike anti-aircraft missiles depended on combining vacuum tubes in encoding and decoding stages with over 200 transistors in logic circuits.[20] Vacuum tube-solid state hybrids were necessary because solid state devices had not advanced far enough to completely replace some vacuum tube applications. Military needs for computation speed accompanied reliability, pushing solid state electronics development.

Impelled by the urgency of the threat to the United States, driven by a belief that technological solutions held the key to the future, and encouraged by the rapid advance of solid state componentry, the military fostered the solid state electronic parts industry. The Air Force's official report on ballistic missiles put it this way:

> *The ballistic missile program, superimposed on an already rapidly expanding electronic industry, had created an added impetus to expansion in specific fields. . . . Great strides were required almost immediately in the state of the art . . . to meet the meet the environmental conditions, high reliability standards, and accuracy requirements.[21]*

Pressure from military systems had collateral benefits. As Shockley, Bardeen, and Brattain earned the Nobel Prize in Physics in 1956,[22] TI began consumer electronics by manufacturing junction transistors for an advanced portable transistor radio.[23]

Solid state electronics were clearly not a panacea because not all electronic parts were created equal. In fact, the new technology's process controls were very difficult. In the mid-1950s, transistor production lot yields were about 10%.[24] Even at that, and considering the cost of finding which of the parts produced were good or bad, the industry very quickly began making significant profits. TI was making 1,000 semiconductors a day, and 10,000 other devices per month.[25] By 1957, annual U.S. production exceeded 30 million transistors, driving the individual piece cost drastically lower, but still providing annual sales greater than $100 million.[26]

---

[20] Michael Riordan and Lillian Hoddeson, *Crystal Fire: The Invention of the Transistor and the Birth of the Information Age* (New York: W.W. Norton, 1997), 202–203. The hybrid device was designated the AN/TSQ.

[21] Gantz, *The United States Air Force Report*, 166–167.

[22] Riordan and Hoddeson, *Crystal Fire*, 6.

[23] Ibid., 7.

[24] Ibid., 211.

[25] Ibid., 230.

[26] Ibid., 254.

In 1957 and 1958, the electronics industry mushroomed, and the overall demand by the missile programs became minor. Total market penetration, though, hardly tells the whole story. For some part types, the third generation ICBM program was actually the dominant buyer of high technology extreme performance parts. This also allowed advanced technology processes to be fed back to commercial products, affecting the larger industry. For instance, the ballistic missile report went on to say, "Possibly the greatest challenge to the industry is the need for developing and manufacturing electronic equipment capable of continuous, accurate performance under the extremely harsh environmental conditions in ballistic missile operations. Acceleration, altitude, speed, vibration, heat and other environmental requirements are radically different than those we have had to cope with in other military programs." However, for certain kinds of parts, that was not the case (as the demand for solid state componentry would soon show).[27]

Of all the parts of the missiles that raised concern for reliability, the guidance systems had the greatest impact. When enumerating the sources of problems bearing on the mission effectiveness of early long range missiles, SAC Commander-in-Chief General Thomas S. Power enumerated eight factors ranging from human error to inaccurate targeting information. Most of them related to trajectory calculation and "malfunctioning during flight of one or more of the thousands of delicate components."[28] Continued dependence on delicate electronics was therefore out of the question. Robust, reliable, fast solid state electronics held vast potential for long range missiles. That segment of the overall market, though small for the moment, was crucially important for national security and national defense.

[27] Gantz, *The United States Air Force Report*, 167–168; Thomas J. Misa, "Military Needs, Commercial Radios, and the Development of the Transistor, 1948–1958," in *Military Enterprise and Technological Change, Perspectives on the American Experience*, ed. Merritt Roe Smith (Cambridge, Massachusetts: MIT Press, 1987), 284; Stuart W. Leslie "How the West Was Won—The Military and the Making of Silicon Valley," in *Technological Competitiveness: Contemporary and Historical Perspectives on the Electrical, Electronics, and Computer Industries*, ed. William Aspray (New York: Institute of Electrical and Electronics Engineers IEEE Press, 1993), 80. By contrast to the total number of transistors are numbers on the part of Western Electric, the leading manufacturer through the mid-1950s, which was produced 1.3 million transistors in 1958, of which more than half, 693,000, were for military application. Western Electric's proportion of sales to the military was typical of the whole electronics industry. The greatest impact by 1958 was not in the components themselves, but in the demands on test equipment.

[28] Gantz, *The United States Air Force Report*, 183.

Chapter 4

The Final Frontiers

Satellites were about to become the heralds of the opening of a new frontier. However, this was not the only new frontier. Strategic missile systems were opening new frontiers in electronics reliability. These new frontiers were related in important and fundamental ways.

A new cause of great demand for highly reliable electronics arrived in 1957, but before the Soviet Union launched the first artificial satellite. By the spring of 1957, Air Force research showed a solid-fuel ICBM was possible. Such a third generation missile would have tremendous potential beyond the liquid-fueled Atlas and Titan. A solid fuel missile, for instance, had no need for fueling prior to launch. Neither did it require ground handling nor have any of a litany of other shortcomings that made Atlas merely an interim ICBM. With solid state electronics, such a third generation missile would literally be ready to go on a moment's notice.

In August 1957, the Air Force Ballistic Missile Division's (AFBMD, successor to WDD) Col. Edward N. Hall, Chief of Propulsion Development, was asked to develop a medium-range, solid-fuel missile to be the land-based counterpart to the Navy's submarine-launched solid-fuel Polaris.[1] Within two weeks, Hall drew up specifications for a remarkable three-stage solid-fuel missile that would be inexpensive to build and maintain.[2] Learning from the Atlas and Titan, the new

---

[1] Robert L. Perry, *The Ballistic Missile Decisions*, Rand Pamphlet P-3686 (Santa Monica, California: RAND, October 1967), 17–18; Robert L. Perry, "The Atlas, Thor, Titan and Minuteman," in *The History of Rocket Technology*, ed. Eugene M. Emme (Detroit, Michigan: Wayne State University Press, 1964), 155; Neil Sheehan, *A Fiery Peace in a Cold War* (New York: Random House, 2009), 409–415; Joseph Albright and Marcia Kunstel, *Bombshell: The Secret Story of America's Unknown Atomic Spy Conspiracy* (New York: Times Books, 1997), 248. Perry described Hall as a "near fanatic" about solid fueled rockets, and gives extensive credit to Hall and the work he promoted for solving the problems associated with such rockets. The story of Hall's dismissal from developing the Thor and how he went about developing the Minuteman are covered in detail in Sheehan. As a sidelight, Edward Hall's brother, Theodore Alvin Hall, was one of the most damaging, and uncaught, atomic spies who penetrated the Manhattan Project.

[2] Major George A. Reed, "U.S. Defense Policy, U.S. Air Force Doctrine and Strategic Nuclear Weapon Systems, 1958–1964: The Case of the Minuteman ICBM," dissertation, George Washington University, SAF/PAS Document 87-0202, 1986, p. 55.

Implosion: Lessons from National Security, High Reliability Spacecraft, Electronics, and the Forces Which Changed Them, First Edition. L. Parker Temple III.

missile's maintenance depended on high reliability and a "remove and replace" approach to achieve a near 100% alert rate.[3] The deadline imposed on Schriever's group meant the missile had to be mass produced, tapping into the American ability to gear up manufacturing capabilities on a broad basis. Hall chose to designate the as-yet-unnamed missile Weapon System Q, because it was one of the few letters of the alphabet unused for weapon system designations. Weapon System Q "was the first strategic weapon capable of true mass production."[4]

At the same time, WDD had responsibilities for an ICBM byproduct. Any missile sufficiently powerful to deliver a nuclear warhead to the other side of the world could also launch a small satellite into orbit. The possibilities for artificial satellites had been under study in the Services since the end of World War II but were made feasible by the ICBM. Both the Soviet Union and the United States had announced intentions to launch satellites in 1958, during the eighteen-month International Geophysical Year.

Schriever could not dilute the effort dedicated to developing the ICBMs without jeopardizing the urgent necessity of an operational missile in 1960. Thus, in 1957, he expanded the role of Ramo Wooldridge to include systems engineering support to the satellite programs. To accommodate this, Ramo Wooldridge created the independent Space Technologies Laboratory (STL). STL had the Air Force's trust, and received access to both government information and other contractors' proprietary data (as they were not in competition with anyone). Within the year, Ramo Wooldridge and STL merged with Thompson Products to form TRW, Incorporated.[5]

The U-2's intelligence created a disparity in public and government awareness of the world situation. Until the U-2, Eisenhower had pushed hard to advance U.S. strategic capabilities, IW and EW. Preventing a nuclear Pearl Harbor meant, in the absence of better information, and in light of the dire U.S. intelligence estimates, he had to assume that the Soviets were either equal to or possibly pulling ahead of the United States. The U-2 dramatically changed the perceptions of the President and his top advisors. Eisenhower had used the U-2 photography to moderate the rate of the U.S. strategic arms buildup. Not only were the Soviets not ahead, the United States had seized the initiative during World War II, and never relinquished it. Not even close. What the President and his advisors knew, though, could not be mentioned outside the highly classified circles aware of the U-2.

The disparity became painfully obvious on 4 October 1957, when the Soviet Union launched Sputnik, the world's first artificial satellite. The fact of the Soviet feat was not a surprise to the American Intelligence Community, which had actually expected it for several months. Intelligence from the highly classified U-2

---

[3] Ibid., 58.

[4] Ibid., 58; Sheehan, *A Fiery Peace*, 411.

[5] Ivan A. Getting, *All in a Lifetime: Science in the Defense of Democracy* (Los Angeles, California: Vantage Press, 1989), 386; U.S. Congress, Office of Technology Assessment, *A History of the Department of Defense Federally Funded Research and Development Centers* (Washington, D.C.: U.S. Government Printing Office, June 1995), 25. STL had a separate board of directors from TRW as a result of efforts to ensure their independence, and therefore the protection of other contractors' proprietary information.

reconnaissance aircraft had alerted top U.S. officials to the Soviets' progress. The only unknown was the specific date on which the Soviet Union would launch their satellite. The American public's reaction to Sputnik was another thing altogether.[6]

Sputnik put into sharp relief an apparent slow pace of U.S. missile developments. Overnight, a sense of urgency seized the American psyche—we could not allow the Soviets to be ahead of us in any sense. Strategic missiles became the subject of an intense push to surpass the perceived Soviet lead. Although 1954, the year of maximum danger, had uneventfully come and gone, the danger of a surprise attack suddenly seemed much greater. Missiles of the size that had launched Sputnik, and its much larger successor, Sputnik II, clearly had the ability to rain nuclear weapons on the United States.

This sense of urgency aided Hall's and Schriever's missile efforts. On 8 February 1958, they presented Weapon System Q to the Secretaries of the Air Force and Defense. Within 48 hours, they got approval to develop the new missile. They immediately renamed the project. On 28 February 1958, the New York Times reported the Air Force would "produce an advanced type of ballistic missile . . . called Minute Man."[7]

In July 1958, AFBMD began selecting contractors. In October 1958, Boeing was under contract to assemble and test the new missile.[8] Minuteman's guidance and control systems went to the Autonetics Division of North American Aviation in Downey, California.[9] Although the final production total was not decided until 1964, the production run of these missiles would be enormous by previous standards, eventually reaching 1,000. That translated into a large demand for electronic parts meeting challenging requirements for readiness, responsiveness, and nuclear survivability. But it was Minuteman's compulsion to obtain the highest attainable reliability that affected the electronics of that day and decades thereafter.

Understanding the impact of Minuteman on the electronics industry requires an understanding of the reliability movement.

*   *   *

Sputnik and its repercussions fueled a national angst over the quality of American education. In addition to the already heavy demand for scientists and engineers driven by the Cold War's strategic arms buildup, the space race's new sense of urgency created an acute need for more scientists and engineers. In order to open the spigot producing more scientists and engineers, a new means of education had to be found. The result was a nationally-revised curriculum in science and math. Whereas the strategic buildup and the space race initially rested on the shoulders

---

[6] Temple, *Shades of Gray*, 95, 97–99.

[7] Sheehan, *A Fiery Peace*, 415; Brig Gen Schriever to Prof. Dwight F. Gunder, 15 June 1955, in Reed, *U.S. Defense Policy*, 53. Schriever had been unexcited about solid rocket motors as recently as 1955, insisting that they might be useful for intermediate range missiles and smaller warheads. Hall's brilliant success had caused Schriever to reverse his position by 1957.

[8] "History of the Minuteman Missile," National Park Service.

[9] Ibid.

of the engineers and scientists from World War II, the long term space race would be sustained by this new influx. Despite projections of needs, in a free society, individuals get to choose what careers they will follow. Unanswered, for the moment, was how many engineers and scientists this emphasis would generate, and whether they could all be gainfully absorbed into and sustained by existing and future programs.

The fifteen years between World War II and the arms buildup and space race put in motion a feast-or-famine effect for scientists and engineers. They learned by doing, or as apprentices to those who had learned through experience. As long as scientists and engineers from World War II and immediately post-War were still plentiful, no urgent need existed to document the lessons they learned. But the feast would inevitably wane through retirements and other attrition. Without these skilled people, apprenticing would no longer be possible. Since these scientists and engineers possessed the core of understanding and executing high reliability systems, the famine that would inevitably follow would affect retaining their knowledge. Some means had to be found to document what they had learned. A cyclical forgetting effect between feasts fed the need for new standards, not so much to tell people who already knew how to do things correctly, but to inform technically trained though inexperienced people. For the moment, documenting lessons learned did not have the priority to gain resources. That would shortly change in response to industry needs, program costs, rising system complexity, and proliferation of highly specialized knowledge.

Reliability was one emerging specialized knowledge area. The large missile programs attracted much of the best talent as they got under way, and it was there that the pressure for improving reliability became most apparent. Experience quickly rose, and with experience came new understanding of reliability. Part of knowing how to do things correctly to meet high reliability requirements depended on clear, shared, and technically correct measures of what reliability meant.

Consequently, in 1957, Edward J. Nucci, from the Office of Electronics in the Office of the Assistant Secretary of Defense (Research and Engineering), took on the task of determining solid state electronics reliability. Although rudimentary by current standards, the level of complexity of electronic systems was growing, and reliability testing was a significant unknown. Nucci had started the Advisory Group on the Reliability of Electronic Equipment (AGREE) on 21 August 1952 under a charter from the Research and Development Board (RDB) (as an agent of the Committee on Electronics). AGREE pulled together an approach to specifying reliability requirements, testing, and verification procedures.[10]

AGREE's reporting chain modified as the DoD reorganized and shuffled functions, but it remained a high level, visible agency and kept true to its purpose "to monitor and stimulate interest in reliability matters and recommend measures which would result in more reliable electronic equipment."[11]

---

[10] *Reliability of Military Electronic Equipment*, Office of the Assistant Secretary of Defense (Research and Engineering), June 1957, p. iii.

[11] Ibid.

By 1955, AGREE determined sufficient knowledge was available, and sufficient interest aroused, to take steps "toward quantifying reliability requirements and toward developing suitable tests to verify that such requirements are met." This was the beginning of the discipline of reliability engineering, as the group organized nine task groups covering numerical requirements, tests, design procedures, components, procurement, packaging and transportation, storage, operations, and maintenance.[12]

Long before consideration of high reliability could begin, reliability itself had to be quantified meaningfully and consistently. Indeed, at the time, the later understanding of high reliability would have been completely mysterious due to the lack of a consistent approach to reliability determination. Reliability calculations for failure rates owed much of their heritage at the time to R.R. Carhart of RAND Corporation. In 1953, he developed a straightforward formula allowing reliability calculations for entities from components to systems. This work was applied to aircraft systems for the most part, but was ideally suited for the emerging electronic systems in missiles. Calculation methods were necessary but insufficient, because calculations require data. Thus, AGREE was to develop "minimum acceptability figures for reliability of the various types of military electronic equipment . . . expressed as 'time between failures' or some other truly quantitative measurement." This included the basic requirements for reliability testing.[13]

In 1955, RADC's Joseph J. Maresky took a major step towards a cohesive reliability methodology when he published *Reliability Factors for Ground Electronic Equipment*.[14] This was one of the first publications establishing the discipline of reliability engineering.

AGREE's task groups generally settled on determining reliability at the equipment (or box) level, because that was where any meaningful data existed. Much of their information derived from field return reports, or the data supplied from a unit pulled from its weapon system due to some failure. Even then, complexity was an issue. Task Group 1, for instance, excluded the Semi-Automatic Ground Environment (SAGE) and the Naval Tactical Data System from consideration. SAGE used large computers to coordinate air defenses over a wide area, and the Tactical Data System was similarly based on large computers.[15] Complexity made them too hard to handle.

---

[12] Ibid.; William Denson, "The History of Reliability Prediction," *IEEE Transactions on Reliability* 47, no. 3-SP (Sept. 1998): SP-321.

[13] *Reliability of Military Electronic Equipment*, iii, vii.

[14] "Pioneering Reliability Laboratory Addresses Information Technology," *Journal of the Reliability Analysis Center* 6, no. 1 (1998): 1.

[15] Robert Buderi, *The Invention That Changed the World: How a Small Group of Radar Pioneers Won the Second World War and Launched a Technical Revolution* (New York: Simon and Schuster, 1996), 381–406 *passim*; *Reliability of Military Electronic Equipment*, 4. Buderi reports each initial SAGE site contained two IBM FSQ-7 digital computers, weighing 275 tons apiece, with 50,000 vacuum tubes and over 700,000 solid state resistors and diodes, over a thousand miles of cables.

Not all the task groups decided boxes were the proper level of analysis, however. Task Group 4 was to "investigate and recommend methods of specifying development procedures to ensure that equipment designs have the required inherent reliability." They decided, in addition to looking at equipment, to include "selection, qualification and application of components for specific circuit and environment requirements."[16] They recognized that real reliability began at the component (or piece part) level. While piece part reliability data might be more difficult to collect (because much of the data was reported at the box level), their scope included specifying how to make or buy the best parts.

The next task group had responsibility for quantifying piece part performance. Task Group 5, in addition, was to establish

> *criteria and methods for specifying the reliability of equipment parts and tubes in terms of failure rate as a function of time and environment, since this is considered essential to a determination of the amount of improvement demanded to meet the over-all reliability requirements of electronic equipments.*[17]

Task Group 5 was also notable because they went on to explain that government regulations were inadequate to meet their proposals, and would need redrafting. Their assessment was not as negative as it appears. They were blazing a new trail, driven by new demands for reliability across the board, but especially in missiles. Their path combined both new theoretical approaches and hard, pragmatic engineering. So,

> *[t]o be helpful to the equipment designer, information on failure rates of components must include (1) the percent that fails per unit time, (2) the critical failure mode, or modes, or the parameter change to which this failure rate applies and (3) the relationship of this failure rate to the environment.*

To organize and keep track of this, Task Group 5 recommended a new permanent DoD organization comprising personnel from industry and the Services. Membership qualifications covered "the fields of research, development, standardization, procurement and quality assurance." The group had the task "of developing military equipment specifications, testing component parts for design capability and developing inspection methods." It was also to accumulate reliability data over time from usage and failure reports.[18]

Although some task groups had gone to the piece part level, they could agree on reliability definitions only for equipment. Quantitative reliability requirements were understood and defined as "the numerical probability of failure-free system or equipment operation during a described time interval under described conditions of performance and environment." Focus on that level drove qualification and acceptance of the equipment to reside at the box level, not piece part.[19] They clearly knew

---

[16] *Reliability of Military Electronic Equipment*, 4.

[17] Ibid.

[18] Ibid., 5; Denson, "The History of Reliability Prediction," SP-321.

[19] *Reliability of Military Electronic Equipment*, 10.

this was not adequate for the long term since reliability starts at the individual piece part. However, practicality and feasibility forced their immediate concentration to the box level.

Their *Reliability of Military Electronic Equipment* was published in June 1957. This provided the basis on which common reliability calculation could be used to support adequate spares provisioning, comparison of prospective new equipment reliability, and many similar uses. These in turn supported fielding weapon systems whose maintenance could be anticipated and provided for, improving operational readiness.

*   *   *

For Eisenhower and Defense Secretary Charles E. Wilson, ending inter-Service rivalry was a high priority. During the AGREE study's efforts, Wilson dealt a severe (but not decisive) blow to the Services' intense inter-Service rivalry. His November 1956 memorandum, "Clarification of Roles and Missions to Improve the Effectiveness of Operation of the Department of Defense," included some hard decisions regarding responsibilities for missiles.[20] Air Force primacy in ICBMs was not an issue. Multiple IRBM programs were his main concern. Specifically, the Air Force's Thor and the Army's Jupiter each had similar performance requirements. His most important decisions assigned longer range missiles, including all land-based IRBMs, to the Air Force, while ". . . enjoining the Army from planning operational employment of missiles with ranges beyond 200 miles." The Navy had neatly dodged the issue, since Submarine Launched Ballistic Missiles had not been categorized as either IRBMs or ICBMs, and in any case represented a unique component of the new nuclear triad (strategic bombers, land-, and sea-based missiles). His decision also meant very large missiles were not long-range artillery for the Army. At least for missiles, the reason for inter-Service rivalry had been removed.

The long-range ballistic missiles topped the National Security Council's January, 1958, ranking of DoD's priorities for research and development:

1. Atlas;
2. Titan (a second-generation Air Force ICBM);
3. Thor-Jupiter IRBM;
4. Polaris;
5. Anti-missile missile weapon system;
6. Vanguard, Jupiter-C (the first U.S. space launch vehicles);
7. Reconnaissance and early warning satellites.[21]

---

[20] Charles E. Wilson, "Memorandum for Members of the Armed Forces Policy Council: *Clarification of Roles and Missions to Improve the Effectiveness of Operation of the Department of Defense*," 26 Nov. 1956.

[21] Robert Cutler, "Memorandum: *Scientific and Reconnaissance Satellites*," 20 Jan. 1958; NSC Action 1846, "Priorities for Ballistic Missiles and Satellite Programs," Ann Whitman File, Administrative Series 17, Eisenhower Library; NSC 1840, JCS 1731/252.

Grouping the Thor and Jupiter missiles together reflected Wilson's decisions. But the sixth item on the list reflected a Cold War consideration left over from before Sputnik. Vanguard was a scaled-up sounding rocket with essentially no direct military ancestry. This was a vestige of Eisenhower's determination that the first U.S. satellite, if it beat the Soviets, would have to be clearly non-military. The very troubled Vanguard's inability to "beat the Russians" to the first satellite led to the rapid and expedient choice of the Army's Jupiter-C as the first U.S. rocket to launch a satellite. It did so on 31 January 1958, eleven days after the prioritized list was published. Despite its misleading name, the Jupiter-C was a version of the Army's Redstone tactical missile. At the time, no one seemed willing to ask how a limited-range Army missile, given the Wilson limitations, could possibly perform a space launch.

None of the first three priority missiles were yet linked to launching satellites (though planning for exactly that was well under way). In effect, though without intent, Wilson had also decided which Service would have the most powerful role in providing access to space. Vanguard's failure and the Soviet use of an ICBM to launch Sputnik meant that long-range missiles were soon to become the main U.S. launch vehicles.

At the same time, Eisenhower took direct aim at other elements of inter-Service rivalry. His far-reaching DoD reorganization proposals were working through Congress. The Service chiefs knew what he was doing and with considerable effort, the Services began coming around to his view that inter-Service rivalry was to be minimized. He knew that the real national defense and national security satellites were being developed for launch aboard the Air Force's large missiles. However, he faced an unexpected and unwanted organizational quandary.

As a result of the public (and, hence, congressional) furor over Sputnik, a governmental reorganization was taking place to create a new space agency, much against Eisenhower's wishes. All the work on satellites and launch vehicles was in the military. To Eisenhower, space was too new to warrant a new agency that he considered unnecessarily duplicative and fiscally irresponsible. Moreover, with the reality of so much effort already ongoing in the military, he envisioned a new agency as one more source of rivalry and destructive competition. Nevertheless, he bowed to political realities. The result was the National Aeronautics and Space Administration (NASA), called into being when Eisenhower signed the Space Act into law on 29 July 1958.

As though to underscore the point of his concerns, a week later, his plans to reorganize the DoD became reality when the Defense Reorganization Act of 1958, Public Law 85-599, was signed on 6 August 1958.[22] The 1958 Reorganization clarified civilian lines of control and direction, further strengthening the Office of the Secretary of Defense. Eisenhower was not completely satisfied with some of the

---

[22] Ronald H. Cole et al., *The History of the Unified Command Plan, 1946–1993* (Washington, D.C.: Joint History Office, Office of the Chairman of the Joint Chiefs of Staff, 1995), 187; *New York Times* (24 July 1958): 1:1, *New York Times* (25 July 1958): 7:5; *New York Times* (7 Aug. 1958); PL 85-599, Department of Defense Reorganization Act of 1958, 85th Congress, 63 Stat 579.

congressional compromises, as he saw that features of the bill kept the Services from moving toward being "fully unified." He had intended to strengthen the operational role of the Joint Chiefs of Staff and the unified commands at the expense of the individual Services. The role of the Services was to organize, train, and equip forces. War fighting resided with the operational commands. Eisenhower wanted to dull inter-Service rivalry stemming from weapon system acquisition budgets by giving operational military commanders a greater say in the equipment they would be called upon to use in times of crisis or conflict.

He never intended to do away with the individual Services. He wanted to subordinate them more effectively to civilian and Joint Chiefs of Staff control. In his view, the more "joint" or "unified" the Services acted, the more they would learn to work together and eliminate destructive rivalry. Forcing greater cooperation in the "equipping" mission of the Services would allow benefits through economy of scale in buying items used commonly across the Services. A collateral benefit of Eisenhower's reorganization was to commonality, an important basis and justification for standards dating back to JAN-1 in World War II. Eisenhower's call for more unity was having an effect, and being felt by the AGREE in an important way.

*   *   *

AGREE, an important step toward high reliability electronics, was inadequate by itself. AGREE's strengths included comprehensive treatment of reliability, setting the stage for what was to come. Its weakness, due to practical limitations in theory and data availability, left open the topic of piece part quality and reliability.

Piece part reliability was handled mainly after the fact, which is to say, reliability was not designed into parts. The reason was straightforward: no data was available to accurately predict a part's reliability behavior. "Using traditional methods of measuring part-reliability, the results of testing one part could not be extended to newer parts and environmental conditions."[23] Solid state electronics' new materials, phenomena, and physics exacerbated problems with traditional reliability methods. Traditional methods were not up to the challenge of supporting the reliability demands of the coming range of strategic weapons. On that point, there was wide agreement. What was lacking was consensus on how to proceed.

In November 1957, however, the lack of consensus began to change. Robert Lusser of the Army's Ordnance Missile Laboratories at Redstone Arsenal in Huntsville, Alabama, published a seminal work entitled "Unreliability of Electronics— Cause and Cure." In one Army missile system, 60% of the failures were due to electronic parts. He made a strong case that traditional concepts for electronics reliability were inadequate. The value of the paper was the effect it had on the thinking of the reliability community, whose focus began to shift to methods

---

[23] George H. Ebel, "Reliability Physics in Electronics: A Historical View," *IEEE Transactions on Reliability* 47, no. 3-SP (Sept. 1998): SP-379.

depending on the physics (to include chemistry) of the parts, rather than on field-return data.[24]

The move toward consensus and the AGREE report in turn spawned the Ad Hoc Study on Electronic Parts Specification Management for Reliability. The new study was co-sponsored by Paul H. Riley, Director of Supply Management for the Assistant Secretary of Defense (ASD) for Supply and Logistics (S&L), and James M. Bridges, Director of Electronics for the Director of Defense Research and Engineering (DDR&E). Bridges had been the AGREE Chairman. The study was created by a 14 July 1958 memorandum of agreement signed by Bridges and the Director for Production Policy in the Office of the ASD(S&L).[25]

This study emphasized six areas grouped into three sets of topics:

(1) Specification-related topics covering reliability of electronic parts and tubes in terms of failure rate over time, and the preparation and coordination of parts and tube specifications;

(2) Testing-related topics covering reliability testing, test procedures for determining compliance with reliability specifications, adequacy of reliability testing requirements, and qualification testing requirements and specifications; and

(3) Methods for exchanging technical characteristics and data on parts and tubes.[26]

The Ad Hoc group met for the first time on 30 October 1958 under the chairmanship of Bell Laboratories' Paul S. Darnell.[27] Darnell had recently published his influential projection of electronic components trends. He projected key parameters of solid state devices versus time, extrapolating into the 1970s. His charts (which proved well off the mark because of unanticipated advances in the underlying physics and materials sciences), covered projections of feature sizes and operating temperatures across several device types, and may have helped to gain attention for his assumption of the chair in the Ad Hoc group. It is also interesting to note the full name of the study group did, at times, include "and Tubes," but this was left out of the final study report. The Ad Hoc Study Group worked over the next year on its report, which would not be published until May 1960.

*   *   *

[24] Ibid., SP-380. Ebel notes another important influence had to do with the failure modes of electrical interconnections. "Electrical Contacts," *Proc. Holm Seminar on Electric Contact Phenomena*, 1955 (first meeting), laid out the majority of failure mechanisms for such interconnections, and served as the industry resource for many years thereafter as the conference continued.

[25] At the time, DDR&E was still Assistant Secretary of Defense (Research and Engineering).

[26] E. Nucci, "Progress Report on Ad Hoc Study on Parts Specifications Management for Reliability," *IRE Transactions on Component Parts* 6, no. 3 (Sept. 1959): 128–143.

[27] Paul S. Darnell, "History, Present Status, and Future Developments of Electronic Components," *IRE Transactions on Component Parts* (Sept. 1958): 124–129.

Meanwhile, competition and profitability spurred rapid technological advancement. As Hall put the finishing touches on Weapon System Q/Minuteman, and production of discrete electronic piece parts (e.g., transistors and diodes) proliferated, weapon system reliability demands inspired the development of "monolithic integrated circuits." "Monolithic" meant that multiple discrete parts were made from single pieces of materials such as silicon or germanium.[28] Integrated circuits, in general, represented miniaturization of larger discrete devices (such as diodes, transistors, rectifiers) using new physical phenomena such as the quantum behavior of electrons described by Wilson in 1931.[29] Integrated circuits eventually became synonymous with integrated electronics, which could be defined as "all the various technologies which are referred to as microelectronics . . . as well as any additional ones that result in electronics functions supplied to the user as irreducible units."[30]

The first such device, a 1 kHz oscillator, was developed by TI's Jack S. Kilby, and was demonstrated on 12 September 1958.[31] This was a landmark event because Kilby demonstrated the integration of a complete circuit containing the equivalent of diodes, transistors, resistors, and capacitors in one piece of silicon. His "solid circuit" was slightly less than half an inch wide.[32] In the fall of 1958, TI announced the "solid circuit" to the Services. Given the realities of the modern age, one might expect the announcement to have been greeted warmly, but that was not the case.

The Services had individually committed themselves to preconceived approaches to integrated circuitry. The mixed reaction reflected the Services' perceptions of their unique computational needs, environments, and a remnant of rivalry.[33] However, in this particular case, the multiplicity of approaches across industry actually facilitated

---

[28] A formal definition for an integrated circuit is "[a] circuit in which all or some of the circuit elements are inseparably associated and electrically interconnected so that it is considered to be indivisible for the purposes of construction and commerce." *Terms, Definitions, and Letter Symbols for Microelectronic Device*, JEDEC standard JESD99B, May 2007.

[29] Specifically defining what a discrete device is can be complicated. A formal definition is "[a] semiconductor device that is specified to perform an elementary electronic function and is not divisible into separate components functional in themselves. . . . Diodes, transistors, rectifiers, thyristors, and multiple versions of these devices are examples. Other semiconductor structures having the physical complexity of integrated circuits but performing elementary electronic functions (e.g., complex Darlington transistors) are usually considered to be discrete semiconductor devices. . . . If a semiconductor device is not considered to be an integrated circuit in both complexity and functionality, it is considered to be a discrete device." *Terms, Definitions, and Letter Symbols for Microelectronic Device*, JEDEC standard JESD99B, May 2007.

[30] Gordon E. Moore, "Cramming More Components onto Integrated Circuits," *Electronics* 38, no. 8 (19 April 1965): 114.

[31] Jack S. Kilby, "Invention of the Integrated Circuit," *IEEE Transactions on Electron Devices* ED-23, no. 7 (July 1976): 650; Riordan and Hoddeson, *Crystal Fire*, 259. The first semiconductor circuit, to prove his theories, was assembled and demonstrated to Kilby's manager, Willis Adcock, on 28 Aug. 1958. However, the circuit was not integrated, so Kilby set about making an oscillator circuit, and had fabricated three by 12 Sept. Kilby demonstrated a circuit called a "flip-flop" a week later.

[32] Patrick E. Hagerty, *Management Philosophies and Practices of Texas Instruments* (Dallas, Texas: Texas Instruments, 1965), 124.

[33] Riordan and Hoddeson, *Crystal Fire*, 255.

rather than interfered with progress. In all, five basic approaches existed to integrating circuitry, with no obvious winner at the time:

(1) modularization;

(2) thin films;

(3) hybrid thin film/discrete circuits;

(4) "molecular electronics"; and

(5) monolithic semiconductor circuits.[34]

There was no clear path to solve the Services' desires. The Navy wanted reliability and sought it in several ways. The Army wanted reliability, ruggedness, and ease of repair from its Micro-Module programs' advanced manufacturing technologies. The Air Force wanted small devices out of its single-crystal Molecular Electronics program. The emergent space programs had to have miniaturization first and foremost, but closely followed by reliability.[35]

The Navy's thin film technologies built on the World War II silkscreen masking techniques of proximity fuzes at the Army's Diamond Ordnance Fuze Labs. For passive devices, the work proceeded rapidly. Using commonly available materials such as glass, ceramics, plastics, and metals, conductive pastes were silk-screened onto a surface to form basic passive devices such as resistors and capacitors. Ceramic wafers with between "one and four passive components were stacked and interconnected with vertical riser wires. A tube socket was mounted above the assembly to that each module was a complete functional unit." The assembly took the name of a popular child's assembly toy, and was known as "Tinkertoy." For a while, the approach was quite successful, with over 5 million modules produced by Illinois Tool Works. The thin film advances stalled, for the time being, with attempts to develop active elements (such as transistors).[36]

Another form of integration approach funded by the Navy was a hybrid circuit variant. It used thin film passive elements and conventional semiconductor active elements. For these devices, the goal was more about high quality at a low price than about miniaturization.[37]

The Army Signal Corps' Micro-Module program was a conservative approach funded from 1957 to 1963. Although RCA was the main contractor for the Micro-Module program, it required "the skills of the entire electronics industry."[38] Micro-Module "involved a multiplicity of large and small microelectronics manufacturers. The compartmentalization of labor and the need to coordinate components' physical

[34] Daniel Holbrook, *Business and Economic History* 24, no. 2 (Winter 1995): 151.

[35] Ibid., 148; Riordan and Hoddeson, *Crystal Fire*, 255.

[36] Holbrook, *Business and Economic History*, 155; Kilby, "Invention of the Integrated Circuit," 648; R.L. Henry and C.C. Rayburn, "Mechanized Production of Electronic Equipment," *Electronics* 26 (Dec. 1953): 160–165.

[37] Holbrook, *Business and Economic History*, 156; Kilby, "Invention of the Integrated Circuit," 648.

[38] W. Walter Watts, "Tools of the Space Age," *RCA Signal* (March 1959): 55.

and performance parameters required participating firms to communicate technical and scientific information rationally and openly." As a consequence, this facilitated the spread of solid state expertise and the advance of technology.[39]

Kilby later recalled the Army Signal Corps was interested in his invention, as long as the TI approach could be fit within the Micro-Module efforts. TI had demonstrated an n-p-n device, but the Army wanted a p-n-p. The manufacturing techniques for p-n-p proved to be harder than expected, so delivery of the device was delayed, during which time the Micro-Module program itself ran into serious problems.[40]

The Army's interests also ranged more widely. Since the number of connections was a source of unreliability, the Army sought methods to address this. Results from hand-soldering, even by skilled technicians, varied considerably. The Signal Corps developed a new process, dubbed Auto-Sembly, which used a copper foil interconnection pattern into which component leads were inserted, and the whole assembly dipped into solder. With refinement, the process eventually became the standard process for making printed circuit boards.[41]

The Air Force's Molecular Electronics stood at the opposite end of the spectrum from the conservative modular technologies. The initial work investigated diverse topics in "semiconductor, magnetic, metallic, and crystalline materials," and their combinations. As the name implies, the hope was to leap to the level of molecules for miniaturizing electronics. The main contractor, Westinghouse, claimed they pursued this path "because the firm had fallen behind in conventional microelectronics technology and so wanted to 'leapfrog' directly to the next generation technology." The Air Force, though, said Molecular Electronics started at the Air Force Materials Laboratory at Wright Patterson AFB in 1953. Molecular Electronics was "a way both to distinguish its efforts from the other services and . . . a conscious move to incubate new approaches to microelectronic circuitry."[42] Molecular electronics also had a champion in James M. Bridges, who had chaired the AGREE. Westinghouse discussions with the Air Force in 1957 and 1958 led to a contract between the two in 1959.[43]

What Kilby had invented was a fifth approach to making an integrated circuit. Eventually, Molecular Electronics' approach would morph into an approach similar to what Kilby had invented.[44] However, when TI announced their "solid circuit," the Air Force, like the Army, was concerned about how it fit into Molecular Electronics. If the "solid circuit" was an example of Molecular Electronics, the Air Force would support it. The majority view held it was not.

---

[39] Holbrook, *Business and Economic History*, 143.

[40] Kilby, "Invention of the Integrated Circuit," 651.

[41] Ibid., 648.

[42] Holbrook, *Business and Economic History*, 153.

[43] Far too advanced for the time, the idea that molecules could be made into useful miniature devices eventually became nanoelectronics several decades later.

[44] Holbrook, *Business and Economic History*, 156.

At nearly the same time as Kilby's invention, and prior to its announcement, Fairchild Semiconductor's Robert Noyce also invented the integrated circuit, but his was a more practical device. His technology, using a refinement of the masking technology of World War II proximity fuzes, used a single crystal of silicon semiconductor etched in such a way as to allow metalized shapes to be embedded. His "planar process" was actually the first key to making integrated circuits ubiquitous devices. With the advent of the planar process, device density accelerated significantly.

Thanks largely to R.D. Alberts of the Air Force's Wright Air Development Center Manufacturing Technology Laboratory, the majority view of solid circuits did not hold. Alberts saw, and was able to convince others, that TI's device would avoid the impending chaos from the different Service approaches, and provided "an orderly transition to a new era." With a systematic design approach, many new devices could be invented. Alberts' group started a series of contracts to take the TI invention and move it forward through research, development, and manufacturing process development and maturation. Other devices followed quickly as TI moved to head off competition from Fairchild. Convinced that the device was viable, in January 1959, TI began work on a patent application, filing for "Miniaturized Electronic Circuits" on 6 February 1959. Kilby's first devices were publicly announced at the Institute of Radio Engineers' conference in March 1959.[45]

Air Force sponsorship would turn the TI patent into practical devices. By 1962, Westinghouse was using "molecular electronics" to mean monolithic integrated semiconductor circuitry.[46]

The integrated circuit was not an isolated case of military investment helping the nascent electronics industry. Just as in the case of transistors, military investment was essential. At the time, smaller and lighter devices than were achievable with vacuum tubes were being demanded for national defense needs. Commercial solid state electronics was getting under way but needed a cash infusion from government laboratories and sales. Well-staffed and -equipped government laboratories provided sustainment and more for the new industry as sales moved toward being self-sustaining. It also allowed commercial companies access to world-class scientists and engineers working in a challenging area of research. Much of that work was on integrated circuits.

The inventions by Kilby and Noyce, followed by a variety of additional improvements, accelerated the rate of electronics miniaturization. At first, miniaturization

[45] Kilby, "Invention of the Integrated Circuit," 651; Riordan and Hoddeson, *Crystal Fire*, 260, 271; Eldon C. Hall, *Journey to the Moon: The History of the Apollo Guidance Computer* (Reston, Virginia: American Institute of Aeronautics and Astronautics, 1996), 16; "Integrated Computer Elements Out," *Aviation Week and Space Technology* (20 March 1961): 87. The monolithic integrated circuit was invented at almost the same time by Robert Noyce of Fairchild Semiconductor Corporation. The U.S. Patent Office finally issued the first such patent to Noyce on 25 April 1961, but in the interim TI had had a competitive advantage.

[46] Holbrook, *Business and Economic History*, 153.

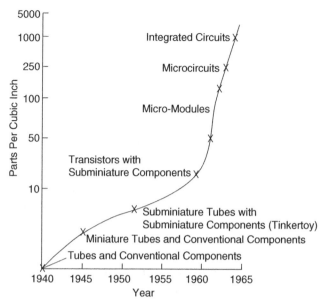

Figure 4.1.   Density of early solid state electronics. Source: After G.W.A. Dummer, "Integrated Electronics," in Ernest Braun and Stuart Macdonald, *Revolution in Miniature* (New York: Cambridge University Press, 1982), 97.

was measured in the number of equivalent discrete parts contained per cubic inch of a device, as shown in Figure 4.1.

Integrated circuits soon outstripped the nomenclature of devices per cubic inch. They would be described by their *level of integration*, which at the time meant the number of embedded devices in the circuit. Thus, initial integrated circuits performing a few simple functions, like the oscillator, were considered Small Scale Integration (SSI).[47] SSI encompassed less than a dozen logic gates containing fewer than 30 transistors. SSI used gate array technology known as uncommitted logic arrays. That is, each logic gate included several transistors, so the total number of embedded devices was a couple dozen. SSI was a tremendous step forward. Much more needed to be learned about how to combine metals into intricate designs functioning as resistors, transistors, and so on. Compatibility of materials had to be well understood, because, for instance, metals and silicon expand or contract at different rates in response to temperature, which could lead to cracking and other problems. Techniques for silicon etching also advanced with experience. The number of logic and

[47] The scale of integration was an early attempt to characterize in compact terminology how densely packed a chip was. The definitions of scales of integration are loose, and attention must be paid to whether the items being counted are transistors, logic gates, circuit elements, and so on. Eventually, the numbers grew so large and so quickly that the size of features became the better referent for the scale of density in a device.

other functions expanded slowly at first, but within a few years began a rapid acceleration as technologies advanced across a broad front.

These developments occurred at a crucially important juncture in the history of the U.S. ICBM developments. On 28 February 1959, days after TI's patent, the highly secret Corona 1 photoreconnaissance satellite, known publicly as Discoverer 1, launched from Vandenberg AFB.[48] The launch was critical for at least two reasons. Obviously, as the first test of a revolutionary new technology—photoreconnaissance from space—it was essential to the future of U.S. national security. It raised hitherto unanswerable questions regarding the extent of national sovereignty of the nations over which a satellite would orbit (whether or not it took a picture of them). Less obviously was what it meant to the U-2, which also remained highly classified. Every U-2 flight was a serious calculated risk, justified by the seriousness of the need for IW intelligence and understanding Soviet capabilities. The Soviets attempted interception of every flight. Time was running out.

Corona was one of two possible follow-up programs. The other was an extremely classified aeronautical *tour de force* for the U-2's designer and developer, Clarence "Kelly" Johnson. Known variously as the Oxcart or A-12, and eventually to be known to the public as the SR-71, the new reconnaissance aircraft would eventually exceed three times the speed of sound. Only time would tell if either follow-up could be ready in time for the inevitable end of the U-2 overflight program.

The criticality of this situation can best be put in perspective by understanding the importance of what the U-2 was finding. The public's awareness of the U.S. missile program up to 1959 had been colored by politically motivated controversy in the wake of the launch of Sputnik in October 1957. A common though erroneous conclusion creating a great swirl of concern held that bomber and missile "gaps" existed between the United States and the Soviet Union, with the Soviet Union in the lead. Key U.S. officials knew no such gaps existed, but the means by which they knew were highly classified. Ever since 5 July 1956, on the U-2's second operational flight, piloted by Carmine Vito, U.S. officials had known there was no gap. On that mission, Vito's aircraft captured the "million dollar photograph" of Moscow's Saratov-Engels airfield. It revealed the Soviets had only a fraction of the bombers claimed by them and accepted without question in the Western press. Yet, in the public mind, Vanguard's repeated failures seemed only to confirm the Soviets were succeeding where the U.S. was falling short.[49]

The visible and repeated failures of the Vanguard launcher created public concern about the United States' technical competence versus the Soviets'. In March 1960, an Atlas missile exploded during fueling, destroying its Vandenberg AFB facilities. That was an especially difficult blow, as the Vandenberg facility was not only for training and testing, but also served as an Atlas operational site. Within a

---

[48] Temple, *Shades of Gray*, 173; Kenneth E. Greer, "Corona," *Studies in Intelligence*, Supplement 17, (Washington, D.C.: Central Intelligence Agency, Spring 1973). Thirteen more attempts were needed until the first satellite reconnaissance image was finally returned to Earth on 18 Aug. 1960.

[49] Temple, *Shades of Gray*, 70. *Shades of Gray* has a more extensive discussion of the controversy and its reality.

couple of months, Atlas suffered two in-flight failures. This was doubly troubling, because not only was the Atlas supposed to be a front-line strategic weapon system, but a version was to be the launcher for the first U.S. human orbital flight—a key event U.S. officials were hoping to use to reclaim public perception of leadership in space. These displays of a lack of reliability were unacceptable on many fronts, but especially for U.S. strategic deterrent credibility.[50]

Gen Power insisted the immediate concern was due to taking shortcuts in testing based on the speed of development and deployment of the missiles.[51] A former ARDC commander, he did not agree with all of Schriever's "concurrency" shortcuts of massive parallelization.[52] But Power's insistence about more testing overcoming reliability problems overlooked a fundamental truth. His viewpoint overlooked the importance of manufacturing processes and piece parts for final reliability. In his view, the object of testing was ensuring each box did what it was supposed to in its expected operating environment. Therefore, testing was to find flaws, replace failed components, retest, and iterate until the box was ready to go.[53] This achieved the objectives of performance and the earliest possible operational date but deferred reliability until the operational phase, where it manifested as maintenance and logistics problems. At the time, there would be plenty of time to upgrade later, but countering the threat of the Soviet Union's ICBMs was paramount. Concurrency succeeded brilliantly. But this made Atlas an interim ICBM solution, since its successors (Titan and Minuteman) were already beginning their development.

The lesson from the initial large missile developments was clear—reliability had to start at the component or piece part level.

The problems of early solid state component reliability became even more complex with the move to critical national capabilities in space. Reliability for space systems was different from the assured readiness for launch driven by missiles. Prior to launch, if a missile component broke, it could be repaired. A satellite was tethered to the Earth by only the slimmest of means—through radio. The ultimate problem, beyond surviving the launch on such missiles, was achieving long term operational reliability needed to support satellites such as Corona, the IW photo-reconnaissance satellite, and Midas, the EW satellite. Although telemetry could be received, and some operations controlled remotely, such early systems were quite limited in their ability to last longer than their nominal two-week missions.

---

[50] Jacob Neufeld, *The Development of Ballistic Missiles in the United States Air Force 1945–1960*, (Washington, D.C.: U.S. Government Printing Office, 1990), 192, 215.

[51] Ibid., 216; Lt.Col. Kenneth F. Gantz, ed., *The United States Air Force Report on the Ballistic Missile: Its Technology, Logistics and Strategy* (New York: Doubleday, 1958), 167.

[52] L. Parker Temple and Peter L. Portanova, "Project Emily and Thor IRBM Readiness in the United Kingdom, 1955–1960," *Air Power History* 56, no. 3 (Fall 2009): 29; Michael H. Gorn, *Vulcan's Forge: The Making of an Air Force Command for Weapons Acquisition (1950–1985)* (Andrews Air Force Base, Maryland: Office of History, HQ Air Force Systems Command, 1989), 40. Schriever had been assigned from the Air Staff to be the head of WDD at the time when Power was ARDC Commander, and in addition to his role as the head of WDD, Schriever was Power's assistant.

[53] This approach, known as "testing in reliability," was thoroughly disproven later, when the effects of extensive rework were found to reduce reliability in microelectronics.

*   *   *

The emerging importance of satellites to support strategic forces made the role of STL more important than ever. The Air Force's national defense space program relied on STL for "the technical competence to meet current and future space age challenges."[54] STL provided continuity and corporate memory when military reassignments moved individuals elsewhere. But with STL's importance came criticism from the emerging aerospace industries, motivating an investigation by the House Committee on Government Operations. The September 1959 House report recommended conversion of STL into a not-for-profit organization along the lines of RAND (formerly the Air Force's Project R and D, operated by Douglas Aircraft Company).[55] Using RAND's model for ensuring independence and avoidance of conflicts of interest seemed logical to all except the parent company of STL. TRW did not want to divest STL. Nevertheless, the Air Force moved to establish STL as a Federal Contract Research Center (FCRC). FCRCs dated back to World War II with the government's need for independent, specialized centers of expertise. TRW was not interested. Discussions with the AFBMD changed things a bit. AFBMD intended to extend a hardware production ban on ICBMs to cover TRW and STL to obtain the necessary independence and lack of conflicts of interest. TRW decided to sell or separate STL by some means.[56]

Further cooperative discussions among TRW, the Air Force, and STL resulted in the Air Force's February 1960 request for an organizing committee for a new not-for-profit corporation with a separate board of trustees. This new corporation became known as the Aerospace Corporation, which would operate an FCRC in facilities across the street from AFBMD. It was established 3 June 1960.[57] Roswell L. Gilpatrick, former Under Secretary of the Air Force and the new Chairman of the Board, announced the new corporation. He said its relationship to the Air Force was "a new approach on the part of the Air Force in the management of its missile and space programs." By 1960, the role of FCRCs had broadened from the earlier emphasis on scientific inquiry to "private support whose objectivity was strengthened by separation from industry."[58]

---

[54] David N. Spires, *Beyond Horizons: A Half Century of Air Force Space Leadership* (Montgomery, Alabama: Air University Press, 1998), 85.

[55] Gorn, *Vulcan's Forge*, 9. Project R and D was one of two seminal projects started by the Air Force's first Deputy Chief of Staff, General Curtis E. LeMay, who had been the commander of the atomic bomb squadrons in the Pacific, and later went on to head Strategic Air Command and the Air Force. The other organization he started was the Air Force Institute of Technology, and together, these supported the highly technological leanings of the Air Force.

[56] Getting, *All in a Lifetime*, 386; U.S. Congress, Office of Technology Assessment. *A History of the Department of Defense Federally Funded Research and Development Centers* (Washington, D.C.: U.S. Government Printing Office, June 1995), 25.

[57] *Aerospace Corporation Annual Report for 1960* (El Segundo, California: Aerospace Corporation, 1961), iii; U.S. Congress, Office of Technology Assessment, *A History of the Department of Defense Federally Funded Research and Development Centers* (Washington, D.C.: U.S. Government Printing Office, June 1995), 25; Spires, *Beyond Horizons*, 86.

[58] U.S. Congress, *A History of the Department*, 11; Spires, *Beyond Horizons*, 86.

There was plenty of work for the new FCRC. Its initial task list, established one month later, rank-ordered their efforts as shown in Table 4.1.

What appeared to be a full plate grew, by November 1960, to include the list shown in Table 4.2. The list included studies and analyses, programs in advanced concept stage, and actual development efforts. All of this activity also quickly raised the potential demand for devices that would work in space.

Table 4.1.   Initial Task List for the Aerospace Corporation

Order of Priority	Identification	Description
1	Program II	Classified Space Program
2	Discoverer	Unclassified name of Corona photoreconnaissance satellite
3	Mercury	NASA human spaceflight program
4	Nike-Zeus and Penetration Aids	Anti-ballistic missile program
5	Midas	Infrared early warning system
6	Agena B	Standard restartable upper stage, used in Corona and its successors
7	Project West Wing	Technology analysis project
8	Hard Base Consulting Panel	Missile basing alternative analysis
9	Test and Instrumentation Support	
10	Communication Support to Ballistic Missiles	
11	Dyna-Soar	Strategic bomber and reconnaissance space plane
12	Orbital Interceptor	Anti-satellite
13	SAINT (SATIN)	Anti-satellite
14	Advanced Planning and Analyses	
15	Transit III	Communications satellite

Source: *Aerospace Corporation Annual Report for 1960* (El Segundo, California: Aerospace Corporation, 1961), viii.

Table 4.2.   Additions to the Aerospace Corporation's Initial Tasking

Advent	Communications satellite
Mid-Range Ballistic Missile	Initial description of the THOR IRBM
BAMBI	Boost-phase antiballistic missile interceptor
Vela	Series of treaty monitoring satellites
Advanced Re-Entry Concept	Missile warheads and other re-entry devices.

Source: *Aerospace Corporation Annual Report for 1960* (El Segundo, California: Aerospace Corporation, 1961), ix.

Military investments in solid state technologies led to the necessity for standard-izing specifications for these parts. The Navy established the first standard for discrete devices (diodes, resistors, bipolar transistors, field effect transistors, etc.). Their initial approach was updating the vacuum tube standard, MIL-E-1, adding a section on discrete devices (initially only diodes).[59] Drawbacks of their approach were the very different natures of the vacuum tube devices and solid state discrete devices. Consequently, in 1959, the Navy's Bureau of Ships consolidated discrete device specifications into the first edition of MIL-S-19500, "General Specification for Semiconductor Devices." MIL-S-19500 was more than a consolidation and fusion of solid state specifications. It incorporated results from Paul Darnell's Ad Hoc study, which was about to be released. It also brought together the standardization work being done jointly with industry through the Joint Electron Device Engineering Councils (JEDEC).[60] JEDEC was the electronics manufacturers' forum for coming together in common cause to deal with issues of standardization, working cooperatively with the government. Its importance went beyond simple technical sharing. Standardization documents' wording was occasionally unclear or did not cover a specific case as a consequence of advancing technology, and JEDEC addressed these and similar issues. Coalescing the Darnell and JEDEC approaches made sense for two reasons. First, the JEDEC system provided a comprehensive system for registering the various kinds of specifications. Second, package types needed some kind of common lexicon and designation system. The Navy's MIL-S-19500 built a framework to provide a single solid state semiconductor device production and testing standard.[61]

* * *

In May 1960, the need for successful satellite reconnaissance assumed considerably greater importance. Eisenhower decided the risk was too great, and no more U-2 overflights would take place after the 26th one, which he authorized to fly that month. Eisenhower's decision was prophetic. Francis Gary Powers' U-2 was shot down over the central Soviet Union by a Soviet SA-2 surface-to-air missile. The Oxcart triple-sonic reconnaissance aircraft would not fly before April 1962, so

---

[59] C.W. Martel and J.W. Greer, "Joint Army-Navy Tube Standardization Program," *Proceedings of the I.R.E.* (July 1944), p. 430. Readers unfamiliar with electronics and military standards will find them a confusing jumble later if not paid attention to. Note that the middle letter, E, meant that the standard stood for electron tubes. In later chapters, MIL-I-38535 will be seen, and the I stands for integrated circuits. Similarly, M stood for microcircuits, S for semiconductor devices, and H for hybrid devices. In the end, there will also be PRF, for Performance. These devices will help sort out much of what follows.

[60] JEDEC was founded in 1958 as a joint activity between the Electronic Industries Association and the National Electronic Manufacturers Association. Since that time, JEDEC as an acronym has been dropped, and now stands as the name of the organization. NEMA stopped participation in 1979, but the cooperation of JEDEC and EIA (as TechAmerica) continues the government-industry standards development collaboration.

[61] *The Reliability Handbook: An Introduction to Military/Aerospace Semiconductor Reliability Concerns, Screening Techniques, and Standardization Programs* (Santa Clara, California: National Semiconductor Corporation, 1979), 69; Nucci, "Progress Report on Ad Hoc Study," 132.

Corona was the only way to see inside the Soviet Union.[62] However, Corona had failed to return any results 13 times in a row. Direct support to the President and his advisors was crucial at this point in the Cold War, yet Corona was not ready when Powers' flight was lost.

*   *   *

Also in May 1960, Darnell's Ad Hoc Study on Electronic Parts Specification Management for Reliability made a lasting impact on the electronics industry. Known as the Darnell Report, its significant findings would structure quality programs for decades to come. Not all of the recommendations were immediately effective for the new long range missile or space programs, but all would eventually be adopted.[63]

The Darnell group recognized that

> *[t]he fast changing state of the electronics art, together with increasingly complex equipments demanding high reliability, has created the need for (1) additional requirements in electronic parts and tubes specifications, (2) faster coordination of parts specifications, (3) the establishment of technical characteristics data for dissemination to design and logistics personnel and (4) a complete review of the parts specifications program to ensure compatibility with the reliability program.*[64]

Compared to the later rate of technology change, of course, the development pace of the late 1950s was leisurely, but their three major recommendations were precisely right.

First, the group recommended establishing "An Advisory Group on Management of Electronic Parts Specifications" within OASD (S&L), to be broadly coordinated with other related groups such as DDR&E. Its management responsibilities included recommending "1. the most efficient implementing actions, 2. the most effective program and 3. changes to any proposed programs that will ensure the implementation" of the report's recommendations.[65]

Second, the Darnell report called for reorganizing some activities. These included assigning the Armed Services Electro-Standards Agency to the OASD (S&L), requiring it to report to the Assistant Secretary's office responsible for standardization. This would improve efficiency by creating a single organization for electronics standards.[66]

The third, and perhaps most far reaching, recommendation stemmed from the study group's consensus on the Office of the Secretary of Defense's poor handling of parts specification management. Comprehensive, detailed, and extensive changes to the processes for development, management, and control of parts specifications

---

[62] Temple, *Shades of Gray*, 385.

[63] *Parts Specification Management for Reliability, Vols. 1 and 2, Report of the Ad Hoc Study Group on Parts Specification Management for Reliability*, Offices of the Director of Defense Research and Engineering and Assistant Secretary of Defense (Supply and Logistics), May 1960.

[64] Ibid., iii.

[65] Ibid., vol. 1, p. 4.

[66] Ibid.

were necessary. They wanted the standardization office to "prepare and recommend DOD policy and program-implementation plans concerning the development of specifications, standards and handbooks for electronic parts and tubes." New specifications and standards development was a resource-constrained activity. It could become a black hole of resources if not tightly controlled. So they included recommendations for controlling, prioritizing, and scheduling the standards effort. The changes mandated collection, analysis, and retention of "parts characteristics data, including test data," supporting the development of new specifications or revisions of existing specifications.[67]

The Darnell group's third recommendation took the earlier AGREE intent (which at the time had addressed reliability only at the box level) by taking reliability to the next level: the root of reliability. With the limited extent of adequate parts data, full realization of the recommendation would take some years. They knew there would be a transition during which reliability would still have to be addressed at the box level, however unsatisfactory that was.

The group's recommendations clearly meant existing guidance was considerably overtaken. DoD's source of assured quality parts, Manual M204, *Military Qualified Products List* (MQPL), was to be "a list of approved sources of supply for qualified electronic parts." Darnell's group found "subquality" parts listed because the MQPL processes lacked both technical backing and necessary management oversight to ensure compliance. Hence, they included revisions of the MQPL qualification procedures. Initial parts qualification data could then be coupled with field failure reporting data. That enabled feedback from problems to the writers of specifications and standards, eventually resulting in updated or improved parts.

Qualification required establishing a reliability level for the part or tube as a necessary first step. Then, before submitting samples for qualification approval, the manufacturer had to "show evidence that he has the test equipment he needs to conduct all tests required in the specification and that he has satisfactory in-plant process control." Qualification was contingent on the manufacturer's "submission of sufficient data representative of production quality." Moreover, all qualification tests were to be conducted "under the surveillance of the appropriate government inspection service." Qualification would have to be reevaluated by periodic audits, or after changes in product design or manufacturing processes.[68] So, after a cycle of problem-to-specification/standard-to-updated parts, the manufacturer was obliged to be reevaluated. The feedback cycle, given the expected technology volatility, meant audits would occur about every two years, unless a manufacturer made substantial changes in the interim.

Parts could be put on the MQPL only after a manufacturer had qualified its processes and test equipment and sample parts had been submitted to an approving agency for verification. Any significant changes to manufacturing processes, such as increases in production rate, required a review and possible reapproval.[69]

---

[67] Ibid., vol. 1, p. 5.

[68] Ibid., vol. 1, p. 7.

[69] Nucci, "Progress Report on Ad Hoc Study," 133.

At the heart of the Darnell revisions was establishing failure-rate levels as the basis for parts specification and qualification. The new concept covered "qualification inspection and re-evaluation, acceptance inspection and failure-rate testing." They were not so naïve to think this radical change was of little or no cost impact. They adopted an initially low threshold to ease the transition. The failure-rate specification was to a predetermined confidence level—a statistical measure that determined lot sample sizes and tests. The group suggested a low confidence level at 80% or less. They also allowed for use of simplifying assumptions and acceleration factors (such as raising the temperature of the part to see what the effects of extended use would do to the part's performance), in an effort to reduce sample sizes to an economical level.[70]

Darnell's group also recognized the crucial importance of an information sharing process. At the early stage of solid state electronics in 1959, parts quality varied considerably from process to process, even within a single plant, and certainly across the industry. Information on problems with electronic components could come from many places, including the manufacturers, companies using the parts in larger assemblies, and from the users of the equipment in the field. Information sharing among all interested parties could have a significant impact on improving the quality of parts being made, supplied, and used.[71] In fact, it could also increase competitiveness as all learned from problems and lessons learned even before standards were updated. A secondary effect of this was a (gentle) industry-wide acceleration of technology development.

A number of more limited data sharing efforts already existed. A Battelle Memorial Institute analysis of parts test data had also examined a number of these. Since missiles and their launch vehicle counterparts produced extreme environments in terms of acceleration, shock, and vibro-acoustics, they would benefit as more was learned and shared about how to deal with these factors. The Titan ICBM's information sharing covered the exchange of report data between subcontractors, and was considered an excellent example. The Navy's ballistic missile program support by the Naval Ordnance Laboratory (Corona) was also excellent.[72] The Navy program's strength was providing guidance concerning component choice to contractors involved in reliability. The benefit of information sharing, then, had already been recognized and implemented on some nationally important programs.[73]

With these and other efforts, a comprehensive process began to take shape. However, the missile and ordnance information feedback and sharing programs' very attractive features did not cover all the most desired elements. Other ideas ranged from including information in parts specifications from the manufacturers,

---

[70] *Parts Specification Management*, vol. 2, p. 1. The temperature-conductivity relationship is known as the Boltzmann or Arrhenius equation, and allows correlation of behavior at higher temperatures to lifetime performance.

[71] Nucci, "Progress Report on Ad Hoc Study," 133.

[72] NOL (Corona) or NOLCorona was named for its location in Corona, California, not for the then highly classified photoreconnaissance satellite program.

[73] Nucci, "Progress Report on Ad Hoc Study," 133.

to publishing "failure-rate data, such as the RCA 'TR-1100,' Vitro Corporation Report No. 98, and others in an unedited but bound handbook."[74]

Two pertinent questions were whether a given part would work as intended, and what tests determined that. Although reliability calculations had, of necessity, focused at the box level, qualification and suitability questions could be answered only at the part level. Component level testing was the only way to know how the electronic parts would react to their operational environment, such as in long term, power-on situations like sitting on alert. As the Services soon realized, considerable duplication existed in their approaches to suitability determination testing. Sharing testing data reduced duplication, allowing more tests to be conducted at lower individual cost. Thus, based on the findings of the Ad Hoc study, the Air Force, Navy, and Army agreed to establish the Interservice Data Exchange Program (IDEP). Initially, IDEP covered only missile program parts, but would soon expand as the value to the nation of unimpeded data sharing became obvious.

The Ad Hoc study's agreement on specifications and standards, common means of calculating reliability, parts qualification, and problem information sharing indicated, among many other things, that Eisenhower's push to end inter-Service rivalry had, at least to this extent, succeeded. The agreements made in May 1960 were just in time to support the agenda of a new Administration's Secretary of Defense the following January.

* * *

An advantage of monolithic integrated circuits (IC) was once the manufacturing process started, there was no need for human intervention to solder other components into an assembly and no wire leads to be broken off. One advantage of integrated circuits was that they reduced complexity and connections. A driving factor behind ICs was that conventional circuit complexity increased the number of components. All these components had to be assembled and their connections hand soldered (usually). Failure rates correlated to the number of components and connections, so reducing these by integrating them into a monolithic circuit favored increased reliability. "Though the transistor may have removed some of the heat and power limitations of tube circuitry, it did little to resolve this problem, which J.A. Morton of Bell Telephone Labs came to call the 'tyranny of numbers.' Integration offered the possibility of alleviating this difficulty, as well as miniaturizing the circuitry."[75]

ICs meant smaller packages, more reliable operation, and lower manufacturing costs, and immediately became attractive to the military. ICs could potentially eliminate hybrid systems comprising tubes and solid state devices. Such hybrids were expensive to design, build, and test. Also, ICs opened the possibility of faster computation speeds than vacuum-tube systems.

---

[74] Ibid. Although not immediately adopted, the idea of a single handbook was attractive, and one notable such product was National Semicondutor's *The Reliability Handbook: An Introduction to Military/Aerospace Semiconductor Reliability Concerns, Screening Techniques, and Standardization Programs* (1979), which remains a very popular source.

[75] Holbrook, *Business and Economic History*, 150.

A 1962 survey by *Electronic Industries* magazine found 35 of 59 respondents were engaged in modular packaging activities while 34 were pursuing thin film approaches. Only 15 were looking into monolithic approaches.[76] By 1965, monolithic integrated circuits were clearly superior, and thin film technologies lost out for the time being, though not completely in some specific device areas.

The bright light in this fragmented approach must surely have been the differing approaches to ICs broadened the scope and range of available technology. The industry was well under way; having exceeded $100 million only a few years earlier, sales passed $1 billion in 1961.[77] Despite the commercial success, the influence of the military's needs cannot be overlooked. Much of this early production had military destinations, but the important factor was not market share. The lasting effects had to do with early cash infusions and pushing a broad range of advanced technologies to meet military needs feeding back into the commercial industry.

The "newness" of space programs pitted considerable fundamental engineering against a background of extreme expectations. Hard engineering produced, painfully, initial capabilities in photoreconnaissance with Corona, early warning from MIDAS, and communications using a number of different frequencies in different programs.

Right alongside these were the X-20 Dyna-Soar space plane, and the Satellite Interceptor (SAINT), both of which embodied fantasy and the desire for military human spaceflight. In many ways, both SAINT and Dyna-Soar were far ahead of their times, especially in terms of national policy and their technical achievability. Both, therefore, raised some important political and technical questions.

Dyna-Soar was the Air Force's premier human spaceflight program at the time, and was to have a "hot" skin without heat resistant tiles. Its rapid launch and dynamic maneuvering would provide weapons delivery from space and allow it to do its own post-strike damage assessments. SAINT would carry humans to rendezvous with a potentially hostile satellite to determine whether it was indeed hostile, and, if so, to negate it. Determining the reliability of the SAINT final stage was a difficult process, and held some key insights to the future directions for standards and high reliability practices.

Despite their differences, a synergy existed in the high reliability demanded for strategic missiles and for human spaceflight. The Martin Titan III booster for Dyna-Soar was "the first booster system to be specifically designed for man-rated applications."[78] Typical of the day, man-rating the Titan III for Dyna-Soar focused quality efforts on the design above the component level.[79] However, Martin made no distinction between piece parts, units, or subassemblies in quality inspection. The program was blazing new trails in reliability and quality assurance, as no one had experience

---

[76] Ibid., 152.

[77] Riordan and Hoddeson, *Crystal Fire*, 275.

[78] *Program 624A Reliability/Quality Assurance/Maintainability Quarterly Status Review*, Aerospace TOR-269(4116-21)-1 (October 1963), 1.

[79] SSD Exhibit 62-117, "Reliability and Quality Assurance Program Requirements, Program 624A," 1 November 1962.

with qualification for human spaceflight. Protection of human life drove a clear need for the highest attainable levels. Having said that, however, what were the highest attainable levels, and how could they be achieved? There were no clear answers.

> *The Martin Company indicated that applicable requirements for reliability, quality assurance and maintainability were being included in contract documents to the supplier (i.e., procurement drawings, procurement specifications and/or purchase orders with special note requirements). It was noted, however, that interpretation of these requirements ran from the "sublime to the ridiculous." The area of contamination control, level of clean room (and definition of clean room) type of information required on certifications were indicated as problems. Too many suppliers are not used to the stringent 624A requirements and do not understand the reasons for such controls on the same or similar components to those they have previously supplied on other programs.*[80]

So procedures from similar programs (that is, the Titan II ICBMs) were not held to be sufficient for man-rating the Titan III. Such differences led to confusion by suppliers.

The reason why differing practices were a problem for the suppliers has to do with training and skills in a technical manufacturing workforce. When similar parts or devices are required by their customer to be (unnecessarily) different, problems of negative transfer, excess training, additional break down and setup, and so on turn into cost, reliability, and quality considerations. On long production runs, as was needed for Minuteman I, a machine operator would develop skills and understanding that might not apply to a very similar run of parts for Polaris. Inconsistency, confusion, extra cost, time, and related problems begged for more standardized requirements and manufacturing.

In 1961, the SAINT program noted:

> *It sometimes occurs that complex equipment is built according to approved manufacturing instructions (drawings, specifications, etc.) and yet exhibits a failure rate much higher than the population from which it was drawn. This is due to a number of factors, among them design tolerances, quality control levels, inspection criteria, etc., but the result is the equipment is not truly representative of the design.*

So, although tests could theoretically determine the level of reliability, even when the design was adequate, it was difficult to realize in hardware. Eventually, each of the areas would need to have standards established to preserve what was learned through successes and failures. Without capturing these hard-won lessons in current and new standards, complex systems such as Dyna-Soar and SAINT would always remain unachievable or would be plagued by crippling problems of unreliability. Capturing the lessons from high reliability systems would also lead to successive development of other standards beyond MIL-S-19500 to "tamp down" problems as they arose and became common.[81]

---

[80] *Program 624A Reliability,* 39.

[81] *Requirements, Objectives, and Statistical Methods for SAINT Reliability Demonstration Program,* Aerospace TOR-594(1112) RP-1 (1 May 1961), 3.

*   *   *

In January, 1961, President John F. Kennedy arrived with an entirely new agenda. Though still focused on the national security and national defense space programs, the reform approach of the Kennedy Administration was strongly colored by the new Defense Secretary, Robert S. McNamara. A testimony about the effectiveness of Eisenhower's achieving his reorganization goals was that the issue of inter-Service rivalry was much less on McNamara's mind. Whereas Eisenhower aimed his organization reforms at the military commanders, McNamara intended to improve the efficiency and effectiveness of the DoD.

Within four days of assuming office, McNamara launched a series of reforms to make the DoD operate more like a big business, bringing order to what he perceived as chaos. First, proving that inter-Service rivalry was not completely off of his agenda, he mandated such things as the Air Force and Navy development of a common next generation fighter, or TFX. This became the General Dynamics F-111. Second, he imposed a disciplined budgeting process, common across all the Services, known as the Planning Programming Budgeting System. Finally, he also imposed an approach on systems acquisition to make contractors more responsible for their products.

He called the approach the Total Package Procurement Concept (TPPC). TPPC sharpened a contrast in Cold War acquisition of weapons. Air Force Systems Command (AFSC) had recently been activated out of the former ARDC and a part of Air Materiel Command, based on LtGen Bernard Schriever's ideas. Schriever took his pioneering concepts in the development of ballistic missiles and space systems and turned them into a weapon systems acquisition command. His approach entailed decentralization into product centers, and separating acquisition from operations. In that way, Western Development Division had evolved into Space Systems Division, acquiring national defense space systems. Schriever's goal was to get capable weapons into the field quickly, well, and to keep pace with the perceived rate of Soviet weapons advances. The Cold War, after all, was about survival of the United States, and while cost was important, it was far from the most important thing. McNamara agreed with most of that, with one significant difference. Weapon performance was important, but of lesser importance than cost and efficiency. Cost and efficiency, in turn, meant centralization.

Schriever later assessed the situation:

> We arrived at the right arrangement organizationally and management-wise in the creation of the Systems Command and the Logistics Command. Unfortunately, the implementation became impossible under the policies and procedures as prescribed by Secretary of Defense McNamara.[82]

TPPC, in a sense, extended what Schriever had pioneered, and took it to the extreme. Specifically, McNamara exaggerated the initial planning portion to weapon acquisition.

---

[82] Gorn, *Vulcan's Forge*, 76. Gorn was citing an interview of Gen Schriever.

He separated systems acquisition conceptual design and planning phases from the development and production phases. His purposes were manifold. Separating conceptual concerns from production forced better upfront planning and design, thinking through problems at their earliest point. Though this did not eliminate later unforeseen or unforeseeable problems, the cleavage of the phases allowed handling distinctly different kinds of activities on either side in different contractual terms. What he apparently thought he was doing was drawing a line beyond which the technologists had to stop tinkering, and force manufacturing to get on with the development of the end product.[83]

With conceptual work taken to the extreme, the result was a procurement package to be competed comprising detailed information on "mission areas, weapons performance, possible technical approaches, the limits of current technology, and credible cost and schedule estimates." This Concept Formulation Phase led to a competition in which the bidders detailed "development activities, facilities design, plans, tests, production, schedules, logistics support, and cost." Ultimately, the selected contractor would be obligated to provide:

"**1.** Realistic and fixed performance specifications;

2. Realistic and fixed cost and schedule standards for the development phase;

3. Cost and schedule estimates for production, operation and logistics support adequate for planning activities;

4. Justifications based on cost-effectiveness."[84]

Problems arose with TPPC, but assessing its effectiveness has been hampered by the political environment of the time. TPPC was acknowledged to be flawed, leading to overly optimistic government and contractor cost estimates. Perhaps the most serious flaw was the immensely detailed paperwork TPPC required. "Overly detailed performance standards were specified before existing technology had been tested under life-like conditions. This resulted either in test programs that required more technical capability than necessary; or in test programs that demonstrated deficiencies which, in light of the concurrent nature of the development process, were very costly to rectify."[85]

A consequence of the required paperwork was very high administrative costs in a time before modern office automation. "Huge proposals and contracts, sometimes tens of thousands of pages, required oversight by armies of government employees. Additionally, DoD demands for massive contractor documentation

---

[83] LtCol. Steven C. Suddarth, "Solving the Great Air Force Systems Irony," *Air Power Journal* 16, no. 1 (Spring 2002): 8–9.

[84] Combined ARDC/AFSC History, 1 Jan.–30 June 1961, 30–36; Interview of General Bernard A. Schriever by Major Lyn Officer and Dr. James C. Hasdorff, USAF Oral History Interview 676, Air Force Historical Research Center, 20 June 1973, pp. 22–25; W.D. Putnam, "The Evolution of Air Force System Acquisition Management," Rand Report, Aug. 1972, pp. 10–22, 37; "Air Force Reorganizes Itself," *Business Week* (25 March 1961): 26; *Management Principles of AFSC* (AFSCM 25-1), 25 Sept. 1961, pp. 5–6; all cited in Gorn, *Vulcan's Forge*, 74–75.

[85] Gorn, *Vulcan's Forge*, 76.

persisted even though the data often ceased to be useful by the time it had been collected."[86]

Former Deputy Secretary of Defense for Analysis Alain C. Enthoven defended TPPC, though, pointing out some of the politically-motivated criticisms of TTPC were on systems whose cost performance was actually better than acquisitions a decade earlier. Whichever the case, TPPC was a foundational concept which would, over a decade later, morph into a key acquisition reform called Total System Performance Responsibility, which will be discussed later.[87]

[86] Ibid.

[87] Alain C. Enthoven and K. Wayne Smith, *How Much Is Enough? Shaping the Defense Program 1961–1969* (New York: Harper & Row, 1971), 239–240.

Chapter 5

Minuteman Means Reliability

With a running start in 1958, the short time allowed for Minuteman manufacturing impacted the U.S. electronics industry. To arrive in time, the Minuteman had to set precedents, remove barriers, and push technology—and do all of this on a breakneck concurrent development schedule. The sudden demand meant rapid expansion for some manufacturing lines or their almost exclusive use supporting the missile programs at some point.

There is so much more to the story than that. Histories of the evolution and growth of the semiconductor industry usually give a passing nod to the impact of Minuteman. However, such histories consider only the total number of parts ordered versus overall industry output. The focus of this history has to do with the *quality* of the parts more than the gross numbers. The Minuteman program demanded a great number of electronic parts with the highest attainable reliability.

For instance, Minuteman I (formally designated LGM-30 A and B) briefly absorbed over 90% of all transistor production in the United States.[1] For high-end ICs, military sales were more significant in later years, even after the end of Minuteman II. The final version of Minuteman III, LGM-30 G, bought 65% of ICs in the years when these missiles were built. Such statistics are impressive, of course, but obscure the more important consideration of what the effect was of the push for the highest reliability.

Processes and materials used to achieve the high quality needed by Minuteman affected the quality on commercial lines. Minuteman infused money to solid state industries and raised the overall quality of parts for all users.

Minuteman's revolutionary concept and extraordinary technical achievement created a nationwide industry: "Tens of thousands of industrial and Air Force managers, engineers, and workers [had] to be trained. New machine tools and test facilities [had to] come into being." These efforts changed "the face of America, the make-up

---

[1] It is no longer clear whether the figures for LGM-30 A, B, and G include the ground systems (WS-133A and WS-133A/M respectively), but regardless, the percentages speak for themselves about the impact the MINUTEMAN program had on electronic parts.

Implosion: Lessons from National Security, High Reliability Spacecraft, Electronics, and the Forces Which Changed Them, First Edition. L. Parker Temple III.

of the Armed Forces and the industries which support them."[2] America had no concept of a throw-away culture at the time. Electronic products were intended to last, backed by warranties to repair or replace defective products. Higher reliability parts actually lowered costs to the manufacturers because they incurred fewer warranty returns. The impact of improved reliability tracing back to the demands from the Minuteman and related developments included a collateral boost to commercial electronics reliability.

In 1961, TI developed the first complete equipment application of semiconductor ICs in a data processor for the Air Force Manufacturing Technology Laboratory. The processor was directly due to TI's earlier IC patent and led quickly to their 1962 use in the Minuteman II Guidance System (developed by North American Aviation's Autonetics Division), and the Apollo Guidance Computer (developed by MIT and Raytheon). ICs had become critical military and space applications.[3]

Urgency underscored the second generation ICBM's development and fielding. On 1 February 1961, less than three years after approval, the first Minuteman launched from the Air Force Missile Test Center in Florida. Working perfectly, it hit its target after a flight of 4,600 miles.[4] Within six weeks, construction began on the first Minuteman missile field, and this was ready when SAC placed the first flight of 10 missiles on operational alert on 22 October 1962, during the height of the Cuban Missile Crisis.[5] The timing was propitious. When President Kennedy countered the Soviet Union's placing ballistic missiles in Cuba with the promise that use of their missiles against any nation in the Western Hemisphere would be regarded as an attack on the United States, he was backed by Minuteman. Such was the nature of the Cold War that the Soviet actions were considered to be a response to the U.S. placement of Thor and Jupiter IRBMs around the periphery of the western Soviet Union. The Cuban Missile Crisis resolution was later linked to the decision to remove these U.S. missiles, but McNamara committed to remove them based on the Minuteman's availability. The President's and Secretary's actions would have been less confidently done had the Minuteman been unable to demonstrate its reliability and capabilities right from the very start.

The relentless pace of deployment did not abate. By July 1963, 150 were on alert. These turned into 300 by October, 450 by March 1964, and the 800th missile went in place in June 1965.[6] The last, the 1,000th, was deployed in 1975. The rapid fielding of these large missiles suggests the extensive demand they placed on the national output of some of the earliest solid state devices.

In addition to its other firsts of a kind, Minuteman I was the "first mass production application of semiconductors to high reliability military electronics." The

---

[2] Roy Neal, *Ace in the Hole* (Garden City, N.J.: Doubleday, 1962), 72–73.

[3] Patrick E. Hagerty, *Management Philosophies and Practices of Texas Instruments* (Dallas, Texas: Texas Instruments, 1965), 127.

[4] Neil Sheehan, *A Fiery Peace in a Cold War* (New York: Random House, 2009), 418.

[5] "History of the Minuteman Missile," National Park Service, http://www.nps.gov/mimi/index.htm.

[6] Http://www.strategic-air-command.com/missiles/Minuteman/Minuteman_Missile_History.htm.

reasons for the move to solid state were manifold: solid state components were more rugged and more compact, required less power, generated less heat than vacuum tubes, and represented better reliability because they drove down the number of connections needed within a circuit.[7]

The Minuteman I specification mandated every component—each piece part— "to have its own individual progress chart on which production, installation, checking, and rechecking . . . [t]esting, retesting, and re-retesting more than doubled the cost of each electronic part."[8] Minuteman demands for reliability caused the two most pressing reliability requirements for piece parts to be: (1) orders of magnitude increases in piece part lifetimes, and (2) valid and versatile reliability data soon after the part becomes available for application.[9] The reliability data would fold back into the next available purchase of parts, thus accumulating higher reliability specifications over time.

The first generations of ICBMs (Atlas and Titan) used analog autopilot technology for stability.[10] Guidance equipment only provided steering commands. Minuteman I introduced an integrated digital guidance and flight control system. When it flew for the first time in February 1961, this was the first flight of its kind anywhere in the world.[11] The guidance computer comprised about 15,000 components, and they had worked flawlessly the very first time, as demanded. This Minuteman I success allowed NASA to consider using a similar, even more advanced approach in the Apollo Guidance Computer. The Apollo Guidance Computer was recognized to be so challenging that it had a very long lead time. So, in August 1961, NASA let the first contract of the Apollo program when it contracted with MIT's Instrumentation Laboratory to study the feasibility of an integrated digital guidance computer.[12]

Minuteman I reliability did not rest on the initial design or the high selectivity of parts, but continued as part of a program of reliability improvement.

The Air Force and its prime contractor, Autonetics, carefully considered three alternative strategies to achieve size and weight reductions in the Minuteman guidance package. One involved repackaging discrete components. The second

---

[7] J.M. Wuerth, *The Evolution of Minuteman Guidance and Control* (Anaheim, California: Rockwell International Autonetics Group, 11 June 1975), DTIC Accession # ADB011836, p. 8.

[8] T.R. Reid, *The Chip: How Two Americans Invented the Microchip and Launched a Revolution* (New York: Random House Trade Paperbacks, 2001), 147.

[9] Joseph Vaccaro, "Reliability and the Physics of Failure Program at RADC," *First Annual Symposium on the Physics of Failure in Electronics*, September 1962, p. 5.

[10] This statement also applied to the Navy's first submarine-launched ballistic missiles, POLARIS A1, which began in 1957.

[11] Wuerth, *The Evolution of Minuteman Guidance*, 1, 3.

[12] Paul E. Ceruzzi, *Beyond the Limits: Flight Enters the Computer Age* (Cambridge, Massachusetts: MIT Press, 1989), 92; Linda Neuman Ezell, *NASA Historical Data Book*, vol. 2 (Washington, D.C.: Scientific and Technical Information Division, National Aeronautics and Space Administration, 1988), 172n. Some sources say the contract was let in April of that year, which may be when NASA requested the proposal, but the NASA Historical Data Book is clear that it was in place in August.

entailed use of thin-film hybrid circuits, similar to the approach taken by IBM in its 360 computer. The third was to use integrated circuits. Of the three, the integrated circuit route was understood to promise the greatest potential for size and weight reductions, but the technological distance to be traveled in order to achieve necessary levels of reliability was the farthest. In short, the Air Force chose the high-payoff, high-risk approach over more conservative options.[13]

Minuteman II was to be an impressive improvement over the earlier Minuteman. Strategic doctrine had changed, and demanded a range increase from 6,300 miles to 8,000 miles, but that was only the least of the major improvements. Improvements in the size of Soviet warheads, and expected improvements in their accuracies, meant Minuteman II had to be able to withstand a much more severe nuclear environment than its predecessors. The requirements for survivability were strict and, after the first 60 of the 350 Minuteman IIs, included operation through environments with X-rays, neutrons, prompt dose ionization, and electromagnetic pulse.[14]

Size and weight were important to the missile. The rule of thumb for such a missile was that every extra pound of weight cost $100,000 in extra fuel.[15] So, Minuteman II aggressively pursued the use of integrated circuits to reduce the complexity, improve reliability, and reduce size, weight, and power requirements.

The years 1962–1964 were important for the electronics industry due to these major government purchases. Minuteman II awarded two large development contracts to TI and Westinghouse. In 1963, further contracts were awarded to these two, and to RCA as well. Each Minuteman II used about 2,600 microcircuits, and there were to be 350 of these missiles. The TI contract forced the company to develop and produce 18 new ICs within six months—more than the total number of ICs then in production in the entire industry. During the time Minuteman II was buying components, estimates are it took 20% of the entire industry's annual sales, not to mention how it affected any single manufacturer. TI produced a device that put 22 existing discrete component circuits into a single IC for Minuteman II, producing 100,000 ICs by the end of 1964 and delivering them to Autonetics, the guidance program developer. By July 1965, Minuteman II was the largest single consumer of microcircuits, as microcircuit purchases had surpassed 500,000, with an estimated 600,000 more requisitioned. The delivery rate was about 15,000 per week.[16]

[13] Richard C. Levin, "The Semiconductor Industry," in *Government and Technical Progress: A Cross-Industry Analysis*, ed. Richard R. Nelson (New York: Pergamon Press, 1982), 62.

[14] Wuerth, *The Evolution of Minuteman Guidance*, 6, 10; "Minuteman Is Top Semiconductor User," *Aviation Week and Space Technology* (26 July 1965): 83.

[15] Reid, *The Chip*, 147.

[16] Eldon C. Hall, *Journey to the Moon: The History of the Apollo Guidance Computer* (Reston, Virginia: American Institute of Aeronautics and Astronautics, 1996), 19; R. Pay, "Circuit Weight Slashed in Advanced Minuteman," *Missiles and Rockets* 14 (2 March 1964): 35; R.C. Platzek and Jack S. Kilby, "Minuteman Integrated Circuits: A Study in Combined Operations," *IEEE Proceedings* 52 (December 1964): 1678; "Minuteman Is Top Semiconductor User," 83. Platzek and Noyce noted that the original requirement for new ICs was 20, but this was subsequently reduced.

Meanwhile, NASA's Apollo procurements were growing quickly. Fairchild had emphasized development and sales of its commercial Micrologic line of microelectronics throughout 1961. MIT's Instrumentation Laboratory bought samples, and in 1962, they recommended NASA use these devices as the baseline for the Apollo Guidance Computer. The first block of the computers took "about 200,000 Micrologic elements by the summer of 1964."[17] Some important work went to Fairchild Semiconductor Corporation, supported by Bell Labs and Motorola, for the Apollo Guidance Computer.[18] The demands for high performance and very high reliability were daunting for the new solid state industry, but not out of reach. Fairchild Semiconductor Corporation invented some commercial high-frequency transistors, and was the first to the marketplace with these. Minuteman reliability requirements at first precluded their use in that program as well. However, additional innovation and improved process controls made in cooperation with the Minuteman program allowed Fairchild a role in the military high reliability market.[19]

The "commonality of interest among the contractors to the federal government," wrote two scientists closely associated with the industry, "promoted the high diffusion rate of new information in semiconductor electronics." Everyone was learning from everyone else. Unlike most military developments, microcircuits were easily adaptable to commercial use.[20]

Because Apollo Guidance Computer development and manufacturing was sustained over multiple years through much of the decade, the Apollo program was claimed to be the largest single consumer of integrated circuits between 1961 and 1965. Apollo and Minuteman II accounted for virtually all sales of integrated circuits in 1963 and 1964. In 1965, the average microcircuit cost (for the industry) dropped from $15 to $12 apiece. The heavy influx of funding helped the existing microcircuit manufacturers to expand and improve, but also facilitated market entry for other firms.[21]

Histories of the semiconductor industry rarely appreciate the significance of the impact of this large a proportion going to Minuteman and Apollo. For instance, over the span of the Minuteman acquisition, the percentage of military and aerospace sales declined in market share, as shown in Table 5.1.

---

[17] Hall, *Journey to the Moon*, 19, 141; Norman J. Asher and Leland D. Strom, *The Role of the Department of Defense in the Development of Integrated Circuits* (Washington, D.C.: Institute for Defense Analysis, May 1977), paper P-1271, pp. 10–23. NASA and MIT proceeded with a more advanced design, the Block II Apollo Computer, in the spring of 1964. By the end of the decade, "Block II production had increased the total consumption of integrated circuits to approximately 1 million."

[18] Asher and Strom, *The Role of the Department*, 10–23.

[19] Chong-Moon Lee et al., *The Silicon Valley Edge: A Habitat for Innovation and Entrepreneurship* (Stanford, California: Stanford University Press, 2000), 159.

[20] John G. Linville and C. Lester Hogan, "Intellectual and Economic Fuel for the Electronics Revolution," *Science* 174 (18 March 1977): 23–29; Reid, *The Chip*, 151.

[21] Levin, "The Semiconductor Industry," 36, 62–63; "Minuteman Is Top Semiconductor User," 83. Levin's data shows an even more dramatic drop in digital ICs, dropping from $17.35 apiece in 1964 to $7.28 in 1965.

Table 5.1.   Distribution of U.S. Semiconductor Sales by End Use

End Use	Total Semiconductor Sales Percentages			
	1960	1965	1970[c]	1974[d]
Computers	30%	24%	31%	30%
Consumer Products[a]	5%	14%	14%	23%
Industrial Products[b]	15%	26%	25%	32%
Military and Aerospace	50%	36%	30%	15%

[a] Calculators, watches, automobiles, etc.

[b] Process controls, test equipment, office, and telecommunications equipment.

[c] Average of figures from 1968 to 1972.

[d] Average of multiple sources for 1974.

Source: Richard C. Levin, "The Semiconductor Industry," in *Government and Technical Progress: A Cross-Industry Analysis,* ed. Richard R. Nelson (New York: Pergamon Press, 1982), 19.

The decreasing share reflects the rapid growth of computers and consumer products despite the increasingly large, though inconsistent, military sales. The important point is *not how much* was bought, but *what* was bought. The military was buying the highest-end, highest reliability products, pushing the state-of-the-art. The common lack of discernment between all electronic parts and high reliability parts is a recurring theme in this history, and one that we will see eventually contributed to serious missteps in acquisition policies.

The Minuteman story is more than just about total piece parts produced, though that cannot be overlooked. With the transistor, the Army Signal Corps' central role as "an institutional entrepreneur presiding over technological change . . . undoubtedly increased the pace of transistor development."[22] The industry was changing, however, as Minuteman acquisition progressed from block I to block III. The government was buying a rapidly increasing number of integrated circuits, while the market as a whole accelerated away, making the government share decline as a percent of sales, as shown in Table 5.2.

Before the Minuteman II guidance computer project, commercial integrated circuits "were available only for logic, and had been proved only for low voltage, low power applications."[23] Minuteman II's guidance computer substantially reduced the volume of the Minuteman I computer, attaining one quarter the size of the earlier device. Despite the size decrease, however, Minuteman II's logic circuit board had the capability of twelve Minuteman I circuit boards, and more than double the

[22] Thomas J. Misa, "Military Needs, Commercial Radios, and the Development of the Transistor, 1948–1958, in *Military Enterprise and Technological Change, Perspectives on the American Experience,* ed. Merritt Roe Smith (Cambridge: Massachusetts: MIT Press, 1987), 276.

[23] Platzek and Kilby, "Minuteman Integrated Circuits," 1670. Feature sizes, for the curious, were 1 mil, but in some cases they were allowed to be 0.5 mil.

Table 5.2. Government Purchases of Integrated Circuits Through the End of MINUTEMAN II Procurements

	Total IC Shipments ($ Millions)	Shipments to Federal Government[a] ($ Millions)	Government Share of Total Shipments (%)
1962	4[b]	4[b]	100[b]
1963	16	15[b]	94[b]
1964	41	35[b]	85[b]
1965	79	57	72
1966	148	78	53
1967	228	98	43
1968	312	115	37

[a] Includes ICs produced for Department of Defense, Atomic Energy Commission, Central Intelligence Agency, Federal Aviation Agency, and National Aeronautics and Space Administration.

[b] Estimated by John Tilton, *International Diffusion of Technology* (Washington, D.C.: Brookings Institution, 1971), 91.

Source: Richard C. Levin, "The Semiconductor Industry," in *Government and Technical Progress: A Cross-Industry Analysis*, ed. Richard R. Neson (New York: Pergamon Press, 1982), 63.

memory capacity. In Minuteman III, it would double again.[24] Such advances seem expected in an industry where miniaturization was the dominant focus, but as industry-wide sales numbers obscure important details, so do the increases in capability. Manufacturing technology and circuit design technology were pushed by the project.

Reliability and quality impacted Minuteman costs as well. By 1964, Minuteman I's cost for five years of ownership was less than one-third that of any other U.S. strategic delivery system. That cost was figured only on the cost of keeping the missiles ready to go within their silos. Had the cost of flight reliability been included, the savings would have been even more dramatic.[25] Achieving that level of reliability had not been easy.

The prevailing view at the time was each lot of parts had a normal, or Gaussian, distribution of quality. Some parts in a lot would simply be better than others. An open question was the significance of the disparity between the weakest performing parts and the best parts in any given lot. Put another way: were lots homogeneous, or did wide variations exist?

To find out, the Minuteman I reliability effort drew on the best thinking at the time, emphasizing life testing of components, process controls, and extensive failure analysis. Testing, retesting, and re-retesting weeded out the weakest performing parts. But more was needed to achieve the reliability requirements.

[24] Wuerth, *The Evolution of Minuteman Guidance*, 25.

[25] Ibid., 10.

Minuteman II continued the practices of tight process controls and failure analysis. However, to surpass Minuteman I reliability performance, a new approach to life testing was needed. The life testing that served Minuteman I well was going to be impractical for Minuteman II "because of the extremely large sample sizes that would be required to get statistically significant data within a reasonable time."[26] Also, although fewer missiles were being acquired, they were considerably more complex electronically, and that meant greater difficulty and time in testing the parts. So, component manufacturers were asked to perform various stressing tests, "such as temperature, voltage, heat, vibration, etc." These tests discriminated for problems in device construction and fabrication techniques. The results pointed to areas needing improvement in design, materials, and processes. This approach was made a contractual requirement under the Component Quality Assurance Program (CQAP).

The RADC Physics of Failure work provided an essential complement to the CQAP, because it provided rational causes and mechanisms for the failures themselves. Testing alone pointed only to the failure, but not necessarily the cause or the mechanism. CQAP requirements mandated only that corrective actions be implemented, and if subsequent tests revealed no further failures like the one previously encountered, things moved on. The Physics of Failure completed the understanding necessary to capture lessons learned, improve processes, and instantiate such changes in military standards. RADC's involvement in CQAP for technical direction helped ensure that the two efforts worked smoothly together.

Raw sales numbers such as in Tables 5.1 and 5.2 simply obscure the full story about the impact on the market. Most recent histories point to the early industry being supported by the government funding. That hides the quality and reliability efforts and effects. Minuteman was not simply *buying* advanced technology parts. They were *teaching* the industry to make high reliability parts. And this was paying off. Figure 5.1 illustrates the effectiveness of the reliability improvement initiatives across all the Minuteman blocks. In each case, significant improvements were made, building on lessons learned.

All three versions of the Minuteman specified performance and reliability that surpassed the current capability of manufacturers. This pushed innovation and advanced processes.[27] The CQAP program forced manufacturers to do intense internal process analysis and understanding. Rigor, process discipline, and continuous improvement achieved and then surpassed the reliability goals. All of these paid off for the electronics industry at large.

In June 1970, as it achieved operational status, Minuteman III became the first ICBM to have multiple independently-targeted reentry vehicles. Despite such significant increases in capabilities and complexity, Minuteman III also added more

[26] Captain J.F. Wiesner, "Opening Remarks—Minuteman II Physics of Failure Program," *Physics of Failure in Electronics* 4, June 1966, DTIC Accession # AD 637529, p. 423.

[27] Levin, "The Semiconductor Industry," 64.

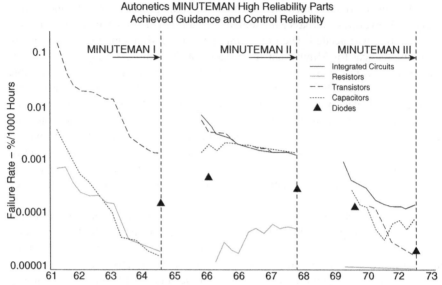

Figure 5.1.   Minuteman program reliability improvement effectiveness. Source: After J.M. Wuerth, *The Evolution of Minuteman Guidance and Control* (Anaheim, California: Rockwell International Autonetics Group, 11 June 1975), DTIC Accession # ADB011836, p. 9; Christopher Lecuyer, "Fairchild Semiconductor and Its Influence," in Chong-Moon Lee et al., *The Silicon Valley Edge: A Habitat for Innovation and Entrepreneurship* (Stanford, California: Stanford University Press, 2000), 179.

reliability improvements. By that time, Fairchild Semiconductor's transistors were showing a failure rate equivalent to one failure in 10,000 years.[28]

Over a decade after the last Minuteman III was deployed in 1975, transistor and diode manufacturers that had been part of the program's acquisition still advertised their products had "Minuteman reliability" with great pride at the accomplishment.

*   *   *

By the early 1980s, two major reasons were accepted as the basis for improved electrical component reliability. The first was the increased use of ICs, which brought with it a significant decrease in the number of connectors. Second was the wider use of solder to assemble ICs, rather than inserting them into sockets.[29] Such assessments overlooked deeper history, reflecting only more recent events.

---

[28] Christopher Lecuyer, "Fairchild Semiconductor and Its Influence," in *The Silicon Valley Edge: A Habitat for Innovation and Entrepreneurship*, ed. Chong-Moon Lee et al. (Stanford, California: Stanford University Press, 2000), 179. In Figure 5.1, that is equivalent to the 0.00009 percent failures in 1,000 hours.

[29] Ralph A. Evans, "Electronics Reliability: A Personal View," *IEEE Transactions on Reliability* 47, no. 3-SP (September 1998): SP-332.

Minuteman was acquired by the Space and Missile Systems Organization (SAMSO, the latest successor to WDD). As the name suggests, SAMSO's dual responsibilities allowed lessons learned from all the blocks of the Minuteman program to feed directly into the quality and reliability of parts for the national defense and national security space programs.

To understand the next step in the story of high reliability parts, we need to retrace some footsteps, and go back in time to events happening at the same time as the early Minuteman program.

Chapter 6

Skinning Cats

Minuteman emphasized and advanced the state of the art of reliability. However, like skinning cats, calculating reliability can be done in a number of ways. In some cases, the differences have to do with fundamental assumptions. More often, the differences have to do with the kinds of data available to perform the calculations.

The traditional approach was "measuring, predicting and testing reliability." The methods in place at the start of the Minuteman program indicated "the early 1950 parts, when used in large quantities in proposed weapon systems, would result in failure rates that would seriously impair the operational readiness of these systems."[1] Consequently, the Minuteman I brought a great emphasis to this specific set of issues in pushing the reliability state of the art. But Minuteman had to build on expertise and techniques that went beyond traditional approaches.

* * *

As a reminder to take nothing for granted, the space industry as a whole was to receive a shock, though not unexpected. The X-20 Dyna-Soar, several decades ahead of its technological time, had extensive cost overruns, launch delays, and significant technological problems. These proved too much for McNamara, who had no use for expensive human spaceflight programs. He summarily cancelled the X-20 on 10 December 1963. While such a major cancellation might have had severe repercussions throughout the industry as contracts were terminated and highly skilled professionals laid off, this was not so much the case this time. As the Secretary of Defense taketh away, so he giveth. McNamara replaced the X-20 program with a much less forward-leaning program called the Manned Orbiting Laboratory (MOL). The MOL was to use the Dyna-Soar's Titan III booster, so the launch vehicle program stayed on track. The electronic parts industry impact was delays of large high reliability electronics purchases, as the Dyna-Soar's building phase was replaced with

---

[1] George H. Ebel, "Reliability Physics in Electronics: A Historical View," *IEEE Transactions on Reliability* 47, no. 3-SP (September 1998): SP-380.

---

the MOL's design study phase. Consequently, the former large purchases of high reliability parts stemming from a few satellites and a large number of missiles wound down over time.

But that was far from the end of the relationship between the military and the high reliability parts industry.

* * *

After Maresky published his *Reliability Factors for Ground Electronic Equipment* in 1955,[2] RCA compiled and released a set of mathematical models for estimating component failure rates. RADC folded the RCA TR-1100 into the *RADC Reliability Notebook* in 1955. The *RADC Reliability Notebook*, and the series of subsequent notebooks, were other predecessors of what would become the primary military standardization document.[3] Revised again in 1959, the RADC book mainly considered Air Force ground equipment until a companion document on cooling ground electronics added the effects of thermal stress.[4] The premise of these documents was that components comprising a device are the primary determinants of failure rates.[5]

In addition to determining expected in-service rates, reliability estimates were used to anticipate maintenance requirements, particularly the level of spare parts needed. This latter consideration was crucial to the new ballistic missile submarine forces of the Navy, since reliability calculations were the basis for determining the kinds of spares to take on long submerged cruises. In some respects, the submarines faced issues not unlike those of spacecraft once out on patrol. Thus, it is not surprising that the Navy in 1962, building on the prior works, released MIL-HDBK-217, "Reliability Prediction of Electronic Equipment." Although competing approaches existed, MIL-HDBK-217's wide use within the DoD as a contractual compliance document quickly drove it to define the traditional reliability approach.

The MIL-HDBK-217 approach required quantitative data sources such as field returns. In a sense, then, it was predictive as much as giving a snapshot in time at current technologies. Reliability predictions needed a more fundamental level of information having to do with the way a part actually worked, the materials with which it was made, and how these would fail. This alternative path began with the physical processes involved in a component failure.

At the time, part failure rate data was collected from observations made during statistical testing, but did not rely on the underlying actual failure modes. To the

---

[2] "Pioneering Reliability Laboratory Addresses Information Technology," *Journal of the Reliability Analysis Center* 6, no. 1 (1998): 1.

[3] Ibid.; William Denson, "The History of Reliability Prediction," *IEEE Transactions on Reliability* 47, no. 3-SP (September 1998): SP-322.

[4] *RADC Reliability Notebook, Second Revision* (Washington, D.C.: Office of Technical Services, Department of Commerce, 31 Dec. 1961), RADC-TR-58-111; *Handbook of Methods of Cooling Air Force Ground Electronic Equipment*, (Griffiss AFB, New York: Rome Air Development Center, June 1959), RADC-TR-58-126, DTIC Accession # AD-148907.

[5] Denson, "The History of Reliability Prediction," SP-323.

time of the Minuteman program, the absence of the underlying failure mode information had not appreciably hurt reliability efforts.

> *Little, if any, effort has . . . been expended to clarify the relationship between total part failure rate and its underlying physical cause at the molecular level.... [I]t is becoming increasingly more difficult to escape a more fundamental approach to reliability improvement. Complex military equipment demands inherent, predetermined reliability in its components today, and yet cannot wait for it.*[6]

From the data available to them, officials at RADC could diagram the kinds of failures being experienced. The thing that was escaping them and preventing their ability to predict how devices would fail, had to do with the underlying causes. They knew the how of failures, but did not have a comprehensive understanding of the why.

For that reason, RADC began the Physics of Failure program in 1961. Joseph Vaccaro had Joseph B. Brauer create the Reliability Physics Branch and its program. From that time until the early 1990s,

> *RADC shaped the course of reliability physics through key study programs to enhance the corporate knowledge of physical properties and failure mechanisms in electronic parts. Until this time, there had been many discussions on various approaches to reliability physics, but the funding to maintain a focused program had been lacking.*[7]

After setting in place contracts with many of the major manufacturers of electronics, RADC set about investigating such things as bulk effects in silicon, noise correlation, high energy radiation effects, ceramic dielectrics, surfaces, interfaces, and more. Their objectives were to detect, identify, and measure individual failure mechanisms and, when possible, use the information to improve the reliability of the type of parts. Also, they wanted to "place the study of the physics of failure on a sound and workable basis," in the hope that this would provide better predictions of individual failure mechanisms.[8]

To meet the sound and workable need, RADC held a symposium on the physics of failure on 26–27 September 1962 in Chicago. This was the first of a series of annual meetings known at the time as the "Physics of Failure in Electronics." The RADC Applied Research Laboratory hosted it jointly with the Armour Research Foundation of Illinois Institute of Technology (later Illinois Institute of Technology Research Institute, IITRI). That first symposium was well attended by 350 people across many technical disciplines.

They had their work cut out for them because, as Figure 6.1 shows, there were very different kinds of failure mechanisms across the many different part types, and as the number of different kinds of parts continued to expand, the extent of the unknowns would only increase.

---

[6] Joseph Vaccaro, "Reliability and the Physics of Failure Program at RADC," *First Annual Symposium on the Physics of Failure in Electronics*, September 1962, p. 6.

[7] Ebel, "Reliability Physics in Electronics," SP-380.

[8] Vaccaro, "Reliability and the Physics of Failure," 9.

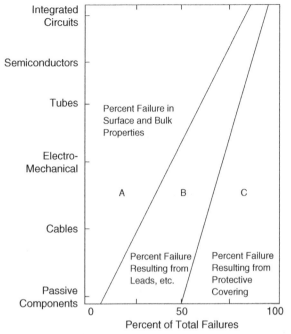

Figure 6.1. The 1962 view of sources of failures. Areas A and B represent mechanical failures which are generally caused by nonuniformity in manufacturing processes. Source: After Vaccaro, Joseph. "Reliability and the Physics of Failure Program at RADC," *First Annual Symposium on the Physics of Failure in Electronics*, September 1962, p. 10.

The purpose of the first symposium was "to relate the fundamental physical and chemical behavior of materials to reliability parameters."[9] The introductory address was given by Harry Davis, who was the Deputy for Research in the Office of the Assistant Secretary of Defense for Research and Development. He told the audience:

> *The Physics of Failure in my opinion is the very heart of the reliability effort, and a thorough understanding of mechanisms and physics of failures will be the principal means by which extreme reliability requirements posed by advanced systems will be met. Our failure to understand the basic mechanisms which lead to short and long term changes in properties of electronic parts and materials is without a doubt the most important cause of the very serious reliability problems which face the nation today. As a result, improvement in part reliability during recent years has been incremental compared to the rapid rate of other technological progress. Procedures for reliability testing and design control continue to be time-consuming and expensive.*[10]

[9] Ebel, "Reliability Physics in Electronics," SP-380.
[10] Ibid.

He also pointed to the magnitude of the Minuteman program's financial backing for reliability physics.

> *As evidence to these requirements, we have the Minuteman System where due to its constant alert characteristic and complexity, the Air Force was forced into a program of some $30,000,000 for parts improvement through control of production processing methods for reliability testing to obtain parts with this increased life characteristic; parts which are not available through normal procurement methods. By comparison, the total annual budget for Air Force research, development, testing and evaluation for electronic components is only $16,000,000.*[11]

The result of the body of technical information gathered by the Minuteman program "provided the nucleus from which the rest of the industry developed."[12]

Setting the precedent, this first symposium was specifically designed to generate information exchange and discussions to maximize the understanding of current problems facing the participants. Unlike many professional symposia, the papers were purposely not refereed, "to allow all new ideas, testing and opinions to be aired. In many cases the work was so new that an objective, knowledgeable referee would be difficult to find."[13]

One of the milestones in the first symposium was the proof of the use of the Arrhenius model of temperature stress for semiconductors. The model, which provided the basis for accelerated life testing of parts, had been used before this, but until the paper by G.A. Dodson and B.T. Howard of Bell Labs, there had not been a justification.[14]

In 1962, complexity was clearly a major source of unreliability. Joseph Vaccaro, introducing the first symposium on the Physics of Failure in Electronics, noted "*no matter how high the quality of individual parts*, the more of them we put into our equipment, the poorer will be the total reliability of the equipment. The implication for highly complex equipment is clear—we must have a very high order of component reliability."[15]

* * *

As the Minuteman missiles on alert reached the 150 mark, the Navy concluded MIL-S-19500 was unwieldy. Piece part quality and reliability testing requirements had matured significantly. Instead of being borne within MIL-S-19500, the testing information could be made more widely applicable if made into a separate standard. This was attractive, since it meant better ability to learn test procedures and retain

---

[11] Ibid., SP-381.

[12] Ibid.

[13] Ibid.

[14] G.A. Dodson and B.T. Howard, "High Stress Aging to Failure of Semiconductor Devices," *Proc. Seventh National Symp. Reliability and Quality Control*, Jan 1961, 262–272; Ebel, "Reliability Physics in Electronics," SP-381.

[15] Vaccaro, "Reliability and the Physics of Failure," 5. Emphasis in the original.

skills, and reduced the time required in setting up and tearing down test equipment. MIL-S-19500 covered a wide range of devices: transistors and diodes of all types, voltage regulators, and rectifiers, all coming under the heading of "discrete" devices. In 1963 the Navy developed MIL-STD-750, "Test Methods for Semiconductor Devices," for testing the kinds of parts manufactured in accordance with MIL-S-19500.[16]

* * *

The success of the first Physics of Failure in Electronics symposium encouraged RADC to hold a second, in 1963, also in Chicago. A larger number of papers were presented and included some noteworthy firsts for that forum. One covered the first real emphases on accelerated life testing, paving the way for many more such papers in the subsequent years. Other firsts included a paper on failure analysis from the users' viewpoint, and on failure mechanisms in gallium arsenide, a semiconductor material alternative to silicon.[17]

Brigadier General A.T. Culbertson, RADC Commander, recalling the start of the Physics of Failure work, told the group that effort was a "concerted program to place increased emphasis on the fundamental physical and chemical aspects of the reliability program. Motivation for the program arose from recognition of the very superficial knowledge available on the basic building block of the military weapon system, the component part."[18] By the third symposium in September 1964, Culbertson noted the statistical approaches were still well ahead of the efforts in the basic physics and chemistry of failure. Consequently, extrapolating the life test data from one kind of device to a similar, but different, one was "largely unrealized." He continued to urge progress at the subsequent symposium, saying "a fundamental physical approach represents the only rational basis for a systematic understanding of part failures."[19] Such encouragement from the RADC Commander was important, because it reflected where he and his organization were trying to go.

Consequently, the Minuteman CQAP announcement at that third symposium proved to be a watershed of other papers by spurring other, similar investigations. Indicative of the increasing relevance of the symposium's papers, the participants heard such topics as the results of an investigation into an industry-wide problem with aluminum electrolytic capacitors. The following year, some of the first papers to take advantage of the prior, but recent, physics of failure work were presented at

---

[16] MIL-S-19500 established four levels of quality assurance for the discrete electronic parts. These came under the designations of JAN, JANTX, JANTXV, and JANS in increasing levels of required quality, the last denoting the addition of space-level requirements.

[17] H.S. Dodge, "Failure Mechanisms in Semiconductors," and R.L. Anderson, "On the Degradation of Gallium Arsenide Tunnel Diodes," *Physics of Failure in Electronics* 2 (Sept. 1963): 328–337.

[18] BrigGen A.T. Culbertson, "Opening Address," *Physics of Failure in Electronics* 3 (April 1965).

[19] BrigGen A.T. Culbertson, "Opening Address," *Physics of Failure in Electronics* 4 (June 1966): xiii.

the fourth symposium. These papers covered active and passive discrete parts, and included one of the first on ICs.[20] Tracking the papers that were first of their kind throughout such symposia provides an insight into the focus of industry, government, and academia regarding the state of the art in high reliability electronics. The papers provide a gauge of the rate of progress in an industry whose underlying technologies continued to advance ever more rapidly.

*   *   *

Meanwhile, the Navy upgraded the IDEP data sharing process. They combined the Navy's Component Reliability History Survey Program (CRHS) with the IDEP. CRHS brought the reliability perspective to the ongoing sharing of testing data on the missile programs. Shortly thereafter, the Navy folded in additional missile testing documentation by adding the Guided Missile Data Exchange Program.

IDEP's success within the DoD soon attracted NASA. NASA's launch vehicles shared great commonality with the large ballistic missile programs (from whence they came). As NASA evolved these missiles for human spaceflight, their reliability became greatly important. So it was that NASA joined the IDEP program in 1965. In 1966, NASA began issuing warnings (ALERTs) to share information about parts not meeting their space program requirements. The ALERT system was eventually adopted by IDEP.

Broadening IDEP's constituency had several benefits. First was the recognition that devices, parts, and materials were becoming common across sources. A problem found by one organization might save impacts to another. Broader membership meant greater likelihood impacts would be caught earlier, translating to savings in program costs and schedules. Second, expansion of IDEP's content recognized the convergence and commonality underlying the technological bases of space, missile, and defense systems.

Strategic missile and space program requirements affected their supporting technologies. For instance, many of the Minuteman III's significant upgrades involved avoiding Soviet anti-ballistic missile systems (first revealed in 1964). The upgrades depended on solid state electronics because of size, weight, and power constraints, but there was a tradeoff with survivability concerns. These electronics had to survive Soviet nuclear anti-ballistic missile attacks at their "destination," but any U.S. missiles attacked before leaving their silos had to be electronically

---

[20] Ibid., xiii; A.J. Borofsky, "Component quality assurance programs for microminiature electronic components for MINUTEMAN II," *Physics of Failure in Electronics* 3 (September 1963): 1–14; R.S. Alwitt and R.G. Hills, "The chemistry of failure of aluminum electrolytic capacitors," *Physics of Failure in Electronics* 3 (September 1963): 93–107; T. Walsh and M. Rocci, "A Technique for Controllable Acceleration and Prediction of Degradation Mechanisms of Electronic Parts," *Physics of Failure in Electronics* 4 (June 1966): 59–73; J. Partridge, E.C. Hall, and L.D. Hanley, "The Application of Failure Analysis in Procuring and Screening of Integrated Circuits," *Physics of Failure in Electronics* 4 (June 1966): 95–139. The first paper on plastic parts was also presented. See S.M. Lee, J.J. Licari, and A. Valles, "Properties of Plastic Materials and How They Relate to Device Mechanisms," *Physics of Failure in Electronics* 4 (June 1966): 464–492.

survivable. Requirements for such nuclear hardness were not very different from the kinds of nuclear effects satellites experienced in various orbits. Vacuum tube-based systems had been relatively immune to electromagnetic pulse and radiation effects, but new solid state devices were not. Solid state devices not only had a lower tolerance to voltage surges, they also reacted in peculiar ways to different radiation types and energies. Understandably, any strategic missile force vulnerability introduced by such devices was a serious concern. Both space and missile systems began specifying their radiation tolerance, or hardness, driving another evolution of solid state electronics.

Whatever the demands on the strategic missiles, adding consideration for radiation tolerance to piece parts meant they would also be well adapted to use in space. In space, the natural radiation in the environment was much more severe than on the surface of the Earth, but much less than needed to deal with in strategic warfare.

No clear evolution path existed for radiation tolerance. The predominant logic circuit component at the time was the bipolar transistor, which appeared to have an inherently higher radiation tolerance compared with other solid state technologies. Such devices were able to withstand the effects from gamma, neutrons, and X-ray radiation. Their radiation tolerance came from their vertical structure, with components buried within the bulk silicon of the microcircuit. That radiation tolerance was at the price of high power consumption, creating problems for spacecraft, where power and heat rejection systems added considerable weight.

Each ounce of satellite weight in any component impacts launch vehicle performance and the satellite's stability, and represents weight that another component cannot have. As a rule of thumb, to get one pound of payload into a 100 nautical mile circular orbit, 100 pounds of launch vehicle fuel and weight are required. Seen in that light, it is easy to see why solid state components were so much more desirable than vacuum tubes. Every bit of saved weight affected multiple aspects of a satellite and its launch vehicle.

Reducing logic circuit power would have important payoffs rippling throughout a satellite's design. Alternative technologies to bipolar transistors existed, though without the technological maturity or radiation tolerance. Such a candidate was the metal oxide semiconductor field effect transistor (MOSFET), which were very effective switches using low power consumption surface effect principles. Field effect transistors were the type of device theorized in the 1930s, but were not the first kind of transistors developed due to difficulties of developing manufacturing techniques for them.

The specific origins of metal oxide semiconductor (MOS) devices are difficult to pin down, because a number of concurrent scientific and technical efforts aimed at solving their unknowns. As early as 1961, device elements for MOS transistors were made in laboratories. Several projects, across manufacturers such as TI, Fairchild, and RCA, received funding from the Air Force and National Security Agency to advance their individual approaches to MOS for use in calculators and computers. The MOS technologies matured quickly (by 1960 standards) with the first commercially available MOS-integrated circuit appearing in 1964.[21] So, what was it about MOS which made it of such interest?

MOS required fewer manufacturing steps than bipolar devices, and that meant lower cost devices. Its low radiation tolerance and somewhat slower speed were, for the time being, important drawbacks. One form of MOS, the complementary MOS (CMOS) technology, introduced by RCA in 1968, seemed most promising, but making such devices radiation tolerant proved a great challenge. By the late 1960s, however, considerable progress had been made, and CMOS devices began to meet the qualification criteria in MIL-M-38510, "General Specification for Microcircuits," about which more will be said shortly.[22]

Much lighter than vacuum tubes and earlier solid state devices, CMOS meant space programs could save weight or opt for more redundancy in backup systems. So, not only was advanced solid state technology making satellites more capable, additional layers of redundancy were beginning to make them more reliable. Reliability translated into greater certainty they would be working when the need arose because it also meant longer lifetimes, which would become increasingly important.

*   *   *

The fifth Physics of Failure symposium was sponsored by RADC and Battelle-Columbus Laboratories, and had grown to 535 people in attendance. This was the last to be sponsored by RADC, who turned over sponsorship the following year to IEEE Electron Devices Group and the IEEE Reliability Group. Classified sessions delved into radiation-related topics, but this was overreaching for the purposes of the symposium. The radiation topics were excluded after that, and left to other forums such as IEEE's annual Nuclear Effects and Space Radiation Conference (NESRC). The differentiation of these conferences and symposia gives a strong indication of the degree of specialization and the depth of the topics being covered. The industry continued to expand, as did the underlying university curricula (which generated an increasing number of the papers presented). Individual engineers could actually make a career of specialization in radiation topics, and even that began to break out into different internal branches. These conferences became important information leveling events.

*   *   *

Despite the solid state electronics industry's youth, some important forecasts appeared in 1965. As in any industry, sales and manufacturing projections or

---

[21] Bo Lojek, *History of Semiconductor Engineering* (New York: Springer-Verlag, 2007), 322–323, 328, 342. Specifically, the kind of MOS was p-type MOS, used for logic circuits. PMOS came first, followed by n-type or NMOS, and finally by complementary or CMOS. While the distinction between p-type MOS and n-type MOS is very important, it exceeds the scope of this present discussion. What is important, however, is that each advanced the technologies of integrated circuits, becoming more efficient and using less power.

[22] Ernest Braun and Stuart Macdonald, *Revolution in Miniature: The History and Impact of Semiconductor Electronics* (New York: Cambridge University Press, 1982), 102; *Reliability Handbook*, p. 121.

technology forecasts were important. By that time, Patrick E. Hagerty had been with TI for twenty years. Hagerty was thinking on a strategic scale about the potentials of the electronics marketplace. In the December 1964 *Proceedings of the IEEE*, he laid out necessary criteria for the industry's future expansion. The following year, he built on these in a book capturing TI's groundbreaking management philosophies and practices.[23]

> *For electronics to be truly pervasive, it must be readily and commonly used by the mechanical engineer, the chemical engineer, the civil engineer, the physicist, the medical doctor, the dentist, the banker, the retail merchant, and by the average citizen in broader ways than just for bringing entertainment to his home.*[24]

Although his vision began with those highly technical consumers, already in the marketplace, who were pushing the advancement of technology, he knew that ultimately the average person had to "need" electronics. This was a vision that went well beyond any sense of national security and defense.

He envisioned stages of electronics marketplace maturity, and considered pervasiveness to represent the terminal phase. That required a "relatively concentrated, highly automated industrial complex" that would supply discrete piece parts and integrated circuits to the rest of the electronics industries and industry in general. It necessitated a "very large number of organizations . . . [utilizing] these basic electronic materials to solve their own and their customers' problems . . . in all sectors of society." He also recognized an important enabling truth. "This integrated circuits industry must have established a common language for the input and output parameters which specify its products." That is, standards were essential to achieving his strategic vision. His tactical view of standards was for "a wide variety of computer programs, which will have replaced conventional engineering handbooks as we know them today and truly allow the user of these basic electronic materials, integrated circuits and compatible discrete components, to design the required electronic functions by computer."[25]

All of these would shortly come to pass, and well ahead of his forecast.

Also in 1965, Fairchild Semiconductor Corporation's Gordon E. Moore turned a seemingly innocuous observation into a forecast that has driven the industry ever since. He noticed "complexity for minimum component costs has increased at a rate of roughly a factor of two per year." He felt certain such a rate of solid state componentry densification would continue for the short term, and was bold enough to extend it into the future. In an April periodical, he expounded what later became known as Moore's Law:

---

[23] Patrick E. Hagerty, *Management Philosophies and Practices of Texas Instruments* (Dallas, Texas: Texas Instruments, 1965), 123. In his book, Hagerty included a reprint of his article from the IEEE Proceedings of December 1964. The IEEE article formed part of the book, but the rest of the book allowed Hagerty to more fully explain how he had helped guide Texas Instruments to be an innovator and leader in this new technology field.

[24] Ibid., 126.

[25] Ibid., 132–133.

Over the longer term, the rate of increase is a bit more uncertain, although there is no reason to believe it will not remain nearly constant for at least ten years. That means by 1975, the number of components per integrated circuit for minimum cost will be 65,000 . . . on a single chip.[26]

In addition, he projected manufacturing costs falling by a factor of ten within five years. Only time would tell if such extreme projections could be achieved, given the extremely uncertain advances that had to be made in manufacturing technology to achieve them.

*   *   *

The Cold War arms race added a new element in the mid-1960s. The Soviet Union had developed, and began deploying, an anti-ballistic missile (ABM) system. President Lyndon Baines Johnson was cool to the need for the United States to respond. However, within the Johnson Administration and Congress, a debate raged over how to respond.

Johnson's attitude puzzled many in the military. Because the Soviet system used a nuclear warhead mounted on a powerful missile, it also had a collateral ability to attack low altitude satellites. Among these low orbiting satellites were some of the most critical capabilities available to the nation. At the time, these most important capabilities were deeply, and thoroughly, hidden. The National Reconnaissance Office (NRO), responsible for all reconnaissance satellites, was so completely classified that its name, its initials, and its existence were not acknowledged. And yet, the importance was discussed by President Johnson when speaking to a group of southern educators in Tennessee the evening of 16 March 1967. What he said underscored the effects the most critical national security space systems were having:

I wouldn't want to be quoted on this, but we've spent between thirty-five and forty billion dollars on the space program. And if nothing else had come out of it except the knowledge that we get from our space photography, it would be worth ten times what the whole program has cost. Because tonight I know how many missiles the enemy has and, it turned out, our guesses were way off. We were doing things that we didn't need to do. We were building things that we didn't need to build. We were harboring fears that we didn't need to have.[27]

Johnson knew where the ABM system was deployed and how extensive it was. His confidence and security in the face of this threat depended on film-return satellite photoreconnaissance. He knew, as he explained to the audience that night, based on what the Soviets were doing, that they were willing to discuss "restriction of

---

[26] Gordon E. Moore, "Cramming More Components onto Integrated Circuits," *Electronics* 38, no. 8 (19 April 1965): 114. At the time, Moore was the Director of Research and Development Laboratories at the Fairchild Semiconductor Division of Fairchild Camera and Instrument Corporation. Three years later, Moore became the co-founder of Intel Corporation.

[27] Evert Clark, "Satellite Spying Cited by Johnson," *New York Times* (17 March 1967); William E. Burrows, *Deep Black: Space Espionage and National Security* (New York: Random House, 1986), vii. The *New York Times* did not report the text, probably in deference to the President's opening words, but paraphrased the content extensively.

offensive and defensive missiles to check the arms race."[28] Standardization, reliability, advanced technologies, and disciplined processes for handling growing complexity all came together to deliver to the President, his advisors, and his strategic planners the knowledge they needed to maintain a properly balanced policy course, and reduce the chance of any Pearl Harbor-like surprise.

*   *   *

In missiles, reliability of piece parts affected availability rates. Once on alert, the strategic missiles needed to be ready at a very high availability rate. Another way to say this is any missile taken off alert status for maintenance was unable, and therefore unavailable, to accomplish its strategic mission. Downtime affected readiness levels, an important operational consideration in the Cold War calculus. Not only was reliability an important element of being ready to go immediately upon the launch command, so was keeping the missiles available on alert. Equipment failure rates had become an important factor in high reliability systems by the mid-1960s. Although missiles could be repaired, downtime for maintenance introduced undesirable operational outages.

A failed satellite, though, was the loss of a critical space capability until a replacement could be launched, which could be a considerable length of time. Robust redundancy helped delay the satellite's inevitable failure. Improved solid state communications devices allowed more extensive telemetry to report, identify, isolate, and possibly correct problems that occurred. Even with failure rates of about 1% per thousand hours of operation, more needed to be done.

Analyses of failure causes indicated a rigorous screening program could find and eliminate many sources of failed piece parts before assembly into equipment. At the time, it was axiomatic that, once a satellite device was turned on, it would either fail within a few hours to a few days, or it would last for a very long time. Early failures, referred to as "infant mortality," were a plague to satellites. Eliminating "infant mortality" would improve satellite reliability, equating to longer lifetimes.[29]

AFSC had a series of specific laboratories supporting its product divisions. RADC's extensive facilities and equipment primarily supported Electronic Systems Division (ESD) at Hanscom AFB in Massachusetts. RADC's primary asset was engineers with close industry relationships, maintaining awareness of the specific state of the art for any current or emerging electronic technology. As the Air Force's center for high reliability electronics knowledge, RADC worked closely with high reliability experts throughout government and industry. They developed the first standardized screening procedures to remove the "infant mortality" failures then plaguing microcircuits.

---

[28] Clark, "Satellite Spying."

[29] T.R. Moss, *The Reliability Data Handbook* (New York: ASME Press, 2005), 10; Bilal M. Ayyub and Richard H. McCuen, *Probability, Statistics, and Reliability for Engineers* (New York: CRC Press, 1997), 396.

RADC's large cadre of Air Force officers and civilians had advanced degrees in electrical engineering and complementary disciplines. More than 50% had PhDs. Many maintained relationships with Syracuse University, expanding their reserves and recruiting pool. They were aware not only of the next generation developments, but of where individual corporations were heading with their own research and development. RADC personnel occupied a privileged position of trust, working directly with industry manufacturers while protecting their technological innovations. In many cases, they were able to help the corporations with additional expertise, avoiding various failure mechanisms. RADC was a central clearing house for issues resulting in IDEP ALERTs. As solid state electronics burgeoned, RADC became a national asset supporting national defense and national security programs. RADC was the one and only place where knowledge of issues and their solutions came together for the government and industry.

At RADC, Richard Nelson developed an in-house thin film and monolithic microcircuit manufacturing facility. This facility allowed the RADC engineers to learn a great deal about these technologies, and how to make screening economically feasible. The knowledge gained resulted in RADC's publication of *Quality and Reliability Assurance Procedures for Monolithic Microcircuits*. That, in turn, became the basis for MIL-STD-883, "Test Methods and Procedures for Microelectronics," in 1968. This was the integrated circuit companion to MIL-STD-750 for semiconductor testing.[30] MIL-STD-883 had two major sections. First, it compiled semiconductor wisdom in a detailed, prescriptive "how to" description of "environmental, mechanical, visual and electrical test methods." The collection of detailed tests was designed to test for "specific reliability and quality concerns" affecting semiconductor devices. Standardization brought with it an important leveling effect, since use of the tests in identical ways across all suppliers helped ensure consistency of quality across manufacturers, increasing the ability to have a competitive marketplace. Furthermore, standardization eliminated differences in electronic specifications for the same parts among the Services and NASA.

The second part of MIL-STD-883 was just as important, covering screening and conformance testing requirements. Recognizing the need to achieve cost-effectiveness, not all semiconductor devices needed to be of equal reliability. Parts for a deep space probe necessarily needed greater reliability than a routine ground-based (and repairable) system. Thus, MIL-STD-883 established different classes of semiconductor parts, each with a tailored set of screening flows appropriate to their intended end use. This was an important step forward, as it recognized space qualification as something different from other kinds of high reliability concerns. Class A parts were intended for use in applications where reliability was a critical concern. In terms of high reliability parts, Class A parts, used in NASA's human spaceflight missions, were to have failure rates no higher than 0.004% per thousand hours of operation.[31] Class B parts had a different set of screening flows for applications in

---

[30] "Pioneering Reliability Laboratory," 2; *Reliability Handbook*, 18.

[31] *Reliability Handbook*, 18.

aircraft or ground systems where reliability was essential, but removal and replacement were possible. Finally, Class C parts had the least reliability concerns and were for applications that were easily repaired in case of failure, and those for which occasional outages were not critical.[32]

MIL-STD-883 also introduced a flexible and expandable numbering system, because no one naïvely thought standards would not evolve as technology advanced.

RADC remained in control of test method development or improvement. Using industry reliability data, RADC could identify where concerns existed. Based on their extensive technical expertise, working with the manufacturers, methods could evolve. New methods required coordination among the major agencies such as SAMSO (supported by the Aerospace Corporation, which had recently been designated a Federally Funded Research and Development Center or FFRDC), NASA, and the Naval Ammunition Depot (NAD) in Crane, Indiana.[33] RADC then handed the result to the Defense Supply Agency (DSA) for coordination among the affected manufacturers, government agencies, and others such as JEDEC. With RADC providing the technical expertise, then, a rigorous process for updating the military standards was in place. The process was highly conservative of wisdom, backed by a long term, stable, civilian (and uniformed Air Force) workforce.

Thus, RADC's convergence of many different facets of electronics expertise resulted in its unique standards role. This was true for a wide variety of terrestrial military systems, but certainly also for space systems. They had either complete or extensive possession of three of the key reasons for standards. Requirements were the directive language of standards, denoted with the language of contractual compliance by use of "shall." Standards benefited by learning lessons from experience using a process that RADC helped raise to a high art. Each lesson could generate one or more "shall" statements and affect multiple standards. Lessons learned were not the only basis of directive language, however. Just as important was information from the manufacturers about emerging technologies, where the underlying phenomenology required some new handling technique or some new kind of testing. The basic physics of the underlying phenomenology could drive tests and methods. RADC's relationship with industry and various laboratories thus served it well in understanding device behavior at a fundamental level, which translated into appropriate "shall" statements. A third source was known as "field return data," or information derived from defective parts returned to the manufacturers. This could point to shortfalls in processes, materials or part designs—information that could be very sensitive to a manufacturer. RADC's access to such data highlights the sensitive role

---

[32] Ibid., 18–21.

[33] U.S. Congress, Office of Technology Assessment, *A History of the Department of Defense Federally Funded Research and Development Centers* (Washington, D.C.: U.S. Government Printing Office, June 1995), 11. On 1 July 1967, AFSC combined two former divisions, Space Systems Division and Ballistic Systems Division, to form SAMSO. In 1967, the Federal Council for Science and Technology renamed the FCRCs as Federally Funded Research and Development Centers to more accurately reflect their role.

the organization played due to its ability to handle and protect such information. More than that, however, industry benefited by gaining advice from the RADC experts, and the latter used their insights to ensure corrective measures got fed back into appropriate standards. RADC was a repository of deep insight into electronics and the military standards derived from that insight.

RADC's continued emphasis on quality control processes benefited the whole semiconductor industry. These processes, demanded by the military standards, supported the best commercial practices, and were an essential element of the success of the high quality and reliability of U.S. semiconductor devices. RADC were able to find problems when everyone else was at a loss.[34]

* * *

To this point, we began with the initial breakthrough in understanding electron behavior in the 1930s which enabled solid state electronics. America learned an important lesson in World War II: advanced technology was imperative for national security and national defense. Although driven by military investment, technology advancements of the 1940s and early 1950s created the popular view of the future where technology promised better life, prosperity, and security. Soon, responding to the threat of the Soviet Union meant more advanced weaponry. Such weapons challenged the fledgling solid state industry to deliver the reliability demanded to face the threat. Proliferation of different solid state devices in various shapes, sizes, and descriptions led to military standards, improving competitiveness, and providing more uniform quality. Test methods and equipment raced to keep up with the advances in solid state electronics.

Military involvement in solid state electronics and reliability improvements forced the early pace of advancement, focused research, and provided much needed capital. This influence was episodic, driven by the immense number of parts for the MINUTEMAN and the Navy's POLARIS. However episodic the buying was, the first lessons of high reliability and quality were learned quickly. The highest reliability systems raised the industry's level of commercial parts.

With production facilities sized to meet the demands of the late 1950s and early 1960s military buildup, excess capacity drove down unit cost and made solid state parts available for new consumer electronics. Quickly, the commercial market forces took over and drove the end item costs more than the military could, except in the area of the highest reliability parts. In that segment of the market, the military was dominant.

The same time frame spawned the U.S. space programs. These were:

1. The publicly acknowledged civil space program, aimed at space exploration and national prestige;
2. The somewhat classified military space program whose purpose was the defense of the U.S.; and

---

[34] Author interview with John Ingram-Cotton, 25 April 2009.

3. The unacknowledged, highly classified national security space program, whose purpose was reconnaissance for determining the intentions and developments behind the closed borders of the Soviet Union.

All three space programs placed increasing demands on the early solid state electronics industry. Failures taught hard-won lessons, and these lessons needed to be captured, retained, and used to improve future electronic parts. Some things were certain:

1. Commercial parts were not the equivalent of military parts;

2. Military parts needed to be much more reliable to meet the demands of operating in space; and

3. The nation was coming to demand information from space to support the highest levels of decision and policy making within the government.

The new frontier in space was tied inextricably to the new frontier in solid sate electronics.

Part II

Startup Transient (1969–1980)

National defense and national security space programs experienced a nexus in the 1969–1972 timeframe. Major changes occurred, affecting program content, direction, and complexity, as well as control over these elements. The changes shook the foundations of many parts of the space program. Different kinds of space programs emerged as military human spaceflight was out of the picture, despite the continued importance of human presence for the civil space program. Development of a reusable space launch vehicle publicly overshadowed the continued emphasis on the critical national defense and national security uses of space. Military standards continued to advance the quality of electronics, but major changes were occurring there as well.

Demands for greater capabilities in commercial and military electronics drove up complexity. Critical changes had to occur in a variety of complex and interrelated ways that would have effects well outside either space programs or the electronics industry. Within each of these, processes and ways of doing business had been refined to a high art. Advances meant some fundamental changes had to occur. Complexity of electronics and of whole systems demanded attention, or else reliability would suffer. Complexity threatened reliability, and therefore quality, until the means to manage complexity caught up.

Many threads came together from 1969 through the end of the 1970s as national defense space systems faced a critical make-or-break point.

Implosion: Lessons from National Security, High Reliability Spacecraft, Electronics, and the Forces Which Changed Them, First Edition. L. Parker Temple III.
© 2013 the Institute of Electrical and Electronics Engineers. Published 2013 by John Wiley & Sons, Inc.

Their capabilities, crucial national importance, and complexity all warred with their need to demonstrate their viability within modern warfare. Space systems could no longer be laboratory experiments or engineers' hobby shop items. They were becoming societally relevant. Going to the moon was an impressive limited feat, but a critical hurdle had to be leaped in the domains of national security and national defense: reliance. If space systems could not be relied upon in conflict, then they were of no real military value. The time had come to demonstrate more than relevance. Space systems had to demonstrate they were robust enough to be relied upon.

To make space systems mainstream would require vision, revisiting fundamental assumptions, and a new way of doing business that built upon the best aspects of what had gone before.

Chapter 7

Changing the Sea State

The years 1969–1972 were about to bring a major change in direction for space programs in general, with focus on a specific reusable launch vehicle, not a destination or mission. During this time frame, opportunities arose for high reliability parts to take a new and much stronger set of processes to their logical conclusions, namely, with the creation of a new class of parts exclusively for use in space. This was possible because NASA's focus shifted to reusable launch vehicles, away from Class A parts, opening the window for creation of a new class. The change was also needed because of the lack of respect paid to space systems by the other-than-strategic commanders in the Air Force.

As space program capabilities potentially touched every aspect of military operations, this became a problem for some Air Force leaders. Their problem was not always a lack of appreciation for the space systems' capabilities. Instead, some senior military and DoD civilian officials became concerned about over-reliance on space systems. The Soviet Union's new co-orbital anti-satellite program concerned some of them, but there was more to it than this threat.

The lack of respect manifested itself as a lack of confidence in space systems dependability due to their fragility, making them easy targets.[1] Such easy targets, then, might provide useful capabilities in peacetime but could not be relied upon in wartime. Space systems were simply not robust, not organic assets, and seemed easily countered. The perception that they were going to be undependable in wartime operations shaped an ongoing set of discussions among military leaders. Policy and doctrine held overreliance, or dependence beyond where such capabilities could be relied upon, was to be avoided.

---

[1] For clarity, here the term "fragility" refers only to the tenuous ties to a satellite in orbit. If anything happens to such a machine, only the on-board capabilities, exercised through telemetry, are available to figure out the problem and correct or compensate for it. The term does not have anything to do with the ability of a satellite and its components to survive accelerations or the vibro-acoustic environment during the launch phase. Neither does it have anything to do with the health of the satellite or constellation due to long-term operation and aging effects.

Implosion: Lessons from National Security, High Reliability Spacecraft, Electronics, and the Forces Which Changed Them, First Edition. L. Parker Temple III.
© 2013 the Institute of Electrical and Electronics Engineers. Published 2013 by John Wiley & Sons, Inc.

Using that line of thought, it therefore became better not to rely on space systems at all, since the way forces trained in peacetime was the way they would fight in war. This threatened to make space capabilities fascinating peacetime curiosities irrelevant to the future of military operations. Warfighters, unfamiliar with what space was really capable of doing, observed from afar the high number of launches needed to keep operational satellites in orbit, the unpredictability of when failures or outages would occur, and the length of time necessary to make replacements. Externally and superficially, these communicated fragility and unreliability, undermining space system acceptability.

The national security space program then suffered an apparent devastating defeat while the civil space program was ascendant not only at home but abroad as well.

On 10 June 1969, Defense Secretary Melvin Laird abruptly cancelled the Manned Orbiting Laboratory, the largest national defense and national security space program since its inception in the wake of the X-20 Dyna-Soar cancellation in 1963. This was the last serious gasp of the idea that truly operational national defense and national security space systems required human presence. That expectation was a vestige of the disappointingly crude technology of robotic aircraft of World War II and the early 1950s. The idea that humans were required raised great expectations for SAINT and Dyna-Soar. When they were cancelled, partially in favor of MOL, some sentiment remained alive that human presence still had value. MOL's cancellation was almost anticlimactic, though, since human presence was clearly not needed. Missiles and satellites had unquestionably demonstrated important capabilities without humans on board, and performed tasks, such as unblinking surveillance of Soviet missile fields for indications of missile launches, for which humans were ill-suited. Cancellation of MOL without a military human spaceflight follow-up marked some acceptance of the suite of robotic space systems.

The suite of space capabilities ranged widely by the end of the 1960s. Communications satellites had differentiated into many specialized systems supporting a variety of military users, such as the Navy's Fleet Satellite Communications System, Air Force Satellite Communications System, Initial Defense Communication Satellite Program (IDCSP), and satellites for the North Atlantic Treaty Organization. These systems exploited a wide range of different frequencies, with different characteristics and coverages, providing critical worldwide connectivity for deployed forces. The Defense Support Program (DSP), successor to MIDAS, was providing essential worldwide EW for missile attacks. The Defense Meteorological Satellite Program (DMSP) was enhancing military operations. Navigation was emerging as the next great capability, bringing together technology from the Navy's Timation program, and evolving the first glimmer of what would become the Global Positioning System (GPS). National security systems such as Corona and its successors were providing images via film-return to bolster U.S. awareness and preparedness worldwide. Edwin Land, in fact, had just finished another technology examination. He and his colleagues, recognizing the importance of near-real-time intelligence to the war in Southeast Asia, concluded that the technology was available to develop a similar capability from space.

Space was not a niche for engineers, but on the verge of becoming a truly operational capability. The remaining question, after the demise of military human spaceflight and the emergence of the dazzling array of robotic capabilities, was whether these capabilities could be depended upon in a future conflict. Achievement of this status owed much to the refinement and advancement of ever higher reliability electronics.

The United States won the race to the Moon a month after MOL terminated. Apollo 11's successful landing in and safe return from the Sea of Tranquility on 24 July 1969 finally completed President Kennedy's quest. With several more Moon landings in the offing, civil space potentials were ascendant. Apollo was an important technological feat providing scientific benefits. The problem was that it had been a success. The United States had beaten the Soviets and accomplished a daunting, challenging task within a decade from a nearly standing start. The feat was undeniable, but not practical. National defense and security had to make space *routine*— what the military would term "operational."

Apollo's success raised a major policy question: what should the nation's civil space program do next? NASA and the Nixon Administration could not sustain national interest in space exploration. The national reaction to the Apollo feat was a complex mixture reflecting the turbulent times of the 1960s and growing public disillusionment.

Richard M. Nixon's Vice President, Spiro T. Agnew, chartered a seminal effort to answer that question. The 1969 Space Task Group, with a parallel effort within the national defense and national security space programs, led to a number of key decisions. The most visible one was the Space Transportation System (STS). Aside from its promised order of magnitude cut in launch cost was its ability to retrieve old satellites and update or repair them in space. Part and parcel of the overzealous sales of the STS, this overlooked both practical and economic realities. Bolstered by the Apollo success, however, routine and reliable use of space seemed only a short time away.[2]

An alternative approach to achieving reliability was the Air Force's efforts in making aircraft more maintainable. However, this increased their complexity, making aircraft less reliable, needing more repairs. This difficult lesson learned in aircraft maintenance was not lost on the developers of spacecraft. The fifteen years of rapid improvement in solid state component technology poised it for a major leap forward.

Reliability can come from redundancy, as mentioned above, which was the path followed to that point. That remained insufficient. Improving part quality was a more difficult step, but was, by the late 1960s, where reliability emphasis had to head.

* * *

The Physics of Failure Symposia continued to mark the advances in high reliability electronics. For the ninth meeting of the Physics of Failure Symposium, the name was changed to the Reliability Physics Symposium. The focus remained the same,

[2] L. Parker Temple, *Shades of Gray: National Security and the Evolution of Space Reconnaissance* (Reston, Virginia: American Institute of Aeronautics and Astronautics, 2005), p. 469.

as reliability professionals took on accelerated life testing. Proof of the validity of the Arrhenius model of temperature stress for semiconductors had been given at the first of the symposia. But by the ninth symposium, longer part lifetimes were an increased testing concern. "With the rapid development of new parts, one concern was determining the field reliability of these parts with a minimum expenditure of time and money. The most accepted method was that of accelerated testing."[3] Advanced reliability was that parts lived longer, and the longer they lived, the more was desired. That is another way of saying the failures per 1,000 hours in Figure 5.1 had gotten so low that predicting the lifetime of a given lot of parts would take a very long time to verify unless their wearout were accelerated. Generally, this was accomplished by heating the parts under an electric load, using the Arrhenius model. Now the emphasis among professionals was establishing acceptable guidelines for accelerated testing of parts, and understanding what the accelerated testing could and could not do.

The following year held an indicator of a key turning point, when the first paper on building in reliability appeared.[4] This was an important recognition: reliability began before the part was even started. The Minuteman program and all subsequent efforts had pushed testing about as far as it could be pushed before another avenue of reliability improvement became lower hanging fruit. This key insight was to have an important consequence in terms of how parts were to be graded, and what constituted the highest reliability parts. "Minuteman Hi Rel" might still have been an important accolade for a part, but that was about to change.

As a mark of the increasing attention paid to ICs, one of the first papers on wafer testing appeared. This topic eventually became so important that it spawned a series of conferences on its own, starting in 1982, and held annually at Lake Tahoe.[5]

* * *

One way to tell when a capability has become important is the owner stops pushing people to use it because they start needing and demanding more of it. National reconnaissance led the way. Although national defense systems were playing important roles, such as EW in DSP's case, they continued to support essentially the same users as they had when originated. That, in business terms, was not a growth industry. EW was fundamental to national security, so it had to be sustained. EW, as a case in point, would evolve to do better as technology advanced, but EW remained the basic mission of DSP.

---

[3] George H. Ebel, "Reliability Physics in Electronics: A Historical View," *IEEE Transactions on Reliability* 47, no. 3-SP (Sept. 1998): SP-382; D.S. Peck, "The Analysis of Data from Accelerated Stress Tests," *Proc. Ninth Ann. Reliability Physics Symp.*, 1971, 69–78; C.H. Zierdt, Jr., "Some Circuit Considerations and Operating Precautions for Accelerated HTRB and Operating Testing of Semiconductor Devices," *Proc. Ninth Ann. Reliability Physics Symp.*, 1971, 79–83.

[4] W.H. Schroen, J.C. Aiken, and G.A. Brown, "Reliability Improvement by Process Control," *Proc. Tenth Ann. Reliability Physics Symp.*, 1972, pp. 42–48.

[5] George H. Ebel, "High Temperature Reverse Bias Testing at Wafer Level," *Proc. Tenth Ann. Reliability Physics Symp.*, 1972, pp. 82–87.

In Napoleon's time, Clausewitz had used the term "fog of war" to describe the conflicting, confusing, and often disjointed information finding its way to a commander. More information was not better in such circumstances, just more noise. The changes being wrought in intelligence collection and dissemination by modern tactical aircraft were lifting the fog of war in tactical situations, saving lives and materiel.

Such growth comes when someone, not a "traditional" customer, demands more because it has become essential to their way of doing their mission. In 1973, that is exactly what happened in terms of national reconnaissance. The Army had lost out in the early space programs in 1958 and 1959. They became consumers of information derived from NRO and Air Force systems. The Army's chief of staff for intelligence (G-2) came to believe the Army's future battlefield effectiveness depended on more effective use of the NRO's intelligence information. From preparation of the battlefield—those activities devoted to understanding the place where military operations were about to take place—to the conduct and aftermath of battle, national security systems had important capabilities and even greater promise. As the closed circles responsible for deciding the future of national reconnaissance discussed the move to Land's recommended near-real-time capability, the Army's G-2 got ready to take maximum advantage. Starting with a small budget, the Army began programs in 1973 to adapt national security capabilities for use in tactical military operations. The program was entitled Tactical Exploitation of National Capabilities, or TENCAP.[6]

The Army's TENCAP programs were intended to not only take advantage of some of the electronic intelligence sources, but also to pave the way for a coming revolution in the availability of imagery.[7] Soon, in part due to congressional direction, all the Services began TENCAP efforts to leverage systems developed to support strategic indications and warning, early warning, nuclear detonation detection, and communications.[8] Some of the TENCAP efforts aimed at taking the information that was already present and making it useful and available to the tactical warfighting community. Other efforts developed new tactically relevant payloads.

Near-real-time information was not entirely new; it was only new to the collection and dissemination of some kinds of reconnaissance information. The most important efforts came from experience with tactical reconnaissance capabilities of the latter part of the Southeast Asia war. Those changes involved making reconnaissance information, mainly imagery, from fast flying, low altitude aircraft available to tactical units in near-real-time. So, in practical terms, images taken by aircraft

---

[6] E. Mitchell, *Apogee, Perigee and Recovery: Chronology of the Army Exploitation of Space* (Santa Monica, California: RAND Corporation, 1991), 72.

[7] Robert Perry, *NRO History*, vol. V (Washington, D.C.: National Reconnaissance Office, declassified 2002), 45–48. The development of such a radical new reconnaissance system dates from the output of a panel headed by the remarkable Edwin Land in 1968, according to Perry. The Army systems were to be in place and ready to deal with the expected new near-real-time imagery when it went into operation a few years later.

[8] *Department of Defense Appropriations for 1987*, House Committee on Appropriations, 99th Congress, 2nd Session, Part 3, 680–681.

would be available to front line units within minutes to a few hours after the aircraft's flight. The early designation of such a capability was a "see it now" system, which reveals the intent.[9]

Satellite images available days later remained strategically valuable, but tactical warfare's pace marginalized them. Intelligence information from "see it now" systems, electronic processing delays aside, was available immediately, making possible timely responses to fluid warfare changes. Electronically linking aircraft sensors through relay stations directly to the processing and exploitation units in the rear echelons dramatically reduced timelines. As each delay in the chain from collection through delivery was reduced or eliminated, near-real-time became synonymous with instantaneous availability of reconnaissance information. Such capabilities in aircraft changed the pitch and pace of tactical warfare, slowly at first, but eventually to the point where U.S. forces would be able to detect and react to an enemy movement faster and more accurately than the enemy. And *that* made intelligence information not only tactically relevant—it made it revolutionary and a significant information advantage for the United States.

Such changes could be adapted for strategic and national intelligence organizations. By the late 1970s, the first near-real-time space systems were in place, capable of supporting national, strategic, and, initially to a limited extent, tactical capabilities. Surmounting the hurdles to provide satellite information to tactical units was not easy, nor did it happen all at once. The inexorable movement to tactical relevance, and therefore operational support, was gaining momentum by the early 1980s. The availability of very high reliability space-qualified electronic parts contributed to spacecraft robustness, lengthened spacecraft lifetimes, and made these capabilities real.

*   *   *

Space capabilities were being pushed to be more "operational" because they offered vast improvements to solve such traditional military problems. Making them operational, however, meant much better parts had to be used. The NASA Class A parts, the highest reliability parts available, were not good enough. To see what was happening to high reliability parts, we need to step back a couple of years and pick up the evolution of parts technology and requirements.

[9] Perry, *NRO History*, 45–48.

Chapter 8

Space Parts: From A to S

During the tumultuous changes in 1969–1972, high reliability parts and processes made steady improvements in the background.

A large part of improving quality and reliability is the feedback process. That is where failures are analyzed and lessons learned and then retained by some means, feeding the overall experience forward to the next system in line, preventing repetition of past problems. For spacecraft, this process involved certain key military standards and specifications; older versions that continued to be updated as well as new emerging ones. This all took place within a framework of disciplined engineering processes. While the focus here is on electronic parts, standards that addressed every aspect of high reliability systems were developed, for spacecraft alone or for military systems in general. The military standards bureaucracy was rigorously managed and controlled, providing for a "trial by fire" for any proposed changes, supported by extensive technical analysis in laboratories such as RADC's Rome Laboratories, within the Aerospace Corporation, major prime contractors, and electronics industry manufacturers. The advances in reliability spoke for themselves.

The high reliability processes were in dynamic tension with the marketplace, where volume and yield drove down the cost of commercial electronics. Electronics makers balanced continued investment in technology advances (such as smaller integrated circuit feature sizes) against reliability. They were not alone in their interest in reliability; standards had to take such advances into account. Nothing could be left to chance because not every technology was suitable for use in space, so control and testing processes were vitally important. These forces remained in balance for the moment.

* * *

The nation's space programs were about to face other significant changes. The major success of Apollo and the loss of MOL both had implications for a highly skilled workforce. With no immediate need for more Apollo spacecraft and launch vehicles, the program wound down without a successor, and along with the sudden loss of

Implosion: Lessons from National Security, High Reliability Spacecraft, Electronics, and the Forces Which Changed Them, First Edition. L. Parker Temple III.
© 2013 the Institute of Electrical and Electronics Engineers. Published 2013 by John Wiley & Sons, Inc.

the MOL, a large number of skilled space program engineers had no jobs. A major realignment had to occur. For some, whose careers ran through World War II, retirement was an option. But for those new engineers who had come into the space program following *Sputnik*, suddenly there were few alternatives and little corporate memory.

The sudden lull after MOL and Apollo was exacerbated as the Space Task Group's recommendations were politicized, and delayed for three years before a Presidential decision. This was the first major paroxysm in retaining lessons from hard experiences. MOL and Apollo had advanced their respective technologies and learned things about parts, materials, and processes for space that were important to retain.

In Apollo, the most devastating loss of three astronauts in a fire during a launch rehearsal was due to a choice of materials incompatible with a pure oxygen atmosphere. Hard-won lessons like that had to be retained and errors not repeated. Simply writing down what was learned helped, but retaining what was written to enlighten and inform subsequent generations of engineers was a problem. Retention of such wisdom required embedding it in manuals and other documents, hopefully contractual compliance documents, to prevent recurrence of past problems. Documentation was essential to bridge generational experience gaps.

A step in the right direction was codifying processes from the top level down. Every aspect of a spacecraft acquisition needed close attention, and all parts had to fit together, literally and intellectually. AFSC had pioneered a concept known as the weapon system, recognizing a complex weapon had within and behind it many independent and contributing parts, forming a system. Conceptualizing weapon systems as highly interrelated entities addressed some complexity issues. This approach enabled greater advances in complex weaponry, which in turn required advances in systems thinking. By 1969, this conceptualization had broadened its scope to include more than the integration of separate boxes into a functioning whole, and had become comprehensive of the entire process of acquisition through operations. AFSC's 375 series manuals governed weapon system procurement. These covered aircraft and aircraft systems well, but not space and missile systems' peculiar, unique environments and reliability requirements. The 375 series was a standard acquisition guide, but needed additional embellishment to be sufficient for space and missile systems.

Consequently, SAMSO's Reliability, Maintainability, Quality Assurance and Personnel Subsystem Branch published its *Assembly-Through-Flight Test and Evaluation Guide* in June 1969.[1] The guide provided guidance to all system program offices so their test and evaluation programs could attain "100% mission success."[2] The scope (assembly through flight) excluded piece parts, starting *after* qualification at the component level. A component was

---

[1] *Historical Report*, Reliability, Maintainability, Quality Assurance and Personnel Subsystem Branch, 1 Jan. 1969 to 30 June 1969, attachment 1: *Assembly-Through-Flight Test and Evaluation Guide*, HQ SAMSO, June 1969, found in HQ Space and Missile Command History Office.

[2] *Historical Report*, Reliability, Maintainability, Quality Assurance and Personnel Subsystem Branch, 1 Jan. 1969 to 30 June 1969.

an assembly or any combination of parts, subassemblies and assemblies mounted
together which may be functionally independent of, or an independent entity within a
complete operating subsystem, but which provides a self-contained function necessary
for proper subsystem and/or system operation.

While input parts quality was undoubtedly important, qualification came from the
component or "box level." The document mentioned piece parts, but not their qualification or testing. Instead, piece parts were described as the level of assembly
leading to a component. Piece part testing was accomplished mainly at the receiving
inspection to be sure that what was received was what was expected, but no further
discussions describe qualification at the piece part level. The parts were checked for
proper performance, but actual qualification remained at the box level of assembly.[3]
The underlying assumption was that the best possible parts were being acquired, so
then all else followed. This assumption, based on history, was about to be challenged.
The parts being acquired would shortly become a point of the strongest emphasis in
the continuing evolution of highly reliable spacecraft.

The guide set the goal of 100% mission success, which it defined
accordingly.

To successfully accomplish the mission for which the system is designed all
components/equipment and subsystems must perform their portion of the total mission.
To determine that all components will perform their mission objectives, a method of
testing has evolved and is continuing to evolve.[4]

Mission success depended only on functional success at the component level and
above.

Conformance to specification was verified at every level of assembly, but the
lowest level of assembly where meaningful testing occurred was at the box or subsystem level, not the piece parts. However, piece parts were recognized as having a
contribution.

Proper performance of each piecepart, component, subsystem and system is verified
prior to allowing it to become part of a larger assembly. Each succeeding checkout
operation is based upon a progressive growth pattern that maintains the integrity of
previous tests and takes advantage of all applicable data from earlier testing. In the
final analysis it is mandatory that the sum total of all tests performed on the system
exercise the vehicle through all its mission requirements.[5]

The limited nature of piece part testing indicates it was not a mandatory element
of the testing program. Furthermore, the guide also indicated only selected components had "historical jackets" or pedigree information tracking on their piece parts.
Pedigree information included the lot-date codes for the manufacture of the piece

---

[3] *Assembly-Through-Flight Test and Evaluation Guide*, 2–1.

[4] Ibid.

[5] Ibid., 2–4.

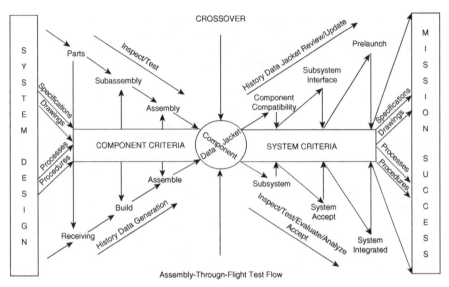

Figure 8.1.   The crossover concept. From the *Assembly-Through-Flight Test and Evaluation Guide*, 2–6 to 2–8: "The left side, component criteria is the source for parts selection, build methods, assembly and in-process test and inspections. It is the fundamental basis for determining the in-process acceptability at the pre-component levels. [That is, at the piece part level.] In the pre-component stage, criteria is [*sic*] a relatively fixed and firm basis for decisions. On the right side of component, the criteria is [*sic*] system oriented. There is a shift in emphasis from criteria as the source to actual demonstrated performance. Performance is evaluated in light of test results and history data to determine if the criteria has [*sic*] been met in more of a qualitative and end-to-end sense. Decisions bearing on the integrity and worthiness of systems on the right become a matter of analysis, evaluation and individual engineering decision rather than mere application of hard and fast criteria. In the center, the component is established as a functional entity to be evaluated by both methods."

parts, and included any significant events in the handling of the parts. Historical jackets only became mandatory for all subsystems and above in this manual.[6]

Building spacecraft with close attention to detail was becoming increasingly complex, covering a wider range of engineering disciplines in the process. Handling all of this increasing complexity required innovative organization and management.

"The crossover concept" (see Fig. 8.1) visualization helped to deal with the intricacy of spacecraft. An orderly, disciplined approach to assembly and testing would ensure every device worked, and as the whole spacecraft was incrementally built, each interim subassembly, subsystem, or system would work as well. The emphasis in testing was detecting devices that did not work as designed, and finding

[6] Ibid., 2–5. It should be noted, however, that this appears to have applied only to satellites. In HQ SAMSO/LV Memorandum for AFSC/CV, Subj: Launch Vehicle Reliability Improvement Program, 20 July 1972, considerable concern existed over the lack of history jackets on the piece parts in the ATLAS E/F launch vehicles.

any workmanship problems. "The crossover concept . . . clearly depicts the component as the critical point in the system growth cycle. Note that the horizontal line represents applicable criteria. On the left of the component crossover, the criteria is [*sic*] oriented toward component control. On the right, it is system oriented."[7] This suggests piece part testing began at receiving inspection, mainly to ensure the parts were what was ordered and undamaged, not necessarily getting rigorous testing until included in higher levels of assembly. Qualifying boxes by ensuring they functioned as expected was sufficient for the level of complexity of these spacecraft, and at the practical limits of theory and test equipment.

The crossover concept covered the "how to" of building a satellite once the system had been designed. That begged the question about how the design came to be and what requirements drove the design. Decomposing requirements from the highest level to the point of specifications sufficient for designing was a critical hierarchy driven by the Federal Acquisition and other regulations. The Federal Acquisition Regulations (FARs) drove the process by codifying all the rules about contracting for goods and services among all the agencies in the Executive Branch of government. The Department of Defense expanded, clarified, or tailored these regulations to its specific acquisition activities in the DoD 5000 series of directives and instructions. Specifically, the DoD's top-tier acquisition directives implementing the FARs were Directive 5000.1, "The Defense Acquisition System," and its accompanying Instruction 5000.2, "Operation of the Defense Acquisition System." AFSC then expanded, clarified, and tailored these to more specific methods applicable to Air Forces acquisitions (in its 375 series of directives). Finally, the unique elements of space and launch vehicle acquisitions were covered by SAMSO's regulations. To these, and in concert with them, the technical aspects of any specific acquisition were added.

The approach was a rigorous process of successively more detailed requirements and specifications, until the system was fully described sufficiently well to be built. Various incarnations of requirements decomposition had used an approach starting with an initial conceptual description to what was known as an A Specification. Usually, the "A Spec" described and decomposed the system at a gross level in terms of overall performance. For an airplane, for instance, the A Spec might require a top speed at a certain altitude, with a minimum range required. In the following "B Spec," the system program office defined the key elements of performance and design sufficiently for a contractor to begin work. In some incarnations, the B Spec formed a part of the request for proposal, which contractors expanded in a companion B Spec detailing their concept and how it would work. Once selected, the contractor would develop and deliver to the government "C Specs," which were a range of engineering discipline-specific documents that would constitute the basis of design.

The crossover concept began where that process left off. By the end of the 1960s, however, this approach was starting to creak under the weight of increasing systems complexity. The problems were, in part, the inflexibility of the process,

---

[7] *Assembly-Through-Flight Test and Evaluation Guide*, 2–6 to 2–8.

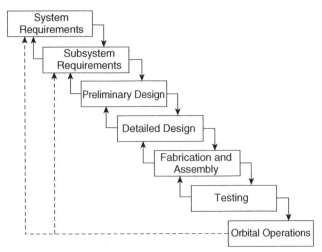

Figure 8.2. Waterfall approach. From conceptual ideas, each step increases the requirements' specificity, devolving higher level statements into one or more lower level statements. As the final design translates into fabrication, manufacture, and testing, the focus changes from understanding the complexity of the problem to applying solutions and testing whether everything works. Once in orbit, however, all that can be done is to learn from problems and feed lessons learned back to any future acquisitions. Adapted for space systems acquisition from Winston W. Royce, "Managing the Development of Large Software Systems," *Proceedings, IEEE WESCON*, Aug. 1970, pp. 1–9.

dealing with simple as well as extremely complex systems. Few inflexible "one size fits all" solutions work well universally. Additionally, depending on the program and acquiring division, the level of specificity in the A and B specifications varied considerably. The upfront requirements decomposition work needed to evolve.

Into that niche came an idea from software development. Like electronic parts, computer software was becoming increasingly complex. In 1970, then, a new approach emerged, known as the "waterfall" approach for graphically obvious reasons (Fig. 8.2).[8]

The waterfall approach was a conservative, disciplined, and risk-averse approach to systems acquisition. Its logical flow down dealt with complexity, and its feedback steps tended to catch any gaps or errors. If it looks like a very high level Gantt schedule, such an impression is not far from the truth. The waterfall's feedback iterations, though, were an advance over the step-wise Gantt approach. The waterfall did not adapt well to highly concurrent or highly risky systems acquisitions. Neither did the waterfall adequately address the need for repeated and extensive testing at all levels of assembly, as this is somewhat hidden inside each of the waterfall blocks.

---

[8] Winston W. Royce, "Managing the Development of Large Software Systems," *Proceedings, IEEE WESCON*, Aug. 1970, pp. 1–9.

The waterfall was not significantly different from the A, B, and C specification approach, although its iterative nature handled complexity issues far better. The first several waterfall blocks were the "design" starting the crossover concept, and the crossover itself simply expanded the "fabrication and assembly" and "testing" blocks. An important problem remained. Where did electronic piece part design and selection occur, and how did these match the requirements and specifications? Parts only entered the crossover at receiving inspection, so they had to have been selected in the large "design" box. The waterfall covered everything nicely, but at too high a conceptual level to understand the intricacies of parts, specifications, and their interactions. Second, the crossover's emphasis on testing assemblies meant that piece parts were sort of part of it, but not really. The only place parts appeared was when the emphasis was on receiving and the testing approach was simply "inspection." Everything really starts with parts, though. Detailed specifications of specific parts were fundamental to each subsystem design, and thus had to tie to unit specifications and system requirements. Complexity had only been dealt with at a level commensurate with the sophistication of the systems and their components for the time being. Parts reliability and quality either lay in two camps, or none at all. As system complexity continued to accelerate, this would have to be dealt with in the fullness of time.

For the time being, what then was the nature of the parts being inspected at receiving? Sources of parts varied, from high reliability parts for military avionics systems to parts built on "captive lines," or manufacturing lines whose output was owned by the government. Such lines were a means of assuring an adequate quality supply for Minuteman and Apollo.[9] NASA established a special category of parts to get first class or Class A treatment. Such parts were used on NASA Class A programs such as Apollo. The Class A parts had additional specifications relating to types of tests performed on the parts prior to delivery to the government. These tests were clearly related to improving the quality of the input parts. Rather than forcing the government to screen and reject parts based on the yield of a manufacturing process, the manufacturer would screen the parts prior to shipping to increase the effective yield to the government. A number of important lessons sprang from NASA's Class A parts program. By the time of Apollo 11's Moon landing, NASA had achieved high reliability with their electronics. Fairchild Semiconductor's national sales manager for aerospace and defense marketing, Gordon Russell, summed up what they had learned. "We implemented internal visual inspections and designed traceability procedures that had never been done before. Our quality control people can trace lots all the way back to wafer processing."[10] These additional steps were

---

[9] To be sure, on the Minuteman program, Autonetics put a great emphasis on parts quality. That program was the exception rather than the rule, however. At Autonetics, the quality effort took as much as 30% of the budget, and this was much higher than other programs were able to afford. Vendors supporting Autonetics were also required to implement early statistical process controls. The result was dramatic, but not widely applicable to the new space programs. Captive lines were also used in Navy missile programs such as Polaris.

[10] Eldon C. Hall, *Journey to the Moon: The History of the Apollo Guidance Computer* (Reston, Virginia: American Institute of Aeronautics and Astronautics, 1996), 147.

important steps on the path to the highest reliability parts, but were not yet the full answer.

Such experiences were making it increasingly clear the highest quality parts began their existence as the highest *reliability* parts right at the start of manufacturing. They came into existence differently than parts of lower reliability. Experience demonstrated quality could not be "tested in"; a lower grade part could not become a higher quality part simply by testing. Screening such lower grade parts could only find the lucky ones that might be suitable for space use. Another truth was clear: testing was a critical element of spacecraft reliability, and the earlier it was performed, the fewer problems were found as parts were assembled into boxes, subsystems, systems, and finally spacecraft. Testing of each level of assembly was expensive, but the cost of early testing paled in comparison to the expense of tearing apart a unit, reworking and then retesting it, as well as the consequent costs in schedule delays or loss of needed capabilities in orbit. Reworking also reduced the subsequent reliability of the unit. At a minimum, then, testing would have to be pushed back to the starting point (parts), and that meant incorporation in military standards.

As illustrated by the crossover, far more items and activities contributed to improving the reliability of spacecraft than military standards and specifications. Yet, for military applications, the increasing importance of solid state electronics to military systems meant more detailed specifications were needed to ensure consistent quality, standardization of part designations, and test methods. Everything starts, after all, with parts. And parts start with materials and manufacturing.

*   *   *

Although begun at about the same time as MIL-STD-883, "Test Methods and Procedures for Microelectronics," the underlying manufacturing specification for microcircuits, proved more difficult to release. MIL-STD-883 had some important shortfalls. For instance, it was missing specific electrical requirements for the devices that were needed for standardization across manufacturers. Also, its recommendations for the sequencing of some processes were subject to a wide range of interpretations. No way existed to provide or enforce a common interpretation within MIL-STD-883.[11] Without prior standardization, and given the wide range, even in 1968, of the different approaches to microcircuit design, the standardization task was not trivial. Not until 20 November 1969 was RADC able to release MIL-M-38510, "General Specification for Microcircuits," to cover all (ten) current integrated circuit categories.

MIL-M-38510 represented a major shift in terms of tightening quality and reliability of parts and manufacturing processes. Like MIL-STD-883 before it, MIL-M-38510 depended heavily on the knowledge gained from Richard Nelson's thin film and monolithic microcircuit manufacturing facility. It closed the holes in MIL-STD-883, bringing forward discrete device testing from MIL-STD-750 and MIL-S-19500. By introducing a qualification and certification process for military electronics, it

[11] *The Reliability Handbook: An Introduction to Military/Aerospace Semiconductor Reliability Concerns, Screening Techniques, and Standardization Programs* (Santa Clara, California: National Semiconductor Corporation, 1979), 27.

provided the control mechanisms to prevent the proliferation of interpretations and ensure continued process integrity. Only by compliance with MIL-M-38510 could a manufacturer put the "JAN" mark on their part. (JAN, which had only stood for "Joint Army Navy" prior to the Air Force's inception, now simply represented parts compliant with standards supported by all of the Services.)[12]

In addition to standardizing microcircuits, it updated the very much earlier concept, left over from JAN-E-1, called the Preferred List of Vacuum Tubes. Qualified suppliers could list their products on a government-controlled list known as the Qualified Parts List (QPL). The QPL differed by having considerably more technical backing and government control. Thus, the QPL corrected the 1959 Darnell Report, finding that many parts on the earlier standard parts lists were actually substandard.[13] A designer could refer to the QPL, find the parts meeting their high reliability needs, and then refer to the device's specifications to incorporate them appropriately into a given system.

MIL-M-38510 became the gatekeeper of the QPL, which was for the moment the gateway to the military microcircuit marketplace. MIL-M-38510 also introduced a novel solution for the lack of microcircuit interchangeability. The standard had attached to it a number of individual, detailed specifications for different microcircuit devices covering additional parameters and electrical tests needed for the device to be listed on the QPL. This allowed the basic portion of the standard to address the parameters common to all microcircuit devices, but to have the flexibility to continue to evolve rapidly as new devices appeared. The additional detailed specifications were initially designated as "/xxx" to indicate they were a subset of the MIL-M-38510, and they received specific numbers. Quickly, these individual specifications became known as "slash sheets" for this method of designation, though their proper designation was Standard Military Drawings. RADC experts identified the most demanded high reliability devices and used these as the basis for the initial slash sheets, which they also wrote. RADC became the most prolific center of production and update for the slash sheets, and continued to supply these even after responsibility for administering the microcircuit program transferred elsewhere. RADC was a national asset for the complete understanding and characterization of microcircuits and other devices, having at times greater insight into the devices than did their manufacturers.

The QPL was not the only list in town. The Defense Supply Agency (DSA) operated the military parts supply system from Cameron Station in Alexandria, Virginia. DSA's military parts and materials responsibilities were very broad, including maintaining an alphabetic list, by product, of manufacturers. DSA established

---

[12] "Pioneering Reliability Laboratory Addresses Information Technology," *Journal of the Reliability Analysis Center* 6, no. 1 (1998): 2. The sequence of MIL-STD-883 to MIL-STD-750 and MIL-S-19500, and then into the MIL-M-38510, MIL-I-38535, and MIL-H-38534, and finally into MIL-PRF-38535 and MIL-PRF-38534, portrays the evolution of technology and how it drove the evolution of testing and qualification approaches.

[13] E. Nucci, "Progress Report on Ad Hoc Study on Parts Specifications Management for Reliability," *IRE Transactions on Component Parts* 6, no. 3 (Sept. 1959): 133.

and ran a quality assurance effort under the name "Mill Run Products Quality Assurance Program."[14] Mill Run products were "produced, stocked or distributed in anticipation of purchase by government activities and contractors for use in U.S. government equipment." Manufacturers could participate, which is to say they could provide products to the military, by submitting the proper forms to DSA, which would ensure the applicant complied with appropriate directives. The DSA list had little specific impact on the high reliability parts of space and missile systems, but that was going to change.

Space programs and large missiles (and their launch vehicle cousins) were about to undergo a major change in the quality of their parts. About as much as could be squeezed out of the rest of the assembly-through-flight process had been done. Lieutenant General (LtGen) Sam C. Phillips, the SAMSO Commander, recognized the greatest gain would come from improving the quality of the parts in the systems comprising national defense space. Although the National Reconnaissance Office was still covert, it closely followed how things were evolving in national defense space, and often hid behind the national defense space programs to ensure the needs for national security space were met. Phillips' activities, then, represented the combined interests of national defense and national security space. He knew very well not all the necessary parts were even available through NASA's Class A parts program. The Class A designation was really tied to MIL-STD-883 testing methods. Class B parts were intended for use in conventional avionics environments, and Class C were expected to be used in ground systems. So, if improvement could be accomplished at any level, this would affect a wide range of missile and space programs. If the parts provided to all these systems were of uniformly high quality, the cost of acquisition might go up slightly, but so would reliability. He recognized that the same measures that could improve the operational reliability of strategic missiles would also increase the lifetime of satellites, reducing the number of satellites and launches. The savings from needing fewer satellites and launches more than offset the cost of better parts, and would have a potentially huge impact on defense and security program budget requirements.[15]

The payoff depended on the details of what changes were needed, how they were to be administered, and the resource impacts of the changes. In early 1970, Phillips asked for a study concentrating on the appropriate alternatives to improve quality in satellites and missiles. The results were presented to him in April 1970.

The quality assurance activities at SAMSO "were at their lowest ebb since the formation of AFSC" when Phillips assumed command of SAMSO in 1969, and had steadily improved since that time.[16] The study showed the next important step was creation of a separate staff organization responsible for the quality assurance function.

---

[14] DSA Regulation 8240.2, 14 June 1971, "DSA Mill Run Products Quality Assurance Program."

[15] SAMSO/CC Memorandum, Subject: Program Review Board, 27 Dec. 1971.

[16] Ibid.

Audits were an important element in quality assurance, and something on which the new staff office would have to focus. Audits of various sorts were conducted to review a part manufacturer's processes or some aspect of a prime contractor's quality program. Other kinds of audits reviewed financial records and bookkeeping practices, or the handling and testing "pedigree" of parts and devices used in spacecraft. All of these were essential to an effective quality assurance function. The emphasis, regardless of the type of audit, was on independent review—a different set of eyes to catch any flaw. With so many different kinds of audits, different programs, and a plethora of contractors, auditing for an agency the size of SAMSO absorbed considerable resources.

Auditing to assure quality had a widespread, uncoordinated infrastructure for examining quality from many perspectives. Thus, it made sense for SAMSO to combine forces with Air Force Contract Management Division (AFCMD) for conducting quality assurance surveys. They signed a memorandum of agreement to that effect in December 1970.[17] A similar arrangement existed with the DSA's Defense Contract Administration Services (DCAS). These arrangements served more purposes than shepherding scarce resources, and allowed some classified national security programs to use audits scheduled by others to cover their interest in some manufacturers' capabilities.

The different roles and responsibilities led to confusion with multiple agendas beyond the principal audit, and began attracting unintended and unwanted attention to the classified national security participants. DCAS accused SAMSO of audits that "were not entirely product oriented and smacked of QA [that is, Quality Assurance] system surveys," not related to contract administration. At the same time, AFCMD had become "overly sensitive . . . concerning a 'buying' function participating in QA surveys of facilities under their administrative jurisdiction." To make matters worse, AFCMD was struggling with its own role in quality assurance and considering reducing the amount of effort expended on it.[18] The growing emphasis on improving quality meant national defense and national security had to encourage the other agencies' continued work in quality assurance.

This growing quality emphasis was shared across the Services and NASA. The Services' Joint Logistics Commanders reached an agreement with IDEP to enhance the types and amount of information being shared, subsequently to be known as the Government-Industry Data Exchange Program (GIDEP). The more widely information was shared across government and industry, the better the overall quality of the available parts, the earlier problems were caught (saving money and schedules), and the more competition invigorated continued innovation and improvement. GIDEP's creation was also supposed to minimize "duplicate testing of parts and materials through the interchange of environmental test data and technical information."[19] The

[17] Memorandum for General Lyon, Subj: Quality Assurance Surveys, 1973, supporting document VIII-18 to *History of Space and Missile Systems Organization, 1 July 1972–30 June 1973*, vol. 7, p. 1.

[18] Ibid., 2–3.

[19] *Military Standard Government/Industry Data Exchange Program (GIDEP) Contractor Participation Requirements*, MIL-STD-1556AUSAF, 29 Feb. 1976.

Joint Logistics Commanders selected the Navy to manage the GIDEP. GIDEP allowed the national defense and national security space programs, contractors, and suppliers to weed out any but the highest quality parts for their systems.

As part of his getting a grasp on the quality of missiles and spacecraft, Phillips established the Permanent Review Board in December 1971, shortly to be renamed the Program Review Board (PRB).[20] Regardless of its final name, Phillips' original name signaled his intent this was to be a long-term way of doing business. The PRB would have wide-ranging scope, but the most critical activities depended on some short-term activity during what was known as Phase 1:

> *In the interest of assuring maximum confidence in mission success involving SAMSO systems . . . a Permanent Review Board [will] assess our program status from the conceptual stages through system launch readiness. . . . The initial phase will consist of a comprehensive review and analysis of SAMSO system failures and anomalies as presented by major system program offices. In addition to identifying specific failures/ anomalies, causes and corrective actions taken, an objective of the board will be to analyze events for indication of trends or similarities. Correlation of success/failure experience with management of functional elements impacting system development and checkout should provide useful lessons learned and aid in the avoidance of repetitive failures.*[21]

The scope went beyond design adequacy, and included such things as production processes, fabrication, workmanship quality, validity of tests, and evaluation efforts, and configuration control. This sweeping review struck at the heart of reliability problems, ensuring the results would address the most important issues.

Phillips' initiative captured several key items. The feedback from failures and anomalies was critical for retaining lessons learned to improve future procedures. This process was about to become substantially more rigorous. Also, it showed the sources of reliability concerns were fairly well understood, being grouped into design, production, fabrication, workmanship quality, and testing. Each of the sub-groups of the PRB looked at one or more of these topics. The result was an overlapping set of results from different but complementary perspectives, where the best ideas would rise to the top. The first PRB, on 6 January 1972, was an all-day affair. It concentrated on launch vehicle failures over the previous two years, but allowed for earlier analyses from other agency boosters, such as NASA's Delta (a launch vehicle derivative of the Thor).[22] The first day included a lengthy session on the past two years' reentry vehicle failures.

Phase 1 of the PRB held three additional full-day sessions, with the final day including an all-afternoon summary review of the prior sessions' failure analyses. The initial PRB was important because its results were a watershed event in the

---

[20] SAMSO/CC Memorandum, Subject: Program Review Board, 27 Dec. 1971.

[21] Ibid.

[22] Delta was the fourth derivative of the Thor IRBM to serve as a launch vehicle. The fourth in line received the fourth letter of the Greek alphabet, and has been known ever since as the Delta, in many further modifications and incarnations.

history of the military and national security space. Its members were the key figures who matured the space programs in the early 1970s. Phase 1's chair was LtGen Phillips, and others included BrigGen Thomas W. Morgan (soon to be SAMSO commander but Vice Commander at the time); BrigGen Lew Allen (soon to be AFSC commander and Air Force Chief of Staff, but then commander of Secretary of the Air Force Special Projects, part of the National Reconnaissance Office); Dr. Ivan Getting, President of the Aerospace Corporation; AFSC Deputy Chief of Staff MajGen Kenneth W. Schultz (shortly to follow Phillips as SAMSO commander); and the commanders of Arnold Engineering Development Center (AEDC, an AFSC test and development center), AFCMD, and ESD (AFSC's electronics counterpart to SAMSO, and parent command of RADC).

Nothing escaped scrutiny. In addition to the obvious emphasis on the hardware in the failures, every office in SAMSO examined its own areas. For instance, the SAMSO Quality Assurance Branch established the Quality Assurance Working Group "to investigate the problems attributed to deficiencies in contract quality provisions and breakdowns in contractor quality assurance programs."[23]

The week following the first PRB meeting saw an agreement between AFSC and DSA on 13 January 1972. DSA had been involved in the PRB through the local DCAS Region. They agreed "to perform joint quality assurance surveys to provide a more comprehensive evaluation of contractor performance and provide Program Directors with the increased visibility necessary to insure product integrity."[24] These joint quality assurance surveys began later in January, with four completed by June. There was good reason for joining together:

> *Defense Supply Agency/Defense Contract Administration Services Regions . . . are required to accomplish QA system surveys of contractors for which they perform contract administration functions to evaluate the effectiveness of the contractor's QA program/system and to assure that contractors comply with contract requirements. Air Force Systems Command (AFSC) Program Offices are required to perform product oriented surveys at facilities under DCAS surveillance to determine the adequacy of contract technical requirements relating to quality and product conformance to design intent. AFSC and DSA/DCAS have determined that it would be beneficial to satisfy these requirements by participation of AFSC personnel in DCAS QA system surveys.*

This was symbiotic. Contract administration inspectors were not qualified to understand the "conformance to design intent" of any system, but neither were the AFSC technical inspectors expected to be conversant with contract administration (except in the broadest terms). The joint arrangement also helped minimize the turmoil in the inspected organizations, since one inspection now served where two were needed previously.

The PRB Phase 1 process emphasized the expectation of quality at the highest levels of command. Phase 2 continued the review and refinement process. The main courses of action identified and established in Phase 1 continued in Phase 2 as the

---

[23] Historical Report, Quality Assurance Branch (DRTQ), 1 Jan.–30 June 1972.
[24] Ibid.

process became institutionalized. Thus, by the PRB of 29 February 1972, barely seven weeks after the PRB's first meeting, the agenda had reduced to considerably detailed content. In other words, they were no longer dealing with high level decisions of policy and organization, but had begun in earnest the necessary and vital detailed work.

By that time, LtGen Morgan, chair of the PRB Executive Committee, started the Electronic Piece Parts Working Group. Another group that was quickly renamed, it became the Working Group for Electronic Piece Parts (WGEPP). The groups' objective was to "[d]etermine the impact of electronic component and subsystem failures on selected SAMSO programs and recommend future courses of action to reduce or eliminate identified problems."[25] Dr. William F. Leverton of the Aerospace Corporation chaired the WGEPP, comprising three other Aerospace members and three government members, working from February through December 1972.

The WGEPP publicly released their summarized findings and conclusions. The first to receive the report were SAMSO's major contractors. The 15 January 1973 report pulled few punches. Their comprehensive results established the agenda for what needed to be done, and included more than the next decade of the U.S. space programs.

The group found that, whatever the case had been during the U.S. strategic buildup through about 1963, military influence over electronic parts was largely gone. Even more disturbing, they found "SAMSO's ability to make its high reliability requirements known and to exert meaningful impact on the piece part vendors . . . [was] negligible." That was all the more troubling in light of the assessment that electronic piece parts lacked reliability and standardization. Such parts "as purchased from vendors did not have consistently adequate reliability for space and missile systems."[26]

All was not lost, however, and successful programs continued to light the way. "[S]ome exceptional major programs . . . had evolved techniques that focused management, engineering and fiscal attention on electronic reliability."[27]

If space programs were ever to achieve full acceptance by military commanders or be considered as operational as any other forces, then change was needed. The WGEPP said the change had to emphasize electronic parts.

Space qualified, high reliability electronic piece parts were the *sine qua non* of high reliability space programs. No matter how much management attention and financial support were applied to the problems of achieving high reliability, the high reliability demanded by space systems could not be attained without space qualified parts. However, the challenges were manifold:

> *Procurement of devices in the past had been fragmented by programs, contractors, and subcontractors; standardization of devices within or between programs was virtually nonexistent; specifications and screening tests, based largely on the*

[25] Program Review Board Meeting Minutes, 27 April 1972, Attachment 4.

[26] *History of SAMSO, 1 July 1972–30 June 1973*, 453.

[27] Ibid. The programs were Poseidon, Apollo, and Minuteman.

inadequate and untimely military specification system, reflected nonuniformity and lacked cohesiveness; close monitoring and control of vendor processes and screens by the contractors and subcontractors ranged from reasonably good to totally absent; and low cost rather than high reliability had often been the dominant consideration in device procurement.[28]

To their credit, the strong support from the highest military and industry leadership in the space programs achieved considerable headway during the course of the study. In a relatively short time, major efforts had been devoted to defining the reliability requirements and calculations to support the imposition of high reliability standards, and the standards themselves had begun to take shape. All of this comprised the broad response needed for the necessary kind of broad management, engineering, and fiscal support constituting the path to achieving high reliability.

The WGEPP had looked into integrated circuits when it began that February:

Because of the problems brought out to the PRB concerning Integrated Circuits (IC's) we need to develop a listing of highly reliable (Hi Rel) IC's for use by all Programs. Where Hi Rel IC's are included in the list, their use would be required or a waiver granted. The IC's would be purchased in lots and provided for program use. Possibly ASD [Aeronautical Systems Division] and ESD could "buy in" on the production line for their use.[29]

To accomplish its broad objective, the group performed detailed examinations of several SAMSO programs and programs of other government agencies. They sought identification of management and technical elements that significantly impacted system success or failure. "Specifically, the group sought to assess the impact of normal versus unique procurement, design, and quality assurance procedures on the ultimate reliability of electronic components."[30] By April 1972, the same group began examining, through MIT's Lincoln Labs, solid state amplifiers as replacements for traveling wave tube assemblies.[31] Whereas some kinds of solid state parts were in mass commercial production, solid state components were still immature—in the sense that they were not offered as standard end items. Examples of where SAMSO wanted high reliability componentry to advance to common offerings were telemetry systems and command destruct systems to support common buys.

One of the first problems the WGEPP addressed was reliability calculations for microelectronics. The available sources were obsolete or inaccurate. The most recent were the December 1965 issue of MIL-HDBK-217-A, "Reliability Prediction of Electronic Equipment," which was undergoing revision at the time, and the September 1963 version of an RADC Technical Report, TR-67-108 (also being revised). The Navy's Failure Rate Data (FARADA) program collected and analyzed reliability, and that information was about to become very important. The upshot was a

---

[28] Ibid., 453–454.

[29] Program Review Board Meeting Minutes, 29 Feb. 1972.

[30] *History of the Space and Missile Systems Organization, 1 July 1972–30 June 1973*, HQ SAMSO History Office, Los Angeles, California (1973), 451.

[31] Program Review Board Meeting Minutes, 27 April 1972.

mélange of methods, confusing by themselves and inconsistent among each other. This created problems for determining system reliability, of course. It also affected source selections and systems comparisons, because comparing reliability calculations based on different sources was hopeless. It also drove industry to distraction, since capturing and reporting reliability data depended on the manufacturers. In short, the wide range of reliability predictions and calculations created a problem for all involved in high reliability electronic componentry.

RADC credited reliability calculation problems to obsolete and inadequate microcircuit data, because the "troublesome problems are parts design and fabrication related."[32] The problems all came down to keeping up with the rate of technological change and the underlying physics of failure.

Before proceeding, understanding the ways failures occur will help in what follows. Earlier, a concept called "infant mortality" was introduced, but it is only one of the ways in which parts and devices fail. Figure 8.3, known as the "bathtub curve," captures the major failure types.

The "infant mortality" phase's characteristic decreasing failure rate can be attributed to manufacturing process errors, flaws, imperfections, or human factors. For testing, electronic parts are powered up and their characteristics monitored for signs they will fail early. "Burning in" screens for weak performers. The last phase, "wearout," is dominated by increasing failures as parts age, and, in the case of spacecraft, radiation begins to shorten their lifetimes. The expected life can be

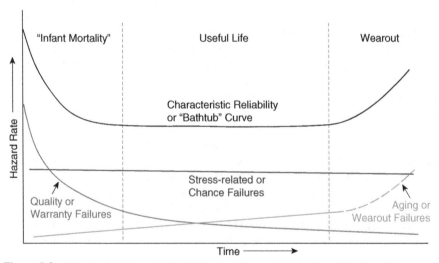

Figure 8.3.   The classic failure rate "bathtub" curve. Sources: Adapted from T.R. Moss, *The Reliability Data Handbook* (New York: ASME Press, 2005), 10; Bilal M. Ayyub and Richard H. McCuen, *Probability, Statistics, and Reliability for Engineers* (New York: CRC Press, 1997), 396.

[32] Ibid.

predicted through accelerated testing or testing at increased temperature levels. In between there is a period of relatively steady performance, with failures mostly due to overstress conditions or shocks. Usually, parts are de-rated or run well below their maximum operating specifications to alleviate stress and reduce stress-related failures. Generally, parts surviving the initial burn in live a long time until aging effects begin to dominate.

In each phase, physics determines the way a part works due to materials selection, feature sizes, operating parameters, and other fundamental factors. Knowing how a part works helps define how the part will act when it fails, but that is not the full story. Parts that operate on certain physical principles fail due to other, related physics and chemistry. Many physical factors cause and affect the mode of failures and collectively are called the physics of failure.

Physics of operating, physics of failure, and testing contribute to the model of behavior captured in the bathtub curve. The exact shape and length of the curve, though, depends on field returns of failed products and related production data. Taken together, these are the basis for failure rate calculations.

RADC's concerns with failure rate calculations being out of date spanned all three types in the bathtub failure rate curve and the supporting data. Without data, the shape of the curve and the length of the stable operating lifetime in the middle were merely notional, and unhelpful in calculating the expected lifetime of a system.

Consequently, inconsistent reliability calculations were well off the mark from reality, serving no one well. The WGEPP illustrated the problem with some contractor experience with common solid state parts (Table 8.1). The wide disparity between the Military Handbook methods, the RADC Technical Report, and actual experience illustrates the problem. The disparity among the contractors emphasized the need for common specifications, and handling processes. While the RADC methods seemed closer to real experience than the Military Handbook, neither produced results sufficiently in agreement with actual field experience. Another important finding jumped out of the data. The more complex solid state integrated circuits were experiencing problems well out of proportion to the other part types.[33] The integrated circuits presented a particularly difficult set of problems for improving reliability. This served as notice that much more attention was about to be given to ICs.

Table 8.1. Experience of Failure Rates versus Calculated Rates for Some Selected Part Types

Fail/10^9 Hrs	TRW	Hughes	Lockheed	MIL-HDBK-217	RADC TR 67-108
Diode	0.8	0.5–44	7–12	140	5
Transistor	4	23	10	300	15
IC	20	40	15–50	NONE	50

Source: Program Review Board Meeting Minutes, 27 April 1972, Attachment 4.

[33] Program Review Board Meeting Minutes, 27 April 1972, Attachment 4.

Table 8.2.   Reliability Predictions and Failure Experience

Fail/10⁹ Hrs	Reliability Prediction	Manufacturing and Test Experience	Orbital Experience
Diodes	10	174	0.1
Transistors	10	590	0.15
IC	15	330	2

Source: Program Review Board Meeting Minutes, 27 April 1972, Attachment 4.

Although contractor experience was obviously variable, there were gaps between predictions and failure rates in testing versus operational experience in space (Table 8.2). This identified disconnects in testing requirements, which were clearly more stressing and stringent than the actual operational environment, and the clear lack of correlation to reliability predictions. Their conclusion at the time was "high failures during manufacturing and test is [*sic*] costly and should be corrected."[34] But the problem was somewhat of their own making. The quality of parts had to have a bearing on very costly rework and schedule slips in manufacturing and testing.

Failures were indeed costly. Using 1972 dollars, a feeling for the impact of actual failures puts the importance of reliability into perspective:

- IC in-flight control electronics failed in a thermal vacuum, incurring a retrofit cost of $150,000;
- six failures in system test required a rerun of tests at a cost of $350,000;
- Transistor failure caused a ten-day impact on flight vehicle;
- Several parts' failures in flight not catastrophic due to redundancy and operational changes, but failures in test cost $6000 each to correct;
- UHF transistor failure retrofit cost $150,000.[35]

System test was near the end of a spacecraft's manufacturing (Figs. 8.1 and 8.2), when the vehicle was assembled and tested as a whole. When the spacecraft is subjected to the simulated space environment in a thermal vacuum chamber, it is almost ready to send to the launch site. Any problems found at these points require expensive disassembling of the spacecraft. Clearly, if part quality problems could be discovered earlier in development, very expensive repairs could be mitigated. There would always be some chance of a part failing in the manufacturing process, or that a mistake in a design, improper handling, or similar issues would not be discovered immediately. The concept, though, was to discover all of these issues as early as possible, avoiding costly failures during system test.

The WGEPP examined additional data related the cost of various grades of parts to their failure rates (Table 8.3). MINUTEMAN's high reliability exceeded NASA

[34] Ibid.
[35] Ibid.

Table 8.3. Cost of Higher Reliability

Part Class	Fail/10^9 Hrs	Cost of Part
Standard Grade	300	$3.60
High Reliability	6	$6.30
Minuteman	2	$30.00

Source: Program Review Board Meeting Minutes, 27 April 1972, Attachment 4.

Class A part types. Clearly, very high reliability parts were available and had significantly lower failure rates. They might cost nearly 10 times as much as a commercial part, but they were more than 100 times more reliable. They had a 50-fold improvement over military avionics grade (the middle row in the table). Avoiding costly, late development cycle problems was within reach when using the highest quality parts available. This meant changing a program's budgeting profile, since it forced higher cost early in a program when the more expensive parts were bought. But since parts constituted only about 5% of the total program cost, the minor shift in the budget profile enabled a program manager to more nearly hit the planned later expenditure profile because there would be fewer late development problems.

All of this led the group to observe "solid state devices as purchased from vendors do not have consistently adequate reliability for space systems."[36] While flight failures had occurred, the cost of rework and retest in terms of dollars and schedule was very high. The factors identified as contributing to the solid state device problems included:

- Fragmentation of purchasing (small lots);
- Inadequate standardization of devices, specifications, or screening;
- Lack of close vendor-contractor relationship;
- Lack of control of subcontractors;
- Government source inspection of limited value in achieving high piece part reliability.[37]

Each also contributed to reliability improvement, and therefore to the agenda of items needing to be addressed.

The fact that the WGEPP found so many fundamental and far-reaching items so quickly is important. Of course, it shows reliability improvement was finally getting the management attention it needed, starting with LtGen Phillips and reaching to the lowest level of SAMSO. It also suggests these items had been known, to some degree, prior to Phillips' emphasis on reliability, but had never gained sufficient support for implementation. That was now changing.

Finally, the WGEPP's ability to pull together these important metrics says something about the availability of program data. Together with GIDEP's recently

[36] Ibid.
[37] Ibid.

incorporated FARADA calculation capability (under the name of the Reliability-Maintainability Data Interchange), these data were essentially readily available for analysis. The Aerospace Corporation analysts could put together this important story. The data were not used to judge the contractors themselves so much as they were necessary to understand the health of the industry and especially the SAMSO enterprise, crossing strategic missiles and satellites. A comprehensive view of reliability issues resulted.

WGEPP recommended several important steps. They called for management oversight charged with developing and implementing policies, procedures, and techniques "to assure that all SAMSO flight programs consistently employed superior electronic parts that demonstrated quality and reliability characteristics which were compatible with USAF mission requirements."[38] To accomplish this, the group recommended establishment of the SAMSO Electronic Parts Management Board. Chaired by a SAMSO director-level individual, and co-chaired by Aerospace, the purpose of the group was to recommend parts policy to the SAMSO commander and establish failure rates for reliability analysis, specifications for parts, and a standard parts list. This was more than a simple bureaucratic solution. By making the government and the FFRDC jointly responsible, it made certain the technical underpinnings for reliability would be part of every decision. Furthermore, by requiring FFRDC signature on all decisions, this provided long-term memory of the issues, even after the military members moved on to other assignments.

The Electronic Parts Management Board reviewed reliability of any new technologies, and centralized failure reporting and analysis. They would also, in some exceptional cases, establish and maintain captive/controlled lines of manufacture. This was, of course, when the military requirements far exceeded those of commercial industry, and the military could advance a manufacturer's business by taking control of an advanced technology line, getting it proved, and then reaping the benefit in support of a critical national program. Eventually, the line would be turned back to the manufacturer for commercial or other military use. Captive lines already existed for NASA's Viking space exploration program, Sandia's (nuclear weapons) devices, and about 25% of SAMSO's IC requirements (many of which were actually bound for national security space systems).

At the time (April 1972), most military and national security space programs were using NASA's Class A parts, and had been since 1969. Class A electronic piece parts were specified in MIL-STD-883, "Test Methods and Procedures for Microelectronics." MIL-STD-883 had little to do with manufacturing processes and more with how to test parts to verify their quality. The surprising thing is, despite longstanding wisdom that quality could not be "tested in," Class A parts usage had improved reliability. In other words, knowledge that testing was not the answer to reliability was demonstrable and essentially unquestioned, but the acquisition culture had not addressed manufacturing of the highest reliability parts. That was shortly to change.

On 14 June 1972, the PRB and WGEPP turned to electrical connectors and a policy on the use of Class A piece parts. Their trend analysis identified a common

[38] *History of the SAMSO, 1 July 1972–30 June 1973*, 454.

problem with contaminated, corroded, and loose connectors, so attention was quickly drawn to these as a matter of priority. When they turned to the Class A parts policy, however, the problem was quite different. The policy was clearly meeting some resistance from the program offices. While program offices would often balk even at measures designed to enhance reliability and mission accomplishment, in this instance that was not the problem. The Class A parts policy was not going to be made mandatory because of an ongoing change in the civil space program.

NASA's Class A parts program had a problem. As 1972 approached, NASA's civil space program was deep in its post-Apollo drawdown. Although recommended by the Space Task Group in 1969, the STS still awaited the President's go-ahead. The NASA Class A "captive" production lines were becoming less available because NASA had little need in this interim. By 1972, almost no Class A parts existed. Any policy mandating Class A parts made little sense. Something akin to the NASA Class A parts was going to be needed but was not in existence.[39]

Top-level support arrived as a regulation from the SAMSO Commander establishing the Commander's Policy Book. This in military parlance made it mandatory.[40] The Policy Book's purpose was

> to provide a means of communicating the Commander's policy to all SAMSO personnel. Policy statements to be included are those not suitable for a SAMSO regulation or other publication and which express the personal desires of the Commander concerning the conduct of the SAMSO mission.

A lieutenant general's stated desires should be understood as orders for those working for him. This was an important part of the overall process of control needed for establishing a true high reliability space piece parts system. As failures and anomalies were resolved, or, as in the case of Class A parts, a policy was needed, they could be added to the Commander's Policy Book. Once in the Policy Book, handling was still under consideration, but the idea was the policies would be parsed out to other directive documents as appropriate. The root causes of failures or anomalies whose corrective actions could be identified would be accumulated, and eventually handed over to the controlling authority as a change to a military standard or specification. Once the new military standard was updated, for instance, the appropriate section of the Commander's Policy Book would be deleted. Annual updates to other documents were required, and in practice often occurred more frequently.

When the WGEPP affirmed prior suspicions that the large number of anomalies (48) over the previous three years failures were due to electronic parts problems, LtGen Schultz released another new Commander's Policy.[41] His careful word choice

---

[39] HQ SAMSO/DR memorandum, "Availability of High Reliability Digital Microcircuits," 29 Oct. 1973.

[40] "Commander's Policy Book," SAMSO Regulation 5–4, 12 July 1972. By 1980, the numbering system changed to SAMSO Regulation 550 series, and numbered 24 different policies covering topics ranging from acquisition management training to reviews and audits, failure modes and effects analyses, and selection and purchase of microcircuits.

[41] *History of SAMSO, 1 July 1972–30 June 1973*, 460. It would be interesting to have more information on the specifics of these 48 cases, but they were merely summarized in the source.

indicated availability problems of the highest quality parts. On 12 December 1972, he issued a directive that microelectronic parts used in SAMSO programs would be "procured and tested (screened) to the most stringent requirements commensurate with the intended applications." Despite this good start, practical considerations meant he could not mandate exclusive use of the best parts (the NASA Class A parts) because of poor availability. So, the policy became, "Microelectronic devices that meet Class B screening requirements (MIL-M-38510) or, when they can be procured, microelectronic devices that meet Class A screening requirements (MIL-M-38510) should be used on all SAMSO systems with long life mission requirements."[42]

Policy has its own subtleties, reflected in this language, since Class A parts "should be used." Stronger language, such as Class A parts "shall be used," would have left no room for misinterpretation, but Schultz's hands were tied, for the moment, by declining Class A parts availability. Consequently, program managers still had discretion to choose less expensive parts even when Class A parts were still in stock. The budget was the wrong criterion for solving reliability issues—the cheapest parts would not solve space program reliability issues.

Furthermore, most manufacturers were not willing to make Class A parts. This unwillingness was not about the small market; it had to do with inadequacies in the specifications. Manufacturers considered the Class A requirements "inadequate to meet their processing needs." Many available parts exceeded the Class B requirements, but were inadequate to meet Class A, resulting in a hodge-podge of screening processes by various manufacturers' own interpretations of the Class A requirements. A clearer, more stringent and comprehensive specification was needed.[43] Only a new specification would meet the Commander's intent to use only the best available parts.

The SAMSO group understood the criticality of the need to prove that space systems offered more than interesting peacetime capabilities. Superior parts were needed for important national defense and national security reasons at the earliest possible time. SAMSO began developing an expanded set of screening requirements filling in the details manufacturers needed to make the highest reliability parts.

Due to the strong time imperative, they went to RADC because of the latter's responsibility for the MIL-M-38510 microelectronics standard. RADC had just released the first revision, MIL-M-38510A, on 3 July 1972. With all the effort required to produce new versions, they were on a rough production cycle of roughly every two years, so another version would not be forthcoming any time soon, as MIL-M-38510A went to the back of a very long queue. SAMSO also approached the Defense Electronic Supply Center (DESC), the DSA organization responsible for qualifying manufacturers to produce the various classes of parts. Both organizations were sympathetic, but neither could support SAMSO's needs in a timely manner.[44]

[42] "Microelectronics for Space and Missile Systems," SAMSO Commander's Policy, 12 Dec. 1972.

[43] *History of SAMSO, 1 July 1972–30 June 1973*, 456.

[44] Ibid.

The time imperative left no choice. SAMSO undertook the task itself through the Aerospace Corporation, starting with investigations of high reliability assembly lines, and then other government agency specifications.[45] Beginning with meetings at the Aerospace Corporation on 5 December 1972, continuing through 1973, they examined how to obtain the highest reliability electronic parts necessary to support space programs.

In doing this, the influence of what had been learned through the entire course of the MINUTEMAN reliability programs (particularly the CQAP) informed the SAMSO and Aerospace people as they adapted this information to apply to space systems.

They looked at existing controlled production lines and held discussions with manufacturers and system contractors used by SAMSO program offices. The WGEPP found the "best parts were being manufactured on 'controlled' lines" because controls over process and manufacturing changes were strictest. They realized their best course of action was not further tightening the screening criteria, but rather starting with the best possible parts. This was just a further recognition that "testing in" quality was not possible, but such entrenched ideas die hard deaths. "It was felt that it would be better to build-in reliability rather than screen out faulty parts."[46] The highest reliability parts had to be made that way from the very start of their manufacture, as papers at the Reliability Physics Symposia had been recording. Put another way, the early declining failure rate in the bathtub curve often reflected unwanted inclusions, or foreign particles, in the basic materials used in the solid state devices. Screening once the parts were made would weed out the most egregious offenders, but did nothing to make the parts better from the start. Preventing the inclusions out the outset, and throughout the manufacturing process, meant fewer parts would fail, screening or no screening. The Minuteman CQAP had demonstrated this was true.

Literally, quality in microcircuits began with the silicon wafer's fabrication. The "wafer" was the metalized silicon starting point for a variety of integrated circuit devices. The way this was done reflected a desire by SAMSO to avoid dictating the methods of manufacture. Instead, they emphasized tight control of documentation and who had the authority to sign off on changes.

Achieving the highest obtainable reliability meant a new class of parts, and it had to be one that could not simply be upgraded from military avionics-grade parts. The new class would in effect be "totally unique, and . . . be handled and processed differently from wafer fabrication onward."[47]

Two particular captive lines proved to be most influential on the new specification. Texas Instruments and Motorola worked with SAMSO to convert their own specifications for electronic piece parts into the military specification format. Using the "internal detailed process specifications" from both companies as a departure

[45] Ibid., 457. The NASA 85MO specification in particular was studied but also found inadequate.
[46] Ibid.
[47] *Reliability Handbook*, 35.

point, with technical assistance from the Aerospace Corporation, the SAMSO group developed MIL-M-33, which quickly changed to MIL-M-0038510B(USAF), "Microcircuits, General Specification For." It was published on 1 October 1973.[48]

The redesignation made the document's intent clear as a direct substitute for MIL-M-38510A. It included much detail in terms of the general design, manufacturing, test, and product assurance requirements for procurement of microcircuits produced on controlled lines for use in Air Force missile and space programs. Most critically, it also established the requirements for a new kind of high reliability device, known as Class S.[49] However, MIL-M-0038510B(USAF) was administered by the DSA as part of the Mill Run Products Quality Assurance Program, not by RADC. On 15 October 1973, Texas Instruments became the first producer of Class S parts when they applied according to the Mill Run procedures, and registered 35 different products.[50]

Class S was distinctly new and different from Class A. Table 8.4 illustrates the extent to which Class S exceeded Class A. Without getting too far into the specifics of the differences, the colors highlight those areas where the new Class S specification exceeded those for Class A, or those areas where Class S required something for which there was no equivalent in the Class A parts. Class A parts had been very good, since they had supported the NASA human spaceflight programs, but the extent of the colored lines indicates Class S was a significantly tighter specification. Clarifications and additions allowed Class S to pave the way for a major increase in spacecraft reliability.

The WGEPP's timely action meant parts quality would not decrease with the loss of Class A. Their actions also showed the value of tying together independent technical experts, industry manufacturers, and government administrators. In ten months, they had created the standard for the highest reliability parts ever, and had a manufacturer lined up to produce the parts. That was the very good and hopeful news, but implementation by programs would soon become a problem for unanticipated reasons.

Without getting too far ahead of the story, it is important at this point to note the supersession of Class A by Class S was neither immediate nor a single step. A

---

[48] *History of SAMSO, 1 July 1972–30 June 1973*, 458. The controlled line specification used was from Texas Instrument's controlled line for the P-95 (Special Projects) line of Lockheed.

[49] *Reliability Handbook*, 35. A close examination of the standards governing Class B parts on the Qualified Parts List with those of Class S showed a great deal of commonality. The same eventually became true when Class B gave way to other classes for military avionics. The *Reliability Handbook* made it very clear, however, "Class S should not be viewed as 'upgraded' Class B devices. Class S devices are totally unique, and . . . must be handled and processed differently from wafer fabrication onward."

For the sake of readability and to avoid clumsy distinctions unnecessary for the point being made, I have chosen to use the term Class S to stand for all classes of high reliability space parts, which include the Classes K (for hybrids) and V (for microcircuits). It is hoped that the knowledgeable reader will forgive this nod to simplicity over precision and pedantry despite appearances of being a Best Commercial Practice for lazy writers.

[50] Historical Report, Engineering Support Branch (DRUE), 1 July–31 December 1973.

Table 8.4. Class A versus Class S Requirements

	M-0038510(USAF) Class S	M-38510 Class A
General		
Qualification of specific supplier line	1	1
Qualification of specific part types	1	1
Qualification conformance tests (re-qual)	1	1
Configuration control, formal	1	1
Resident source inspectors (DCAS)	1	0
Clean rooms for wafer and assembly operations	1	1
Traceability, complete, two way	1	0
Made in USA only	1	0
Design and Construction		
Package, hi-rel only	1	0
Metallization and leads of same material	1	0
Surface passivation	1	0
No internal rework allowed	1	0
100% Screening		
Internal visual by supplier	1*	1
Internal visual by DCAS	1	0
Stabilization bake	1	1
Temperature Cycling	1	1
Mechanical shock testing	1	1
Vibration (particle detection)	1	0
Centrifuge	1*	1
Hermetic seal, fine and gross	1*	1
Power burn in, 240 hours	1	1
Delta drift electrical, 13 mils	1	0
Electrical limit tests	1	1
Radiographic (x-ray) examination	1*	1
External visual examination	1*	1
Sample Screening (destructive)		
Scanning electron microscope	1	0
Wire bond test and control	1*	1
Die bond shear strength	1	0
Destructive physical analysis (de-lid)	1	0

(*continued*)

Table 8.4.   *(continued)*

	M-0038510(USAF) Class S	M-38510 Class A
Miscellaneous		
Accommodation for existing system designs		
Dual marking of parts:		
MIL and contractor part number	1	0
Record storage—years	7	3
Personnel training and qualification	1	1
Manufacturing flow chart	1	1
Organization chart	0	1

1 = YES, 0 = NO, * = Additional requirements above MIL-M-38510 and MIL-STD-883
▨ Class S requirements without Class A equivalent
▨ Class S requirements exceed Class A requirements

Source: HQ SAMSO/DR memorandum, "Availability of High Reliability Digital Microcircuits," 29 Oct. 1973, Attachment 2.

transition period existed because some Class A parts remained in inventories and in production for some time. Also, the initial specifications for Class S were not where they needed to be, as will be seen shortly.

Improving reliability started with improving part quality, as experience showed everything starts with parts. But a standard by itself, no matter how well done, was insufficient. The goal was not the best parts for the sake of the parts, but to make nationally important space capabilities reliable enough to be considered dependable by warfighters. The WGEPP also recognized the most successful acquisition programs had effective parts, materials, and processes *management*.

For that reason, they developed a companion to MIL-M-38510B(USAF), and gave it a title in the arcane language of standards, "Electronic Parts, Materials, and Processes For Space And Missile Applications, Standard Control Program For." SAMSO-STD 73-2A (updating the original release of 15 June 1973) established "the criteria and minimum requirements for the preparation and implementation of a Parts, Materials, and Processes Control Program for use during the design, development, fabrication, and test of spaceborne equipment."

Creating highly reliable spacecraft from the highest reliability parts is more than simply sticking the parts together. Effective management of a parts, materials, and processes (PM&P) program demands detailed inspections and sampling of parts when they are received, attention to special storage and handling requirements, and documentation of every step in the life of the part, through to the final reviews and audits prior to launch readiness of the satellite. PM&P management would evolve as testing improved or as problems pointed to some manufacturing process. At the time, they were addressing collecting the data necessary to pinpoint the cause of any problems detected.

Orbital failures are rarely so obvious and clear-cut that the cause leaps out for the investigators. More often, there are many potential reasons for a failure, each having somewhat different signatures. Sometimes, subsequent failures might be mistaken for the reason all else happened. However, isolating the very start of the failure symptoms is essential. That original failure, called the root cause, is at the most fundamental junction of all subsequent problems that may radiate out from it in a complicated system. This required an in-depth understanding of all components and assemblies regarding the ways in which they could fail, and how each failure mode would look in the reported telemetry from the systems in orbit. Frequently, the symptoms of the failure could mean any number of different causes. Less often, one failure cascades to subsequent failures. Differentiating among possible root causes was based on the different signatures of each of the different possible causes— the physics of failure. To be able to differentiate, investigators depended on data collection, documentation, and full descriptions of all the inputs and outputs of the suspected parts or materials. Root cause determination led to better understanding, and hopefully to lessons to be recorded, shared using GIDEP, and eventually added to the applicable military standards.

SAMSO-STD 73-2 incorporated by reference a number of other standards (such as MIL-M-38510 and MIL-STD-883). The evolution toward a comprehensive space parts program, however, was incomplete. The SAMSO-STD still relied on design criteria in places rather than specifying positive management controls. It allowed for so-called "standard parts" for use in space systems. A standard part was

> *one which meets the engineering requirements of this standard and which is selected from either:*
>
> *a. The parts listed in Table 1 of SAMSO-STD 73-4.*
>
> *b. Specifications references in the applicable General Equipment Specification (MIL-E-4158, MIL-E-8189, MIL-E-8983 & MIL-E-21200).*
>
> *c. An explicit list of parts identified in the contract as program standard parts.*

The 73-2 standard contained a comprehensive set of processes, criteria, screening, and tests to ensure the highest reliability parts entered spacecraft.

The new emphasis on effective parts management called for early warning of problems. Early warning in this sense was not like EW, detecting an attack under way, but more akin to IW, strategic warning. A manufacturer having problems with a specific kind of part that might have a Class S relative, or simply problems with a number of different parts, served as a heads up to parts managers. This did not mean avoiding the manufacturer. Instead, it meant a review had to take place, possibly as part of a program's audit of the manufacturer's facility and processes. Such audits sometimes included specialists from RADC who could consult on technology and process issues for the mutual benefit of the government and the manufacturer. The outcome could mean better parts and better competition in industry, but it depended on data sharing.

Another standard (MIL-STD-1556) implemented an Air Force version of the GIDEP ALERT program known as the Defective Parts and Components Control Program in 1975. GIDEP mainly dealt with military avionics class parts, which DoD was buying for its high reliability applications. Much of the GIDEP information had little to do with high reliability parts for space, but it served as a valuable source of additional information about parts themselves, and about suppliers who were having process problems that might affect space program parts. Thus, GIDEP served an important function beyond the specifics of the information shared, often affecting things far afield.

Most of the individual pieces of the high reliability program were in place when Texas Instruments line "became SAMSO's first control line available to all programs."[51] Only one thing more was needed. The SAMSO office of contracts created the necessary paperwork directing the use of the "List of Preferred Parts for Space and Missile Systems," dated 9 November 1973. Now, not only were the governance, specifications, and standards in place, but they could be made contractual obligations.

The SAMSO Engineering Support Branch issued all the documents for establishment of Class S parts, including the necessary contractual implementation language, as an information package to all System Program Offices (SPOs) on 29 October 1973. The launching of the new Class S parts was also accompanied by the Branch's issuing of three successive newsletters to the SPOs in July, September, and November 1973, as part of what they considered "the continuing information interchange."[52]

Everything was in place for a major leap forward in high reliability. But just like Bill Cosby's routine about his football team getting psyched up at half time and ready for the second half, only as they try to leave the locker room, they find the door locked, the good news stopped just about at the point Texas Instruments signed on as a supplier of Class S parts. The one and only supplier.

Parts manufacturers did not rush into this new market. The reasons were apparently widely varied. Perhaps the dominant reason was that the market size was limited. An important change in the electronics industry was under way. This was the time when consumers discovered hand-held calculators were an indispensible part of living, whereas five years earlier, no one could imagine the need for such a device. The cost for a four function calculator had plummeted from about $200 in 1970 to about $15 in 1975. Higher end, scientific calculators at $395 in 1972 became $295 the following year, and fell to $100 by 1975. The change was in the microcircuits. The price of the devices leveled off as the cost of the microcircuits was no longer a factor in total price. By 1975, most college engineering schools dropped use of the slide rule and converted to the new electronic devices, in a dramatic illustration of a new technology completely swamping an old technology in less than half a dozen years. The consequence was a huge spurt in commercial chip production as planned production line sizes could not meet the demand. One engineering

---

[51] Historical Report, Engineering Support Branch (DRUE), 1 July–31 Dec. 1973.
[52] Ibid.

school's order for its incoming freshman class tied up the Texas Instruments production line of their SR-51 calculators for two weeks.[53]

In 1973, to encourage the new Class S supply chain amidst the rising tide of commercial parts, SAMSO and the Aerospace Corporation began to host an annual Space Parts Working Group, inviting manufacturers, suppliers, and others related to the space parts industry.[54] The group began by informal information sharing on topics of mutual interest, to advance the quality and reliability of the highest reliability parts for spacecraft. Quickly, the meeting became an important annual forum.

*   *   *

More was at stake than the unacceptability of the increasing cost of space systems due to partial or complete mission losses. The major military commanders and their operational counterparts were telling the SAMSO commanders that space systems were simply too unreliable to be counted on in military operations. The only answer to such criticism was to work at reducing the risk of failure wherever it popped up.

General Phillips and his successors honed the high reliability electronics and spacecraft acquisition processes to a fine edge. Through the efforts of the SAMSO commanders and their staffs, a system was in place carrying with it extensive infrastructure in government acquisition offices, FFRDCs, industry, government laboratories, and the standards process.

That might have been the end of the story, since Class S had been defined and all the relevant implementing pieces were in place, including a manufacturer. But that was hardly the end.

[53] Ernest Braun and Stuart Macdonald, *Revolution in Miniature: The History and Impact of Semiconductor Electronics* (New York: Cambridge University Press, 1982), 189–190; "When Slide Rules Ruled the Rockies," *Checkpoints* (USAF Academy, Colorado: Association of Graduates) 37, no. 3 (Dec. 2008): 55; author interview with Thomas Turflinger, former parts engineer with Navy CRANE, July–Aug. 2010.

[54] The Space Parts Working Group begun at that time has continued to the present in an unbroken chain of annual meetings, though the focus has evolved along with the membership.

Chapter 9

There's S, and Then There's S

The U.S. military was about to confront the significant gains in capabilities offered by spacecraft. The late 1960s and early 1970s were characterized by reluctance to rely too heavily on spacecraft. Most military commanders considered satellites fragile research experiments requiring scores of engineers to keep them running. Although satellites' demonstrated capabilities would greatly enhance military operations, their reliability and dependability in warfare were unproven. Other aspects of these concerns surrounded space systems' ownership. The most critical space systems were owned and operated by the NRO, outside of military control except in a national crisis. These national security systems served the National Command Authority and the Intelligence Community, but only loosely supported most military operational commands (other than the strategic commands, with whom the NRO had a long history of IW support). National defense space systems were owned and operated by AFSC—an acquisition command, not an operational one. That seemed further proof the scientists could not let go of their "expensive toys."

SAMSO and NRO work to increase spacecraft reliability countered most such parochial and uninformed views. The payoff was spacecraft that began to achieve the extremely high reliability for which they eventually became known. Advances in high reliability would have multiple consequences in the cost of spacecraft, improved reliability and capability, longer spacecraft lifetimes, and the need to impose a more "operational" point of view on the military space program. Overlaying these concerns was a general interest in making military acquisition processes more efficient.

* * *

Until the early 1970s, the Services had viewed space systems with some skepticism. The fragility of these early satellites outweighed their military utility in the minds of many responsible for planning and executing tactical military operations (such as the war in Vietnam). For instance, the film-return reconnaissance satellites provided information (when it was not cloud covered) well after any given tactical military operations ended. The military, consequently, had adapted several kinds of aircraft

Implosion: Lessons from National Security, High Reliability Spacecraft, Electronics, and the Forces Which Changed Them, First Edition. L. Parker Temple III.
© 2013 the Institute of Electrical and Electronics Engineers. Published 2013 by John Wiley & Sons, Inc.

to provide near-real-time reconnaissance. But not all space capabilities were marginal.

For strategic forces, reliance was a different problem. DSP's early warning of missile attacks against the United States was an important pillar of continental defense, as it would alert the missile and bomber forces for retaliatory strikes in a thermonuclear war. Although the Chiefs of Staff of the Air Force historically had strategic operational experience, space systems were maturing and expanding their ability to serve tactical operations. Weather support from the DMSP had a notable effect on one series of tactical military operations. The landmark was the near-real-time weather support during the Cambodian hijacking and U.S. recovery of the S.S. *Mayaguez* in May 1975. The *Mayaguez*, an unarmed merchant vessel of U.S. registry, became the focus of a major international incident. The incident proved to be highly important in the operational employment of space systems supporting tactical forces. DMSP's timely weather imagery, for instance, was credited with saving lives and aircraft because it allowed on-scene commanders to revise where fighters would rendezvous with aerial refueling tankers. Moving the locations from cloud-covered areas to areas where these operations could be done in clear weather impressed military planners. DMSP had primarily been intended to support strategic forces, but this showed space capabilities could benefit tactical forces as well. In this way, space capabilities began to be used more by tactical commanders, despite continuing concerns about their reliability and availability.[1]

Furthermore, the national security space systems supported Strategic Air Command's (SAC) targeting and gave IW intelligence, especially during such things as Soviet exercises and troop movements. Providing such information helped SAC plan the uses of its strategic forces, and manage the preparedness of these forces.

At the same time, communications satellites were proving critical to the instantaneous worldwide command and control of military forces worldwide. The good and bad consequences provide important insights into the evolution of space capabilities towards acceptance. For strategic forces, communications satellites had been essential to maintaining worldwide positive command and control. But the strategic commanders' acceptance of space capabilities had never been in doubt. The tactical commanders in Southeast Asia and the Philippines had other thoughts.

For instance, first launched in 1966 with a constellation of 28 satellites in total, the IDCSP had enabled a new era in command and control of military forces. A year after the launch of the first eight satellites, they linked the commands in Southeast Asia to Washington, D.C., by a combination of satellite relays and ground links. By 1968, with the full constellation in place, the system's name became the Initial Defense Communication Satellite System. The positive control of globally-deployed forces, which had been so much a necessity for strategic forces, evolved into a new form of micro-management of the tactical commanders. Communications satellites allowed abridging the military commanders' authority to conduct operations for whose success they were responsible. Although they were the ones

---

[1] W.H. Greenhalgh, Jr., *U.S. Air Force Reconnaissance in Southeast Asia, 1960–1975*, Study 135, (Montgomery, Alabama: Air Force Historical Research Agency, 1977), 254.

"on site," instantaneous communications enabled the Pentagon to pull back tactical targeting priorities and decisions. Anecdotally, some tactical targeting decisions were even made by President Johnson. Those responsible for the forces in the actual conflict area saw communications satellites as intrusive and a fundamental change to field responsibility. Some thought such command from afar was irresponsible or ill-considered. These lessons, good and bad, being learned and recorded would be an important consideration in the struggle for acceptance of space capabilities.[2]

Space capabilities were slowly and unevenly becoming operationally important. Serving the National Command Authority and the President's senior security advisors, as well as the strategic commanders, space capabilities made a difference to the highest priority users. Gradually, capabilities were also becoming relevant to tactical military users, despite concerns about their fragility.

*   *   *

Concern about fragility went beyond the technological. The importance of some systems had made them potential targets—something about which tactical commanders knew. The Soviets had anti-satellite weapons. These were not new capabilities. In fact, anti-satellites had existed for some time. However, an important change was in works.

Both the Soviet Union and the United States had operational systems that could attack satellites in low earth orbit. These were direct-ascent systems, using anti-ballistic missiles to loft a nuclear weapon into the path of a satellite. The U.S. Nike Zeus anti-ballistic missile, Project 505, assumed this mission in 1964. Zeus was quickly overtaken by Project 437 in 1967. Project 437 was a Thor IRBM standing alert on Johnston Island in the Pacific. None of these capabilities were particularly useful, because experiments with high altitude nuclear detonations revealed they created more problems than destroying a single satellite warranted. Nevertheless, these systems remained in operation well into the 1970s, with Project 437 becoming inactive in 1975. These systems were replaced because newer approaches could minimize the severe collateral effects.

Beginning with experiments in rendezvous during the 1960s, the Soviets developed, and had put into operational status, a co-orbital anti-satellite by 1972. The interceptor was essentially a large upper-stage of a space launch vehicle. By careful timing and trajectory choices, the interceptor would be launched into space just ahead of its target. It would then blow itself apart, creating a debris field through which the target would have to fly, destroying it without resort to nuclear weapons.

The United States took a somewhat different approach. In the 1950s, several experiments using air-launched missiles demonstrated a possibility of attacking low earth orbiting satellites. Weapon System 119 was an air-to-surface missile intended to give strategic bombers some stand-off capability. After WS-119's evaluation, some residual missiles were launched in an air-to-space mode by Boeing B-47

[2] Donald H. Martin, *Communication Satellites, 1958–1992* (El Segundo, California: Aerospace Corporation, 1991), 95.

Stratojets and Convair B-58 Hustler bombers, designated Bold Orion and High Virgo respectively. On 19 October 1959, Bold Orion's missile passed within about four miles of the U.S. Explorer 6 satellite. The test was not intended to actually intercept Explorer 6. The objective was simply to get close enough to demonstrate the capability which, had it been armed with a nuclear weapon, would have been lethal for the satellite. These tests, and some other experiments run by the Navy using Douglas F-4D Skyray fighters, proved just how technologically challenging an air-launched anti-satellite system would be.

The nuclear anti-satellites' prompt and persistent side effects severely limited their utility. With the advent of the Soviet anti-satellite, the United States needed some equally efficient non-nuclear alternative to hold Soviet low earth orbit satellites at risk. If feasible, an air-launched anti-satellite would have an important potential feature: it could be launched by an operational tactical fighter aircraft. This might appeal to the tactical part of the Air Force and change some of their views about space systems. As challenging as air-launched anti-satellite technology was, by the mid-1970s, several lines of technology maturation began to come together.

*   *   *

As the anti-satellite storm began brewing, around the time of the *Mayaguez* incident, Chief Scientist of the Air Force Michael Yarymovych released an important forecast of technologies which would support the future of the Air Force. In May 1975, he presented the results of *Towards New Horizons II* to the Air Force Chief of Staff. The report contained some radical ideas, but laid out a sound approach to achieving future technological superiority. Five of the top eight recommendations required space and microelectronics technology to achieve them. The study said the Air Force should:

1. Take full advantage of computer technology;
2. Draft a coherent plan to introduce and operate space systems;
3. Develop survivable space systems;
4. Increase research on directed-energy weapons; and
5. Develop a space-based system to defend satellites from attack.

Two more recommendations, a worldwide digitized cartographic database and an all-weather aircraft linked with various battlefield systems, also depended on computers and electronics technology. Finally, the report recommended a heavy-lift global-range airlift aircraft. *Towards New Horizons II* succinctly captured the American way of war from that time forward.[3]

The visionary study came when the military was leaning more heavily on space systems despite lingering doubts about the fragility of the tiny machines out in the far reaches of space. Improved military capabilities provided by spacecraft were

---

[3] Dwayne A. Day, *Lightning Rod: A History of the Air Force Chief Scientist's Office* (Washington, D.C.: Chief Scientist's Office, 2000), 194; Executive Summary, *New Horizons II Study*, June 1975, pp. 29–32 and vol. 5, "The Role of the Air Force in Space."

undeniable and gradually encroaching on routine military operations. If this was truly the way of the future, military and national security space program leaders had to eliminate the issue of fragility—essentially a lack of confidence in satellite reliability. Yarymovych pointed out ways to address the lack of confidence. *Towards New Horizons II* was indeed prescient, as the microprocessor enabling the personal computer was still four years from maturity.

Yarymovych's study said the Air Force's future was tied to advanced technology strongly related to computers and space systems, which meant highly reliable electronics.

*   *   *

Removing the stigma of fragility was important for operational acceptance of space capabilities, but spacecraft were not like aircraft. Progress was uneven. It was only a matter of a few years before all national defense satellites would be considerably more robust, reliable, and long-lived. In the interim, each satellite was unique due to its original technology content, the different kinds of secondary payloads each carried, the number of anomalies and failures it experienced, and the ways these were worked around.

Class S parts and PM&P programs were beginning to have an effect, but hardly universally. The reality of building, launching, and replacing satellites has its own momentum; changing direction is hard. While a new satellite being built might offer some impressive new technologies or capabilities, the capabilities in orbit were what counted. Those in orbit were slowly beginning to last longer due to the continual improvement process of information feedback, as on-orbit experience (especially root causes of failures) directly affected subsequent satellites being built. Improved reliability of parts meant constellations (groups of like satellites) did not advance smoothly, but in definable generations. Prior satellite generations were still working when a new satellite went into orbit, still providing the older capabilities for some additional length of time. So, while some new satellites using Class S parts might offer much longer lifetimes, overall, many lesser satellites remained in operation.

This added up to each individual satellite requiring a deep technical understanding. AFSC commanders had successfully proved the operational commands were unsuited to handle such technological sophistication or deal with problems that only the makers of the systems could handle. For the time being, the Air Force solution was to retain national defense space systems inside AFSC, without an operational command to which these systems would transition, unlike aircraft. That did not sit well.

Behind the scenes lurked the much more difficult problem of control. AFSC only handled the national defense space systems; national security space systems were owned and operated by the NRO, outside the control of the DoD.[4] Military commanders envied the NRO systems and budgets. Reconnaissance satellites were crucially important, innovative, highly capable, and out of the reach of the budget.

---

[4] Provisions existed at higher defense conditions for the NRO systems to come under the command of the DoD, but no such national emergency had ever arisen.

A movement was spawned within the Air Force to move to an operational command for space. This began partially in the belief it would force development of even more robust and responsive spacecraft. No less important was the fact that a space command would partially make the case that space systems were finally "operational" in a meaningful military sense and could be relied upon. Less overt was the intent to whittle away (hopefully) at some of the budgets that had been out of reach until that time. The seed was planted, and would take time to grow, but control of space systems was going to be contested. The underlying enabler was the advances made possible by very high quality, highly reliable parts affecting satellite robustness, capability, and longevity.

Space systems advanced by their underlying microelectronics, but still required a high level of engineering and technology. The *Towards New Horizons II* report's urging the Air Force to take full advantage of computer technology meant microelectronics were a key element of the Air Force's future—the *whole* Air Force, not just space systems. As terrestrial weapon systems began relying on similar advanced technologies, the day was not too far off when space would be as "operational" as the rest of the Air Force.

An indication of things to come was the Air Force's partnership with NASA on the STS, whose most popularly known segment was the Space Shuttle. The Air Force was providing the Western Launch and Landing Site at Vandenberg AFB, California. Development of the ground processing system for the Space Shuttle was in full swing as *Towards New Horizons II* was published.

Not all the things affecting space parts occurred in satellites. For instance, one of the significant technical issues for the Vandenberg launch processing system was how best to use the large mainframe computers necessary to prepare and launch the Space Shuttle. Specifically, how would information be pulled from the mainframes and processed for display to the launch controllers? With IBM, the Air Force worked through either "dumb" terminals, which would do little more than display retrieved information, or "smart" terminals, which would contain some internal processing capability. Known as "Front End Terminals," these smart terminals presaged a computer technology that IBM would release publicly within a couple of years. The IBM Personal Computer would have a significant role in the evolution of the electronic piece parts marketplace.

Curiously, the Shuttle represented a nexus in many ways. It exemplified the major changes about to affect space programs and high reliability parts. The Shuttle, of course, was the chief outcome of the 1969 Space Task Group. As envisioned in President Nixon's final decision to proceed, the STS would start launching 60 times per year in the late 1970s. Forty of these launches were to be from the Kennedy Space Center, and the rest from the West Coast. This launch rate would allow the STS to pay for itself eventually, according to its advocates, while still reducing the cost of launches by an order of magnitude.

The economics of Shuttle advocates did not anticipate a drastic reduction in launch demand. The STS was envisioned before Class S existed, but did not get the President's approval until after. The effects of high reliability parts and improved parts management programs were already being felt as Nixon approved the STS.

The STS never came close to its 60 launches per year (it fell an order of magnitude short) partly because it was never really built to launch at that rate, but also because the need for that many launches evaporated. Some technology advancements incorporated into the STS, such as smart Front End Terminals in the ground processing system, were also dooming its technology to be well out of date by the time it began launching in the early 1980s.

The Shuttle was sold in part by advancing a model of repairing satellites in space, providing an alternative to more reliable, longer-lived satellites. Only a couple of repairable satellites were ever built, as Class S parts eventually allowed satellites to last so long, repairing them to extend their lives made little technological sense.

Class S was a significant step forward, but experience with this new class of parts was initially limited by several factors precluding its universal acceptance. Of course, advancing technology played a part, but the rate of advance, though more rapid than in times previous, still did not compare to the rate of advance in the last decade of the 20th century. As important as high reliability was, by 1975, no program had budgeted for the more costly Class S parts. Those who used Class S parts used them sparingly in critical subsystems, as the higher cost had to be taken out of existing budgets. Also, initially Class S included only a few limited types. Finally, just as Class S reflected lessons learned from prior hard experience, Class S would itself evolve as more was learned from using these new parts. Class S, it seems, was not cast in concrete.

To illustrate the rate of progress, an example in launch vehicles can serve. In late 1973, after the new Class S parts effort was inaugurated and the new standards were brought out, SAMSO's Launch Vehicles Directorate approached Martin Marietta about "improving their parts upgrade program." The cost was too high.

About a year later, however, the company approached SAMSO and the Secretary of the Air Force Special Projects (SAFSP, the unclassified designation of the NRO's Program A, which used its collocation with SAMSO to cover its existence). SAFSP's work to increase Titan III launch vehicle performance to backup the Shuttle in case of failures or delays "highlighted the increase in parts problems during 1974."[5] That led to a Titan III System Effectiveness Review on 30 January 1975 "to improve the reliability of Titan III piece parts through stricter inspection and testing of the parts."[6]

Then, as though emphasizing the need for improved reliability and validating the remaining military commanders' doubts about relying on space programs, launch vehicle failure rates began to climb. Of 39 Air Force launch vehicle firings in 1975, four had failed.[7]

---

[5] MajGen John E. Kulpa, Jr., letter, Subj: Titan III Piece Parts Upgrade Proposal, 12 April 1976.

[6] *History of Space and Missile Systems Organization, 1 July 1975–30 June 1975*, HQ SAMSO History Office, El Segundo, California, 80.

[7] A launch vehicle failure without total loss of mission was a euphemism for failing to reach the proper orbit. If the satellite were placed in the wrong orbit, something of the original mission might be salvaged, but the launch vehicle would still have been deemed to have failed. The original source for these numbers, though not in doubt, could not be verified independently.

LtGen Thomas W. Morgan, by then the SAMSO Commander, reacted to this spate of failures by imposing a "No Change" Commander's Policy on 29 September 1975. Recent performance improvements to launch vehicles were thought to be behind the rise in failures. Lacking definitive answers, Morgan directed "performance improvement changes will not be permitted for proven, successful hardware/software" except under the most strictly controlled conditions. However, spacecraft continued to grow in capabilities (and, hence, weight). The programs most in need of additional performance were the highly classified missions, so Morgan's cap on performance was clearly only an emergency interim measure.[8]

At the time, MajGen John E. Kulpa, Jr., led SAFSP. Reporting directly to the Under Secretary of the Air Force instead of anyone in a uniformed military chain, Kulpa and his organization had a great deal of clout. The NRO at the time was noted for its lack of bureaucracy and prompt actions to resolve technical issues such as this. On 12 April 1976, he advanced the discussions with a letter to Colonel George Murphy, Director of the Launch Vehicles Directorate at SAMSO.[9] Not too subtly, his reference to the 1973 and 1974 identification of problems intimated no excuse for any later launch failures. Kulpa wanted "a detailed analysis of the problem part types and components/subsystems (historic bad performers)." Colonel Murphy's reaction was the letter applied to his ongoing effort to improve reliability through redundant avionics in the Titan III B, C, and D models targeted for launches in late 1979, but that was too limited a response.[10]

Already under way, starting in mid-March 1976, was the Ad Hoc Study Group on Space Launch Vehicles, chaired by MajGen David D. Bradburn, Vice Commander of ESD. By the time Bradburn's study started, eight more launch vehicle firings occurred with one failure. Their approach was to reexamine the results of the launch vehicle failure review boards. His results were to go to the AFSC Commander, General William J. Evans. Bradburn was an ideal choice for the review, since his command owned RADC, and was therefore deeply involved in the overall military electronics work of the Air Force.[11]

What was it exactly that RADC was doing that it could benefit Bradburn's investigation? The RADC laboratories were national assets serving the Air Force, the other Services, DoD, and governmental agencies, as well as commercial advanced technology parts manufacturers. The multiple national responsibilities served by RADC's laboratories ran the gamut of electronics expertise from active to passive and analog to digital devices. The Reliability and Diagnostics Branch was the national center of expertise in testing and analysis of analog devices, having pioneered many of the analysis techniques for these. That branch developed the quality

---

[8] Commander's Policy, *Changes to Proven Launch Vehicle Hardware and Software*, SAMSO Regulation 5-4, 29 Sep. 1975.

[9] MajGen John E. Kulpa, Jr., letter, Subj: Titan III Piece Parts Upgrade Proposal, 12 April 1976.

[10] Ibid. Murphy's reaction is included in his handwritten notes on the letter as he received it, held in the files of the Space and Missile Center History Office.

[11] *Report of the Ad Hoc Study Group on Space Launch Vehicles*, HQ Electronic Systems Division, Hansom AFB, Massachusetts, 12 April 1976, 1.

and reliability procedures for evaluating analog devices.[12] The Design and Diagnostics Branch was the national center of expertise on digital devices. They maintained advanced capabilities to test the most complicated and highest speed integrated circuits and multi-chip modules. The group also designed tools to help manufacturers "design in" reliability especially to avoid the effects of electromigration and hot electron effects in their devices.[13] The group led DoD efforts in the rapid prototyping of signal processing architectures—crucial to the design of advanced systems for air and space platforms.[14]

They did more than just support applied work, because laboratories did considerable fundamental research. The Reliability Physics Branch examined materials, interfaces in materials, and the underlying chemistry and physics of these as microelectronic feature sizes continued to decrease. The latter was extremely important as smaller feature sizes became subject to new fundamental atomic and subatomic physics mechanisms. Just as the development of the first transistor required understanding the electron's quantum behavior before a practical device could be built, the Reliability Physics Branch examined the advanced behaviors of electrons in various materials and over smaller and smaller distances. In so doing, they provided the basis for understanding the reliability of new technologies, enabling commercial manufacturers to better understand what they were creating.[15] Yet another branch provided analytic tools to evaluate the mechanical, thermal, and electronic performance of devices and components before they were built, and to investigate failures after the devices were fielded. The Design Analysis Branch developed simulation tools to evaluate such failures. Perhaps most important of all was the branches' responsibility to operate the DoD Reliability Analysis Center for the entire Defense community.[16] Behind Bradburn, then, stood a unique national asset capable of tackling the most difficult kinds of reliability calculations, design analyses, and failure investigations and determinations.

The May 1975 loss of the 25th Titan IIIC became a watershed event. The failure review board's analysis led to three critical conclusions. The first was recognition that the Class S parts and the attendant standards were "considerable continued advances . . . made during this period in piece part design, manufacture, and screening." Second, "the failure . . . demonstrated the need for a stronger parts program on flight critical hardware." Finally, "the existing parts upgrade program contained

---

[12] Rome Laboratory Command and Control Systems Fax, Subj: Back-up data, Programmatic Impacts of SECDEF Recommendations, 21 April 1995, 2.

[13] Rudolph Ludeke, "Hot-Electron Effects And Oxide Degradation In MOS Structures Studied With Ballistic Electron Emission Microscopy," IBM J. Res. Develop. 44, no. 4 (July 2000): 517–518. Electromigration refers to a problem due to momentum transfer of electrons due to current flow resulting in movement of metal atoms in a conductor. The result is damage to the conductor degrading the device over time.

[14] Rome Laboratory Command and Control Systems Fax, Subj: Back-up data, Programmatic Impacts of SECDEF Recommendations, 21 April 1995, p. 2.

[15] Ibid., 3.

[16] Ibid.

separate standards for piece parts used by Martin Marietta and piece parts used by subcontractors. It was obviously desirable to replace these separate sets of standards with one common set that would apply to all piece parts."[17] This thinking in the Launch Vehicles Directorate directly shaped some of the Bradburn group's attention.

Bradburn's group quickly focused on some specifics. They examined whether (1) the "expected high standard of vigilance" had lowered, (2) "pressures for economical operations" had been a factor, (3) refurbishing missiles as launch vehicles still had merit, and (4) management actions were needed to "assure maximum reliability."

These were all areas where some actions might be needed to improve reliability and assure mission success. The group's review of launch vehicle failures since 1 January 1965 got them a better feel for the trends. The date was somewhat arbitrary, but the configurations of the launch vehicles used in 1965 were close to those used in 1976. They could not attribute all of the failures to electronic piece parts. Of 37 failures examined for that period, the leading causes were design problems (15) and procedures (12). Only three related explicitly to electronic piece parts.

For instance, on 12 April 1975, an Atlas F sustainer engine was damaged by an explosion in the flame bucket (probably a fuel spill), resulting in an early engine shut down and loss of a classified payload. That contrasted with the 20 May 1975 Titan IIIC failure when conductive particles shorted a transistor in the inertial guidance system, losing the Defense Satellite Communications System satellite and launch vehicle.[18]

Had Bradburn simply looked at that breakdown of the 37 failures, his group would have concluded design and procedures improvements would drastically cut the failure rates. While true, he was also highly aware of the industry trends related to electronics complexity. Therefore, he concluded, the *"reliability of* the *simple launch vehicles,* refurbished or new, *remains high,* while the *reliability of* the *complex upper stages,* as a result of electronic piece parts failures, *is troublesomely low."*[19]

He saw the gross numbers averaging out some important effects, hiding the upper stage problems within the larger numbers. He dug deeper, and found the design errors, accounting for 15 of the overall 37 failures, could themselves be traced directly to electronic parts. "New designs applied to previously used upper stages have malfunctioned as a result of electronic piece parts failure rather than design error." That is, the design was not the problem so much as the designer's expectation the new parts being designed in would function as before. The problem related to complexity. "The risk factors for [complex upper stages] are primarily the complex

---

[17] *History of Space and Missile Systems Organization,* 80–81. The history cites as its original source a briefing, HQ SAMSO/LVXS (Lt Giddens), Subj: "Piece Parts Upgrade at MMC," 31 July 1975.

[18] *Report of the Ad Hoc Study,* 1–2; Don Seta, "JAN S Electronic Parts," *Institute of Environmental Sciences Proceedings* (1985): 310; Madeline W. Sherman, ed., *TRW Space Log Twenty-Fifth Anniversary of Space Exploration, 1957–1982* (Redondo Beach, California: TRW Electronics and Defense Sector, 1983), 86.

[19] *Report of the Ad Hoc Study,* 2. Emphasis is in the original.

electronic systems."[20] The Centaur upper stage was the exception because of its "aggressive electronic piece parts program," and some electronic redundancy enabled by proper use of advanced technology.

Bradburn also apparently found a systematic knowledge retention problem. One of the group's recommendations said "To insure that 'lessons learned' stay 'learned,' SAMSO should establish a formal system to transfer failure information, procedures, and data among the ranges, program offices, and contractors." Clearly, the data on the failures was available, as Bradburn had accessed it in the study. The problem, as his group saw it, was relearning the lessons in each generation.

Bradburn's Ad Hoc Study Group proved to be a key juncture in the high reliability space qualified parts efforts. The Air Force responded, reenergizing the efforts that had led to Class S, tightening specifications, qualification, certification, and a range of satellite design activities. These went to the heart of Bradburn's findings about the core problem with design failures. The efforts were very collaborative, involving as major participants: SAMSO; NASA's Marshall Space Flight Center, overseers of the Class A parts program; DESC, responsible for qualification of Classes B and C parts; and the Navy's Naval Weapons Support Center CRANE, successor to the NAD organization, located in Crane, Indiana. Each organization had some important interest in the quality and reliability of the parts to be used in space programs.

By the end of 1977, the Class S space parts standards, specifications, and guidelines were:

- "MIL-M-38510D, 31 August 1977, 'Microcircuits, General Specification for.'
- MIL-STD-976, 31 August 1977, 'Certification Requirement for JAN Microcircuits.'
- DESC-EQE-44, First draft, 18 November 1977, 'Guidelines for Implementation of Class S Microcircuit Certification.'"

Class S testing was in two MIL-STD-883 test methods, 5004 and 5005.[21] Although established several years before, Class S parts are often not considered to have replaced Class A parts until these 1977 events.[22] The cost of upgrading parts to Class S, seen in light of the issues Bradburn examined, was far less than even one of the failures.

In addition to these important changes, other important advances affected hybrid[23] devices. Unlike their solid state and vacuum tube ancestors, these hybrid

---

[20] *Report of the Ad Hoc Study*, 3. The problem was made worse because despite the complexity, due to the desire to minimize upper stage weight, critical functions were single string.

[21] "Minutes of the Meeting for Microcircuit Manufacturers on the New Class S Space Parts Documentation," Aerospace Corporation, SAMSO-TR-78-72, 15 Jan. 1978, p. 1.

[22] *Reliability Handbook*, 29.

[23] The formal definition for a hybrid integrated circuit is "[a]n integrated circuit that contains two or more of a single type or a combination of types of the following elements, with at least one of the elements being active: film microcircuit, monolithic microcircuit, discrete semiconductor device, passive chip, or passive element printed or deposited on a substrate." *Terms, Definitions, and Letter Symbols for Microelectronic Device*, JEDEC standard JESD99B, May 2007.

devices mixed different kinds of solid state devices, or mixed solid state devices with other things such as magnetic devices. The definition of a hybrid device had moved beyond its earlier reference to multichip modules (that is, more than one small integrated circuit chip) and later multichip packages. Hybrids encompassed any combination involving an integrated circuit and some discrete device (the latter being resistors, capacitors, transistors, etc.). For instance, a hybrid would have some kind of integrated circuit combined with some additional device, either as a magnetic coil, some transistors and capacitors, or other additional parts that were not designed into the integrated circuit. The reasons for not designing them into the part varied. In some cases, the external parts improved computation speed or signal quality. Radio frequency devices might have a magnetic coil whose function was unsupported by microelectronics technology. Hybrids became essential as analog-to-digital converters, alternating current-to-direct current converters, direct current-to-direct current converters, and similar devices. Whatever the reason, the combinations of integrated circuits and additional parts tended to be unique custom-designed devices tailored to a specific need. Considered an aberrant form of monolithic integrated circuits because of their combination of distinctly different parts and devices, no existing standard adequately covered their testing. The August 1977 revision to MIL-M-38510 included a new appendix (Appendix G) specifically addressing hybrid construction. To accompany this new appendix, MIL-STD-883 added a new test method (Method 5008) specifically for hybrids.[24]

This large step in the right direction still had to deal with a problem in terms of military and spacecraft usage related to the QPL. The process for listing on the QPL assumed large lot sizes, but hybrids were essentially custom devices of limited quantity. So, testing and qualifying hybrids as though they were very long run parts, such as capacitors or even some standard integrated circuits, simply missed the marks of cost, schedule, and quality. At the time, no suitable alternative approach existed, so this would be the sore thumb of the high reliability industry for some time, awaiting some kind of practical relief. As hybrids gained wider acceptance in the military and on spacecraft, the problems of their qualification only grew worse.

*   *   *

President Jimmy Carter and his advisors took notice of the Soviet co-orbital anti-satellite (ASAT) system's operational status. Despite only a few actual intercept tests, the Soviet ASAT held hostage many critical national security spacecraft. It used the same booster as the majority of their other satellites, so western analysts considered each launch an ASAT dress rehearsal. The U.S. would respond in kind, based on extensive research and development for such a capability. Carter ordered the Prototype Miniature Air-Launched Segment to proceed into full development in 1978. The program began in earnest the following year. Formally designated ASM-135, the weapon system was called the Air Launched Miniature Vehicle (ALMV). The ALMV was a small cylindrical shape, roughly a foot in diameter and a foot

[24] William J. Spears, "MIL-STD-1772 and the Future of Military Hybrids," *IEEE Transactions on Components, Hybrids, and Manufacturing Technology* 12, no. 4 (Dec. 1989): 506–507.

long, and was one of the most complex programs undertaken to that time. Based on information from tracking stations, extensive planning and targeting computation was done on some of the most advanced computers available. Timing and timeliness were absolutely critical to make the system work. Launched from a zooming McDonnell Douglas F-15 fighter, the ALMV got to space in front of its intended target by means of a two stage rocket. Aimed by an extremely sophisticated infrared sensor, the ALMV used small rocket motors to "hover" in front of the oncoming target. The Air Force required 112 ALMVs, and 40 specially modified F-15 aircraft based in Washington and Virginia.[25]

For some, the ALMV was a statement about space systems being operational, with a real weapon in development. To others, it was a bargaining chip to be used to negotiate an end to the Soviet ASAT program. Some, though, considered it an affront to the peaceful uses of outer space. They set out to kill it with a clever legislative trick.

*   *   *

Meanwhile, Morgan took a page from General Sam Phillips' original PRB (on which he had been a charter member). Morgan updated the PRB, calling it the SAMSO/ Aerospace Select Committee on Piece Parts.[26] Continuous improvement of part quality forced Class S parts evolution. Certification to produce had never been trivial, and as experience mounted, so did scrutiny of manufacturers and their processes.

Class S evolved through incorporating changes to MIL-STD-38510, reaching revision D by 1978. The Aerospace Corporation's James Egan explained "each requirement matches up to a problem we have had one or more times in the past." The changes were not arbitrary or theoretical; they were hard fought, hard won lessons.[27] When Bradburn's group found a problem with unlearning formerly learned lessons, the response was to take those lessons, document the next critical areas in the Commanders' Policies, and move them to the appropriate standards.

What they did not do was clearly document the reason for each change to preserve the problem's linkage to the lesson, and then to the paragraph in a standard. As long as the standards discipline remained intact, this would not be a problem. All who used the standards understood and appreciated exactly the point Egan made. It was not so much that the standard was the one and only way to do something. In some cases, alternatives were presented and allowed. The certainty, though, was the methods in the standards were *known* to work. Variance from them, then, had more risk and less certainty of success. Ignoring standards invited problems.

*   *   *

By 1978, the impetus of the early missile days was long past. The only new missile development under way, Peacekeeper, was far more limited in scope than Minute-

---

[25] Paul B. Stares, *Space and National Security* (Washington, D.C.: Brookings Institution, 1987), 101.

[26] "Minutes," 12.

[27] "Minutes," 20.

man. Class S parts mainly supported space programs, but with their stringent requirements and limited market, some electronics manufacturers were grumbling about the measures required for these parts.

Events had not only refined the requirements for Class S, realities also forced changes in their oversight and governance. Whereas NASA, SAMSO, and DESC had previously performed independent audits of the same manufacturing lines, the new Class S unified all of these to "effectively administer and invoke the Class S qualification and certification requirements."[28]

Class S exceeded NASA Class A requirements when originally conceived. By 1978, in addition to the earlier requirements differentiating the two classes, Class S included lot homogeneity (necessary for statistical significance of sampled lots), scanning electron microscope photos (checking such things as interconnect bonding workmanship quality), and additional requirements on product assurance programs.[29] Before assembling any microcircuit using JAN Class S parts, every wafer lot had to pass an acceptance test, called a quality conformance inspection.[30] One of the causes of the early failures in the bathtub curve was unwanted inclusions in the materials. A wafer lot is the beginning of the part—the first step up from raw silicon. Everything starts with parts, it is true, but parts start with materials, which had to be checked for quality.

The Air Force intended to mandate Class S for all new satellite programs, and wherever possible on any buy of new parts for existing programs. Before that could happen, several controls had to be in place. Producing qualified Class S parts meant manufacturers had to be certified, for two reasons. First, certification established the threshold for entry into the high reliability marketplace, weeding out lower quality manufacturers. Second, it leveled the playing field so a part made by one manufacturer would have quality consistent with the same type from another manufacturer.

DESC's Class S manufacturer certification was only a portion of the process to produce space qualified parts. Parts to be offered as Class S could come only from a Class S certified line, and had to be on the Qualified Products List (QPL, successor to the Qualified Parts List).[31] Thus, certification was only one step in a process that ran as follows:

1. The manufacturer requested an audit and submitted required information to DESC;

2. DESC, with SAMSO and NASA, audited the manufacturer's product line(s) and product assurance program;

---

[28] "Minutes," 7. The cooperative nature of these arrangements can be illustrated by the Class S certification requirement that said no Class S certification of a product line could exist without Class B certification. That meant that the Class S certification requirements were not isolated, but were in addition to or exceeded those for Class B, which then became an essential point of departure.

[29] "Minutes," 37–38.

[30] Seta, "JAN S Electronic Parts," 310.

[31] Since hybrids, in particular, were not a "part" but an assembly of parts, the Qualified Parts List had to expand to account for "Products" that were not single parts.

3. Written notification from DESC informed the manufacturer, following a successful audit, the line(s) were *certified*;

4. The manufacturer then had to be *qualified* (in accordance with MIL-M-38510, which required producing and testing parts), after receiving DESC's authorization to test;

5. DESC reviewed the results and issued the QPL-38510 listing;

6. Finally, recertification was required every two years, and maintenance every six months. Periodic quality conformance testing was expected to be performed and documented.[32]

All of these added to Class S part costs in an admittedly already small market. The size of the market meant unique costs were spread across a smaller group. The only thing that could mean, in order to maintain profitability, was Class S parts had to cost significantly more than avionics-grade or commercial-grade parts. This same phenomenon was noted earlier on Minuteman when comparing military avionics grade high reliability parts to Class A parts. For cost conscious program managers and some contractors, though, this translated to pressure to use the less expensive, lower grade parts.

Dr. William F. Leverton, Vice President and General Manager of the Aerospace Corporation's Satellite Systems Division, addressed that very topic with representatives from all the electronics manufacturers in early 1978:

> On the subject of what kind of market it is, I am pleased that we don't so often hear anymore that it is such a tiny market that this company or that company doesn't care. I think many of us have made that argument. I made it a long time ago when I was on the device side of the business. The truth is, it is a market where the manufacturer can make a profit, he can get his new technology into use and proven.[33]

Although the influence of national defense and national security space programs were no longer as obvious as in the Minuteman days, some still existed. National security and national defense programs continued to push for the most advanced technologies, and often paid for some or most of the cost of qualification. Qualification included extensive testing involving manufacturers and independent laboratories, such as those at RADC. The result was a thorough understanding of the physics of operation and physics of failures. Such complete understanding of the underlying physics enhanced the probability root causes could be found even in remote orbits. Furthermore, they were able to affect improvement to design and manufacturing to mitigate some failure modes, eliminating them before the parts entered the marketplace. Manufacturers could then move these lessons out to commercial offering, improving overall quality of their products in the much larger marketplace.

SAMSO could not wait for the military standards review cycle because they needed to address their reliability concerns immediately. Electronic piece parts were

---

[32] "Minutes," 39.

[33] "Minutes," 6.

essential, but were only one aspect of the larger issue of system reliability, which had been set out in MIL-STD-1543, "Reliability Program Requirements for Space and Launch Vehicles," dated 15 July 1974. The original version had not been updated, and technology's advances affecting failure modes and effects could not wait for the slower review cycle.[34] So important was this topic that SAMSO published its own newer version as SAMSO-STD-77-2, on 22 November 1977.

Class S parts enabled spacecraft developers to push for longer mission durations. The longer a satellite worked, the fewer launches were needed to support its capabilities, which (for a time) greatly reduced the cost of space programs. Although the Air Force was still struggling with questions of overdependence on space systems, the advances in solid state electronics had made possible huge advances in capabilities as well. That meant, despite hand-wringing, the Air Force and the other Services were becoming ever more dependent on key capabilities from space. By 1978, space programs designed for eighteen-month lifetimes less than a decade earlier were then designing for five-year lifetimes and exceeding these. One early Defense Satellite Program spacecraft, designed and built for a lifetime of eighteen months, continued operating well past its seventh year.[35]

Emerging long lifetimes were good news. LtGen Richard C. Henry, SAMSO Commander in 1978, added this and other examples to his Command Briefing, the presentation used to educate visitors about the military space program. Unlike Morgan before him, who had wanted to keep SAMSO's profile very low, Henry knew operational acceptance of satellites depended on outreach such as the Command Briefing. He made it the strongest possible message to his peers that his satellites were well designed, robust, and highly capable.

Achieving long lifetimes required more than redundancy. It took the highest input quality parts, testing, and rigorous application of a range of engineering disciplines. Budget realities were an ever present force driving quality down. To save on the cost of parts, some contractors contended screening of lower quality parts was just as effective.

William J. Aston of the Aerospace Corporation put it this way to the electronics manufacturers and prime contractors:

> *Some programs have tried to rely on screening as the means of ensuring quality. We don't feel that screening is the best way to go. Certainly there are advantages and certainly all our specifications have screening. We just cannot accept the concept that you can buy junk and screen it up to high quality.*[36]

The changes in reliability for the programs that had moved to Class S were dramatic. But that demonstration was insufficient to completely put down the argument about equal quality being achieved by proper screening. Doubters were being won over, slowly, by proven performance superiority.

---

[34] *Reliability Program Requirements for Space and Launch Vehicles*, MIL-STD-1543A (USAF), 25 June 1982.

[35] Temple, *Shades of Gray*, 569–575; "Minutes," 10.

[36] "Minutes," 11.

The cost argument was central. Doubters pointed out the cost of lower quality parts, even with the additional costs of more screening tests (called "upscreening"), was still lower than Class S part costs. Unsupported by experience, the argument was that sufficient variability existed in lower quality parts so screening would identify which ones were best, and these could then be considered essentially equivalent to Class S. Nothing could have been further from the truth. However, the upscreening argument was difficult to counter because it did in fact find parts among the lower quality lots that were the best performers. Such parts, however, were still unable to meet the performance specifications demanded of Class S. Those advocating use of upscreened versus Class S parts usually failed to include in their costs those impacts related to late development failures, necessary rework, schedule slips, and lower orbital reliability. Their arguments, as had become increasingly clear by 1978, really only kept costs lower in the first couple of years of development. The majority of a spacecraft's cost was the skilled touch labor. Parts accounted for only 3 to 5% of total cost, so arguments against higher quality parts was the epitome of penny wise and pound foolish.

As a result of the increasing failure rates in the mid-1970s culminating with the Bradburn review, considerable attention went into determining the underlying reasons. Parts quality was a major factor since a poorly manufactured part, even if it could pass screening, did not have all the other requirements derived from hard-won lessons and incorporated into the Class S parts specifications. However, other things affected the reliability and hence the lifetime of spacecraft. A 1977 analysis of recent failures concluded the reliability issues comprised:

1. Reliance on contractor specifications,
2. Not pushing for standardization,
3. Part manufacturers' and contractors' arm-length relationship,
4. Subcontractors lacking the space sophistication of the prime contractors, and
5. Variable parts quality by program (lack of Class S compliance).[37]

It should be clear many of these were systemic and unrelated to Class S qualification, certification, and usage. However, in 1978, the solution to rising failure rates did not cherry-pick from this set of problems (or the inverse, their solutions), rather viewing high reliability as blending all of these.

Nor were the standards complete. Further Class S specifications arrived as part of the latest revision of MIL-S-19500, "General Specification for Semiconductor Devices," for discrete parts in 1977. Microcircuits had been the largest source of problems, and therefore had gotten attention first, but this latest move captured almost all part types. These specifications, furthermore, acknowledged the need for continued revision, since new technologies lay on the horizon in such things as microprocessors, memory devices, and microcircuits with large scale (and shortly very large scale) integration.

---

[37] "Minutes," 11.

The topic of certification came forward as a separate standard in MIL-STD-976, "Certification Requirement for JAN Microcircuits." To gain more leverage in the marketplace, SAMSO and NASA combined forces, creating the joint SAMSO-NASA preferred parts list for space, further tailoring the QPL to pertain to space only.[38]

Standards proliferation in the burgeoning electronics industry was a mixed blessing. On the one hand, it made keeping up on the latest changes and proposals an important, if somewhat onerous, part of doing business. On the other, differentiation in the standards reflected the extent of the unique nature of the various device types and related processes. The latter meant two things. First, the physics of operation among different part types meant they needed to be made, handled, and tested in somewhat different ways. Second, lessons were being learned, and the associated remedies forced different device types apart. Proliferation as adding complexity was only a perspective from the outside. Breaking out a new standard actually reduced complexity and confusion within a device type. Standards organizations and industry repeatedly found combining dissimilar devices into a single standard made the standard overly difficult to wade through, contributing to errors and misinterpretation. Finally, breaking out new standards allowed greater specificity and optimization of the devices themselves.

Microcircuits had been a particular focus in the national security and national defense communities as different types of microcircuits naturally evolved and differentiated. Microcircuits were poised for a major step forward.

---

[38] "Minutes," 13.

Chapter 10

A Little Revolution Now and Then Is Good

Integrated circuits had been moving forward steadily, surpassing Medium Scale Integration (MSI) by the mid-1970s with Large Scale Integration (LSI). MSI had picked up where SSI had left off at the arbitrary break point of 30 transistors on a single chip. MSI technology quickly approached one thousand features on a single chip (usually counted as transistors per chip). As MSI reached about a thousand features in logic functions by 1978, a major advance was possible. LSI was defined in terms of its circuit elements, not simply transistors. With roots starting at about 100 circuit elements as early as 1965, LSI accelerated toward 5,000 circuit elements on a single chip. In comparative terms, Intel Corporation's 1978 release of the 8086 microprocessor chip had about 29,000 transistors. Within a year, another new Intel microprocessor, adding an 8 bit data bus, was released as the 8088. Using the 8088, IBM was able to produce the first widely successful commercial personal computer, with consequences that would ripple through all aspects of modern society for decades thereafter.[1]

Several avenues of effort on ICs began to converge. As the industry had evolved, ICs were manufacturer specific. A special class of ICs was a task-specific integrated circuit, which, unlike a general purpose computer, did only what it was specifically built to do. The industry had been working on such application specific integrated circuits (ASICs), which were essentially integrated circuits custom tailored to efficiently perform a set of specific computational tasks for a specific customer's

---

[1] Bo Lojek, *History of Semiconductor Engineering* (New York: Springer-Verlag, 2007), 340. Lojek includes a cryptic and difficult to follow history of metal oxide semiconductor technology, and includes General Instruments' 1965 campaign asserting leadership in LSI. As with any citation of an official "beginning" of something, the IBM PC with the 8088 chip was *not* the first personal computer. Such devices had existed for over a decade in various kit forms, products for scientists and engineers, in various operating systems, with prices ranging from a couple of hundred dollars to several thousand dollars. The IBM PC was no more the first PC than Columbus was the first to discover North America, but just as in Columbus' case, it is the first to successfully exploit and market the item.

Implosion: Lessons from National Security, High Reliability Spacecraft, Electronics, and the Forces Which Changed Them, First Edition. L. Parker Temple III.
© 2013 the Institute of Electrical and Electronics Engineers. Published 2013 by John Wiley & Sons, Inc.

needs.[2] Both ICs and especially ASICs were designed using libraries comprising standard building blocks usable primarily only by the manufacturer. These off-the-shelf building blocks use LSI logic, which held back overall performance.[3] As ICs became an increasingly important share of the market, this choked the flow from expansion. However, the industry was more focused on continuing up the Moore's Law progression of decreasing feature size.

The Defense Advanced Research Projects Agency (DARPA) is the DoD agency charged with the technological superiority of the U.S. military. DARPA, as it has often done, investigated Very Large Scale Integration (VLSI) to see if there were some activities that DARPA might "sensibly conduct in integrated microcircuit technology."[4] VLSI chips had begun at 5,000 components at the low end, and reached 50,000 components at the high end. The result of DARPA's interest was a study by RAND Corporation in 1976. The authors, Ivan Sutherland, Carver Mead, and Thomas Everhart, highlighted that the problems with advancing the state of the art stemmed from the focus on continued miniaturization, without work devoted to new computer architectures. They concluded: "There is every reason to believe that the integrated circuit revolution has run only half its course."[5] Based on their report, in 1978, DARPA's Robert Kahn began a program to advance VLSI computing architectures, with $15 million in 1979.[6]

For Mead, in particular, the whole set of issues came down to the need for a new way to design ICs. Miniaturization might continue to make circuit elements cheaper, but the utility of such continued miniaturization was going to be limited because of the interconnections between the circuit elements. He and Sutherland began to popularize some of their ideas.[7] They joined forces with Lynn Conway, whose background in supercomputer design and overall computing architectures while at Xerox Palo Alto Research Corporation (PARC) was invaluable for implementing their ideas.[8] To be successful, they had to clear away unnecessary complications and force a new approach that depended on simplicity. Conway explained, "We figured out how to remove tons of unnecessary design rules and optimizations, so

---

[2] "An integrated circuit developed and produced for a specific application or function and for a single customer." *Terms, Definitions, and Letter Symbols for Microelectronic Device*, JEDEC standard JESD99B, May 2007.

[3] "Engineering Hall of Fame," *Engineering Design*, 21 Oct. 2002; Gina Smith, "Unsung Innovators: Lynn Conway and Carver Mead—They Literally Wrote the Book on Chip Design," *Computerworld* (3 December 2007).

[4] Ivan E. Sutherland, Carver A. Mead, and Thomas E. Everhart, *Basic Limitations in Microcircuit Fabrication Technology*, R-1956-ARPA (Santa Monica, California: RAND Corporation, November 1976), iii.

[5] Sutherland, Mead, and Everhart, *Basic Limitations*, iv.

[6] *Funding a Revolution: Government Support for Computing Research*, Committee on Innovations in Computing and Communications: Lessons from History, Computer Science and Telecommunications Board, Commission on Physical Sciences, Mathematics, and Applications, National Research Council (Washington, D.C.: National Academy Press, 1999).

[7] Ibid.

[8] Ibid.

that it all came clear." The result was a radically new, simplified, and standardized VLSI design methodology with accompanying layout design rules for system and circuit design. That meant VLSI IC and ASIC designs were within reach of many more designers than ever before. They had broken the bottleneck in layout and design.[9]

Using the design rules, Conway also came up with an innovative VLSI prototyping service at PARC. Taking advantage of DARPA's internet forerunner, ARPANET, for collaboration, Conway's service allowed designers from anyplace with access to ARPANET to submit design files and obtain low-cost fabrication.[10]

The emphasis on simplicity was crucially important to what would follow. Personal computers were becoming prevalent. Although by modern standards these first computers were limited in computational power, they were sufficiently powerful to implement Mead and Conway's design ideas in emerging computer-aided design tools.[11] This fit perfectly with their efforts to popularize and proliferate the new VLSI approach. But this radical departure from the norm could not rely simply on popular and trade magazine articles.

The path was obvious to them, and led to another sort of innovation. Mead was a professor at the California Institute of Technology, where Sutherland was the chair of computer science. They began to teach their VLSI innovations. Conway taught the first course at MIT in 1978, and during 1978 and 1979, other universities added courses. Soon, California Institute of Technology, Stanford, and the University of California at Berkeley along with 9 other universities were offering courses using Mead and Conway's new textbook and two accompanying guidebooks. Mead and Conway could offer an important new advantage in addition to their textbook. Their courses gave students hands-on design over ARPANET. Use of the ARPANET and the hands-on design had no precedent, as Mead and Conway blazed yet another trail. After publication of their textbook, *Introduction to VLSI Design*, in 1980,[12] 110 universities were using it. DARPA was interested enough in the collaborative use of the ARPANET that they funded the process, which became known in 1981 as the Metal Oxide Silicon Implementation Service, or MOSIS. By 1982, their funding of VLSI initiatives reached $93 million.[13]

DARPA's engagement in VLSI did not end there. Taking a page out of the earliest days of transistor development, as the Services had done with Bell Labs, so DARPA would do with VLSI. DARPA's interest, as always, was to push the state of the art and foster innovation, but they did so in a strategic sense, provoking innovation, not so much in the underlying details by defining the specifics.[14] In the

[9] Smith, "Unsung Innovators."

[10] *Funding a Revolution.*

[11] Smith, "Unsung Innovators."

[12] Carver Mead and Lynn Conway, *Introduction to VLSI Design* (Reading, Massachusetts: Addison-Wesley, 1980). The draft for the textbook actually began in 1977 with preparation for the 1978 course taught by Conway.

[13] *Funding a Revolution.*

[14] Ibid.

case of VLSI, they also saw the need for information sharing, collaboration, and open publication, minimizing the amount of work that was considered classified. DARPA invested in research at a dozen universities, most of which also used the Mead-Conway design methodologies and tools.

In response to continuing concerns about IC costs, performance, reliability, and suitability for military applications, the DoD began the Very High Speed Integrated Circuit (VHSIC) program in March 1980. The impact of the military standards on ICs slowed adoption of the latest technologies into space programs. In fact, at VHSIC's inception, commercial ICs were thought to be a decade or more advanced over their military standard counterparts in terms of feature size and clock speed. (Clock speed was a measure of how fast certain basic computations could take place.)[15] VHSIC was supposed to use silicon technology to reduce or eliminate the commercial lead.[16]

Phase 0 of the program got under way with nine companies developing 1.25 micron feature size, 25 megahertz (MHz) clock speed devices.[17] Phase 1 was intended to keep the same clock speed but put 5×10^{11} gates in a square centimeter. The goals for phase 2 included pushing the density to 1×10^{13} gates per square centimeter, and increasing the speed to 100 MHz.[18] The VHSIC program was the best-supported intervention in the direction of the electronics market since the 1960s. It would have an important impact on the high reliability microcircuit manufacturers as it progressed.

The VHSIC project, which was more focused on defense needs, did not have the same kind of impact as the Mead-Conway approach supported by DARPA. VHSIC's work tended to be more restricted, both in scope and in its dissemination, but that reflected VHSIC's different focus.[19]

These events became known as the Mead-Conway Revolution, or, more commonly, the VLSI Revolution. By empowering creativity, facilitating design, and expediting manufacturing, independent of feature size, VLSI technology leaped forward. How could military standards keep up?

---

[15] The problem with clock speed soon revealed itself. While important in terms of raw comparability of one device with another, clock speed is only one component of the overall processing power required to produce answers from a computer. At the time of VHSIC, however, no better performance benchmarks were recognized.

[16] *Very High Speed Integrated Circuits—VHSIC—Final Program Report 1980–1990*, Office of the Under Secretary of Defense for Acquisition, Deputy Director, Defense Research and Engineering for Research and Advanced Technology, 30 Sept. 1990, 2.

[17] Ibid., 3–4. A micron, or micrometer, is one millionth of a meter.

[18] *Terms, Definitions, and Letter Symbols for Microelectronic Device*, JEDEC standard JESD99B, May 2007, 1–13. Feature size is related but not equivalent to the number of gates per square centimeter.

[19] *Funding a Revolution*. DARPA's VLSI and the DoD VHSIC programs were aimed at different targets, but for an interesting view of VHSIC's impact, see David C. Mowery and Nathan Rosenberg, *Technology and the Pursuit of Economic Growth* (New York: Cambridge University Press, 1989), 149.

Chapter 11

Quality on the Horizon

Stepping back a few months to the last part of Morgan's tenure as the SAMSO commander, quality assurance received a major boost. Quality assurance traditionally meant the set of disciplines aimed at achieving the highest reliability. That remained true, but achieving the highest reliability for space systems demanded a scope that seemed to exceed the normal bounds of quality assurance functions. Morgan's emphasis on reliability continued and expanded to cover the general topic of mission success. High reliability parts were critical in this formula, but mission success was broader, including:

- Design
- Piece parts
- Microelectronics/hybrids
- Manufacturing
- Contractual incentives
- Testing
- Program reviews
- Motivation/training.[1]

While everything might start with parts, parts alone did not make satellite systems as reliable as Morgan knew they needed to be. This list illustrates the necessary breadth of scope. While it would be wrong to assign them all equal importance, neglect of any of them would affect satellite system reliability. So, Morgan met "with key executives of the major SAMSO contractors to discuss ideas and techniques to enhance mission success." This generated an industry-led joint forum on mission assurance held 25–27 April 1978. Each topic was the subject of a separate workshop. By the time the Mission Assurance Conference was over, LtGen Henry had become the SAMSO Commander. Impressed with the major conclusions and

[1] SAMSO/CC memo, Subj: Mission Assurance Seminar, 2 March 1978.

Implosion: Lessons from National Security, High Reliability Spacecraft, Electronics, and the Forces Which Changed Them, First Edition. L. Parker Temple III.
© 2013 the Institute of Electrical and Electronics Engineers. Published 2013 by John Wiley & Sons, Inc.

recommendations from nine workshops, he offered AFSC Commander General Alton D. Slay a briefing on these.[2]

Slay accepted the offer, as he was interested in ideas that had AFSC-wide application. He was already working on a new direction for improving his command, with a greater emphasis on the early planning effort going into systems.[3]

The result came in November 1978, when Slay initiated a study on quality improvement and where AFSC needed to head. The year-long effort, entitled Quality Horizons, started 22 November 1978 with four objectives:

1. Improve end item quality in field use
2. Make contractors more responsible for their products
3. Make more effective use of resources, and
4. Apply appropriate commercial practices.[4]

The 15 November 1979 final report of Quality Horizons leaned heavily on the WGEPP work. Of the study's six major conclusions, three derived from the WGEPP:

1. Attainment of field product quality is a function of the interest in and priority placed on quality by top managers. (WGEPP conclusion)
2. Governmental and industrial organizations that have succeeded in obtaining high product quality levels blend the assurance sciences into one high level organization that can act as a protagonist in causing tradeoff analyses and in assuring a disciplined integration of efforts to obtain product quality. (Consistent with WGEPP)
3. Agreement exists that product assurance cannot be inspected into any product. Nevertheless, AFSC places more emphasis on conformance verification than attempting to influence product quality through design, process control, and test planning early in the program life cycle. (WGEPP conclusion)
4. While various plans and programs have been implemented or proposed in DoD for reduced levels of in-plant surveillance, program managers are generally reluctant to accept reduced in-plant government quality assurance activities, especially conformance inspections. (New conclusion)
5. Commercial contracts are firm fixed price, with limited customer financing, and sole source follow-on buys are common with vendors that deliver a quality product at a reasonable price. (New conclusion)
6. In the commercial sector, warranties are generally offered only as a result of competitive pressures, such as in the commercial aircraft industry. Performance incentives and award fee provisions are almost never used in the commercial environment, either in the U.S. or overseas, nor do foreign governments employ them. Profits are their main incentive due to firm fixed pricing. (New conclusion)

[2] Letter, Lt General Richard C. Henry to General Alton D. Slay, 21 Aug. 1978.
[3] Letter, General Alton D. Slay to Lt. General Richard C. Henry, 9 Sep. 1978.
[4] *Quality Horizons Final Report*, vol. 1, HQ AFSC, 15 Nov. 1979, 1.

Slay's acquisition responsibilities were far broader than the space program's, and the scope of some systems acquisitions included orders of magnitude more piece parts than the entire U.S. space program. Consequently, his recommendations concentrated on the acquisition of the aircraft and ground systems that were far and away the main business of AFSC. In reforming the way such large acquisitions were performed, Slay could affect great sums of money being spent. When serving the budget-conscious President Carter, economy and efficiency were the watchwords of the day. Thus, the last three recommendations aimed at affecting the very large budgets going into aircraft acquisitions, but sadly did not recognize the effect such changes would have on spacecraft acquisitions.

Slay was surely aware that space systems were not like aircraft systems, but apparently was unmoved by the difference. As reformers after him, he apparently felt justified in looking at the larger part of the problem, assuming the space acquisition community would get in line without any serious or long term impacts.

Slay's actions were really part of a very long historical chain of events, dating to before the Civil War, trying with fits and starts to address the need for a better means of weapon procurements. Reforming acquisition affected millions of taxpayer dollars, and new approaches held the promise of great savings. A significant part of Slay's approach had to do with the way in which contracts were structured. To Slay and his staff, cost-plus contracts had not incentivized contractors to be as economically responsible as they might be if they shared more program cost risk. He wanted to use only fixed-price contracts to attain budget certainty. The consequences rippled through the acquisition system as more responsibility was put on the contractors.

He had made a fundamental mistake. Economic responsibility was not the issue in cost-plus contracts. Such contracts were used when technological risk made any estimate of final cost very uncertain. Such a situation could doubtlessly be abused, but for the most part, sharing the risk of developing an advanced technology had paid off well for the United States. Despite the efforts of Gen Phillips and his successors at SAMSO, space systems acquisition remained a technologically risky business for which cost-plus contracts were ideally suited. What Phillips and the others had done was refine and hone the application of such contracts to the space business. Now, that was about to be overturned. One danger of moving a technological development program forward under a fixed-price contract was that the contractors were very motivated to control costs. Cost containment could come at the expense of the best quality parts.

Here is how that worked: As systems' complexity increased, major contractor teaming also increased. The Quality Horizons study team discerned what they believed were variations on some central themes, and they extracted these themes into recommendations for the report. This is where their recommendations for firm fixed-price contracting came from. They equated complexity with the kind of risk in R&D acquisitions, allowing themselves to conclude that since commercial contracts for complex items could be firm fixed-price, therefore research and development contracts could as well. The conclusion that complexity was analogous to risk in research and development overlooked major dissimilarities. Commercial complexity usually did not involve the same degree of uncertainty inherent in highly advanced weaponry, and usually dealt more with complexities of production and manufacturing. Few commercial fixed-price contracts included the same degree of risk and uncertainty of advancing new technologies as did weapon systems.

But what about commercial contracts where quantities to be produced were uncertain, making pricing difficult? Costs per unit produced basically reflect the total of the fixed costs (e.g., facilities and machinery) plus the variable costs (e.g., labor and materials), divided by the number of units. When cost per unit was a major consideration, but the number of units to be made was not predetermined, such pricing per unit was not only uncertain, it was a risk. How was this handled commercially? The report noted "redeterminable contracts" were used in such new product developments. That meant the cost per unit was not fixed, so the terms could be determined as quantities became clear. So, Quality Horizons recognized there were indeed situations where fixed-price contracting was unsuited.

Space systems, though not exactly new products, still embodied a high degree of risk from advancing technology, for which fixed-price contracting was also anathema. But in the case of space systems, a fixed-price contract could mean only one thing. The only reasonable approach to bidding a fixed-price high-risk development contract was to increase the price to cover the uncertainties. This important caveat did not get included in the final recommendations, and was thus kept out of sight of the senior managers.

Although these were serious flaws, the Quality Horizons study team did make some important contributions to understanding the nature of the defense industries. Importantly, they found nothing consistent about "commercial practices."[5] Their investigation turned up as many such practices as there were contractors. Even *if* the contractors could be motivated as Quality Horizons asserted, what certainty could there be of any uniformity within the systems acquisition community? Quality Horizons found any suggestions for relying on commercial practices was little better than a craps shoot.

Slay's Command Policy Letter 22 imposed some of the Quality Horizons recommendations, especially the firm fixed-price contracts and use of warranties. Full responsibility for cost rested on the contractor regardless of technological risk. Space acquisitions were not recognized as qualitatively and quantitatively different from other acquisitions. Firm fixed-price contracts shifting cost responsibility to the contractors and obtaining manufacturer warranties for high technology or high risk developments could have but one outcome: a higher price for space systems. Price was the contractor's expectation of the cost taking into account the degree of risk associated with space systems, which remained in the medium to high risk category in the 1970s.

SAMSO attempted to waive some of these changes, especially as they would increase program costs. That was not acceptable to HQ AFSC, countering the SAMSO argument with their expectation for increased competition, which would drive the costs down. The AFSC response overlooked the predictable impact on parts quality as the contractors were incentivized to keep costs down over other consid-

---

[5] Ibid., 39.

erations. The result was low proposal prices, but extensive contract changes and cost overruns drove the actual government cost up to where it should have been placed originally. Cost-plus contracts had also seen low proposal prices accompanied by contractor expectations of recouping some of the funds needed to actually execute the acquisition as part of the expected way of doing business. These cost-plus contracts often also included either award fees or incentive fees. The criteria for these awards or incentives often included consideration of how well the contractor controlled cost growth, allowing the contractors to trade between costs charged and award or incentive fees earned to arrive at expected profits. By reversing this trend for spacecraft, the Quality Horizons effort actually set in motion a number of contracts that were underfunded from the outset, but without a mechanism to contain the inevitable cost growth at the expense of quality.

This caused economizing in the programs, something which had always been present but not to the same extent. Economizing forced cuts in areas where the least pain could be felt, which included testing, quality assurance, and areas that would not reveal their consequences until later in development or in orbit. This trend led to further examination of where cost growth originated, and this inevitably led to the military standards. The various parts standards forced controls, tests, and processes over and above the processes that supported commercial electronics. The military standards forced quality first, which led to higher expenses.

Another effect of the Quality Horizons report was the AFSC shift in emphasis from *quality* assurance to *product* assurance,[6] the idea being such a shift would "gain emphasis and attention to preventative efforts during design, development and test."[7] Such emphasis shifts are rarely benign. They contain good intentions, but in implementation other ideas that seem to flow logically cause damage.

So, when one of the recommendations reduced the level of in-plant government surveillance, the idea seemed good. The resulting Minimum In-Plant Surveillance (MIPS) program allowed each contractor under Air Force Plant Representative Office (AFPRO) surveillance to apply for reduced government surveillance. In the near term, reduction of the in-plant representatives saved the government money. Quality Horizons intended only to shift resources from in-plant to "product assurance prevention efforts where greater returns from existing resources can be obtained." The report held that this would "motivate contractors to be more responsible for the quality of their products." That would be a good thing, if the motivation worked. However, it contended with the stronger emphasis on cost, which did not favor quality. Cutting back the number of in-plant representatives enabled reductions in processes that would have been overruled previously. This was another pennywise and pound-foolish move, since problems that could have been avoided early on the factory floor were only to be discovered in later phases of acquisition, when they most hurt both cost and schedule.

[6] Ibid., 56. "*Product assurance* is the application of interdisciplinary skills to accomplish the preventative and conformance activities necessary to assure: That requirements are properly specified, that the design will achieve these requirements and that the ultimate product and/or services will perform their intended functions in the operational environment for the period specified."

[7] Ibid., 2.

A less obvious consequence of Quality Horizons' recommendations was it recognized and legitimized ideas about the role of commercial practices and standards that ran counter to everything learned by hard experience in the military space program. Despite the disparity in commercial practices, Quality Horizons began the open discussion of these topics that would soon take hold at much higher levels in the government.[8] The Quality Horizons changes were short lived, as they proved themselves impractical and wrong.

* * *

The Minuteman program affected the high reliability electronics industry long after the major acquisition activity ended. By pushing electronics industry innovation in reliability and quality, Minuteman drove electronics to their highest level of both yet attained. Papers at the Physics of Failure Symposia and their follow-on pointed to the ways to do even better than Minuteman. When NASA wavered on continuing Class A parts, industry, RADC, the Aerospace Corporation, and the national security and national defense space programs moved quickly to improve Class A with lessons learned from Minuteman experience. A series of launch failures produced further detailed insights that refined specifications for high reliability electronics. The result was the highest reliability parts of all, designated Class S.

The Mead-Conway VLSI revolution changed the way ASICs were envisioned, designed, and manufactured. The IC industry restructured itself in a way that facilitated ASICs in a previously unforeseen way. That further enhanced the ability of military acquisitions to push for unique devices, especially for space applications where size, weight, and power were critical. Advancing the state of the art in ICs and ASICs meant space systems would become more complex, but in so doing, would gain greater capabilities and, hopefully, robustness.

The demand for the highest possible reliability was essential for the acceptance of space systems by the combatant elements of the military. As space capabilities increasingly proved themselves, reliability would be the key to breaking down the last barriers of acceptability.

Slay's Quality Horizons portended changes on the way, but missed the mark. It served to demonstrate that changes are not easy, but also that the system being changed was quite complex. The kinds of approaches that might work for acquiring aircraft did not work well for spacecraft. The ideas set in motion by Quality Horizons, though, gave a push to the thinking that would soon move as a juggernaut though all government acquisition.

---

[8] Ibid., 58. Interestingly, the report included important words about Product Assurance that seemed to reflect the influence of the space experience. "Many managers do not recognize that the assurance disciplines have evolved from 'black arts' to engineering sciences with proven tools and techniques for their proper application. These functions are no longer fragmented efforts but complement one another." But in so saying, the report assumed the space systems acquisition was no longer a "black art" and was on par with aircraft acquisitions, which was a mistake that allowed them to overlook the unique highly structured and rigorous processes being brought to bear on space acquisitions that held no applicability to aircraft acquisitions.

Part III

Switching Transient (1980–1989)

Spacecraft capabilities were becoming significant and undeniable. Yet a large and important segment of the military community had resisted adopting space capabilities. For too long, satellites had been too close to science fiction, relying on scientists and engineers far removed from military operational areas. Quality, reliability, and some public relations were essential to military acceptance. The ultimate expression of acceptance, though, depended on an operational entity owning space systems and making them responsive to military commanders in the field. All of these cost money.

As space capabilities worked to gain acceptance and enter the mainstream, they got more funding and visibility. Continued advances in reliability and quality of electronics increased satellite lifetimes and capabilities, but at increased costs. When the government spends more money on something, greater scrutiny is not far behind. Greater expenditures brought greater scrutiny across the entire spectrum of defense acquisitions.

When capabilities become accepted in the mainstream, they also become thought of as routine. Once they are no longer special because they are different, they become subject to shifting political winds.

Implosion: Lessons from National Security, High Reliability Spacecraft, Electronics, and the Forces Which Changed Them, First Edition. L. Parker Temple III.
© 2013 the Institute of Electrical and Electronics Engineers. Published 2013 by John Wiley & Sons, Inc.

At the time, political winds shifted toward containing military systems costs. Increasing costs were no longer acceptable or understandable. During the 1980s, perceptions at the top of the defense and federal acquisition hierarchy diverged increasingly from the realities at the bottom. Fundamental changes to federal systems acquisitions would drag spacecraft acquisitions along into a new and poorly understood area.

All the while, the world was changing in ways that could not be ignored.

Chapter 12

Crossing the Operational Divide

By 1980, nearly a decade of dedicated effort to overturn the perceptions of fragility, unreliability, unavailability, and other criticisms of space systems finally paid off. Just as the years from 1969 to 1973 formed a critical nexus in the evolution of space programs, the nation was about to experience something nearly akin to that at the change to the new decade. Leadership of the Air Force was about to shift from its historical reliance on former SAC commanders-in-chief to former fighter pilots dedicated to tactical airpower. A new presidency was about to square off against the evil empire of Communism, using space power as part of the leverage to grind the Soviet Union into the dust. New horizons were everywhere.

The general specification for microcircuits, MIL-M-38510A, began creaking at the seams as a direct result of the VLSI Revolution. It was clear that IC requirements, especially in terms of manufacturing and testing, were simply inappropriate and inadequate for hybrid ICs. Hybrids combined integrated or active devices with discrete devices. Until requirements appropriate to hybrids were developed, these devices had to be manufactured and tested based on the disparate requirements of both integrated circuits and discrete devices. The impacts on cost and manufacturability drove both industry and the government to recognize the need for a hybrid specification to handle the complexities of combining disparate technologies, promoting high reliability, and still containing costs.

ICs were key enablers of sophisticated and highly capable satellites, and thus drew considerable attention. ICs moved quickly from small scale integration to MSI, whose smaller feature sizes allowed more logic functions and greater capability in smaller packages. Along with the greater capability came even greater demand for ICs. However, generalized ICs were simply not the best approach to repeatedly run specialized computing algorithms. Put another way, a general purpose computer does many things well, but a task-specific computer built and optimized for a few kinds of repeated calculations, and never expected to handle anything else, could be made

Implosion: Lessons from National Security, High Reliability Spacecraft, Electronics, and the Forces Which Changed Them, First Edition. L. Parker Temple III.
© 2013 the Institute of Electrical and Electronics Engineers. Published 2013 by John Wiley & Sons, Inc.

more robust, smaller, and faster. This was even more important in applications where computation speed was important. Running software was inherently slower than running the same algorithms built into an integrated circuit. Although software could be updated and could even change what was being done, ASICs optimized performance of repeated tasks. The Mead-Conway VLSI revolution meant ASICs were about to become more readily available as manufacturer-specific design tools appeared. Such design tools would free spacecraft designers to greatly increase the speed and efficiency of computing, taking advantage of lower power consumption for greatly increased capabilities.

Improved capabilities were important, of course, as space systems continued to move toward military operational acceptability. For spacecraft design concerns, though, lower power consumption affected many subsystems. Lower power needs reduced the size and weight of solar cells and batteries and meant less heat needed to be radiated to keep thermal balance, and these in turn affected satellite design, weight, cost, and design trades. These benefits could translate to greater spacecraft capability by allowing additional or better sensors, greater redundancy to enhance reliability and lifetime, or simply savings on launch costs. To receive these benefits, though, custom ASICs had to overcome important drawbacks of long lead times and very high cost.

As ICs differentiated from hybrids, and both needed to consider ASIC development, work began on developing a standard that would capture the lessons being learned about custom integrated circuits.

* * *

All U.S. space programs were transforming due to the cyclic, complex interdependencies of highly reliable electronics, resultant longer spacecraft lifetimes, increasingly complex spacecraft, and capabilities of both launch vehicles and spacecraft. Since it was a cycle, breaking into it to explain it is a bit of a chicken-and-egg problem, but everything can be traced to the effects of high reliability, Class S parts.

High reliability space qualified electronic piece parts enabled increasingly reliable and capable spacecraft, but at the expense of fewer satellites being built. Clearly the need had decreased as satellites lasted seven years (and longer) as compared with two years a dozen years earlier. But why an expense as opposed to a savings? Both opportunity costs and program costs were involved.

The opportunity cost became apparent in October 1980, when AFSC headquarters solicited annual budget inputs from its product divisions. Space Division (SD, successor to SAMSO) struck an ominous tone about how longer satellite lifetimes were affecting procurement policy. Spacecraft development time had slowed so significantly, satellites were not being bought at a rate that provided options in cases of a catastrophic satellite outage or launch vehicle loss.[1] When satellites had to be replaced every two years, acquisition of satellites resembled a manufacturing line despite variations between successive models. Longer satellite lifetimes had affected this in an interesting combination of the need for replacement and complexity.

---

[1] HQ AFSC MSG 032200Z Oct. 1980, Subj: Request for FY 83 POM Budget Options.

Significant storage costs created strong cost incentives to slow down production of subsequent spacecraft. A satellite that might actually only need two to three years to be built (earlier in the 1970s) had to be stretched to avoid storage costs, while the agency continued to pay for factory equipment, workforce, and other costs. The challenge since the mid-1970s had been to forecast when a satellite outage would require its replacement, timing that replacement's construction and testing so it was not excessively early or late for the need. Longer times between needs favored more capability and complexity. As microelectronics allowed more capability, new processes, and extended redundancy, satellite complexity rose rapidly. More complex satellites took longer to build and test.

Capability and complexity also increased with launch vehicle capability. When a larger satellite with sufficiently high importance forced Expendable Launch Vehicle (ELV) performance improvements, such upgraded performance, in terms of payload weight, then became available to any subsequent satellites. Satellite programs could use the extra performance available to add redundancy, increase capability, or offer hosting to test new sensors and secondary payloads, making integration of the final satellite more time consuming. This process became cyclic, as greater complexity drove greater time to build and test adding weight, pushing launch vehicle capabilities higher, and making more weight allowable for the next cycle of satellites.

More complex satellites needed more Class S parts, of course, but the increase in part count did not affect the lower overall demand for high reliability parts. Lower demand affected part manufacturer interest in providing expensive, high reliability parts. Their interest remained, however, as long as the increases in prices returned good profits.

A different aspect of the electronic piece parts industry's changing nature became worrisome. The Quality Horizons study raised the issue of the U.S. industrial base's erosion, so General Slay was asked to testify before Congress on the topic. Production costs and cost competition were forcing some important changes. Slay's concern was some key manufacturing capabilities moving overseas, losing some domestic industrial capabilities due to competition from Japan and elsewhere. His House Armed Services Committee testimony on 13 November 1980 highlighted industrial base issues and a growing recognition that the United States was losing an essential element of defense capabilities. Slay went so far as to recommend the United States re-industrialize—that is, reestablish industrial capabilities then lost to overseas competitors.[2] The United States needed to do something to encourage and expand domestic industrial capacity, and that pointed, in part, directly at the U.S. electronic piece parts industry, which was slowly moving overseas.

Slay was not alone in his concerns about moving high reliability, military-grade parts production overseas. The International Traffic in Arms Regulations (ITAR) prevented the export of key technologies. After all, the U.S. military depended on its technological superiority as a key determinant in the outcome of any conflict. Sending advanced technology overseas had to be prevented. In a simplistic example,

---

[2] HQ AFSC 102200Z Oct. 1980 Subj: Congressional Testimony, Eroding Industrial Base Issues.

it may be best to think of the issue in the following way. What if, in some future conflict, a U.S. "smart" bomb depended on some military-grade integrated circuit whose only source of supply was an overseas foundry? In some cases, if additional production were needed to support an extended conflict, it could simply be paid for and that would be that. But, what if the manufacturer were in a country that was the ally of the country being bombed? Even worse, what if the manufacturer were actually *in* the country being bombed? The U.S. military would then be hostage to such an overseas provider's political situation. ITAR attempted to prevent this and similar problems. Thus, the issue of overseas production was another layer of complexity viewed in significantly different ways depending on whether one was buying or selling.

Domestic manufacturing requests for military-grade and space-qualified parts had another consequence emerging with the rise of consumer electronics. In addition to controlling costs and other production factors, such overseas production would be competitive in the emerging global economy. The large volume of overseas commercial production would change the equations about yield from a focus only on lots to yield per day or over longer periods of time. A bad production lot might, in theory, be able to be destroyed without perturbing the overall profitability of a manufacturing plant, if the production rate were high enough. That would suit commercial production. However, high reliability parts manufacturing was only a small percentage (about 2%) of overall U.S. electronic parts. Of the high reliability manufacturing, most was military avionics-grade electronics, and a much smaller percentage was space-qualified. Having to maintain a domestic production capability just to produce this small volume of high reliability parts would raise the cost of these parts, because manufacturers must cover their fixed and variable costs. Ultimately, such low volumes would not support multiple domestic vendors, forcing industry consolidation. But necking down to a single domestic manufacturer would potentially bring into play anti-trust and monopoly laws. There seemed no clear way to resolve these contending pressures.

One partial answer to the eroding industrial base was to act in favor of increased profitability for U.S. manufacturers. Henry sent Slay seven issues related to the industrial base. Five of the seven related to multi-year procurement candidates, which were important because SD acquisitions were "typified by high reliability and low volume."[3] There were contracting advantages to moving from multi-year contracts to multiple-year block buys with incremental funding. In multi-year contracts, parts might have to be bought in different years, reducing the lot size and increasing the cost per part. Buying the electronic piece parts for multiple satellites at the same time would provide a short term but very profitable lot buy for the manufacturers.

Under "Long Lead Funding Applications," Henry addressed the issue of the length of time to produce integrated circuit devices:

---

[3] HQ SD CC Letter, Subj: Congressional Testimony, Eroding Industrial base Issues (HQ AFSC Msg 102200 Oct 80), 21 Oct. 1980.

Banking of wafers to produce integrated circuit devices can save 12–14 weeks of the 72 week device lead time. In addition, banking is particularly valuable during development because a wafer is universal and can be used to fabricate a different device required by a design change.[4]

Wafers are the starting point for a variety of integrated circuit devices. This was the level of manufacture at which space qualified parts began. Wafer lot acceptance tests determined if the wafers were of sufficient quality to proceed into Class S parts manufacture. Having high quality wafers in storage saved schedule, and hence money, for IC-based devices, and large quantity purchases of wafers would lower the unit costs.

Finally, Henry offered an even more forward-leaning proposition based on the economies of scale and commonality of electronic parts across multiple satellite programs. The idea was coordinated:

procurement of parts could be made by the prime contractors and stocked at parts vendors. The government would furnish an up front "stocking" fee and guarantee a usage rate but would not take title to the parts. Individual contractors could purchase from the "store" based on their needs. There is an estimated 20% commonality of piece parts between spacecraft. This would be increased if common catalog hi-rel parts were available off-the-shelf. The resulting increase of quantities would induce vendors to compete.

He intended to incorporate this into SD-STD-73-2C, "Electronic Parts, Materials and Processes for Space and Missile Applications."[5] SD-STD-73-2 was undergoing a major improvement and consolidation from the Commander's Policy Book. It would shortly be retired in lieu of two new military standards specifically intended to govern the management of parts, materials, and processes programs, and ensure the quality of the parts themselves for space programs.

Before that could happen, however, the 1980 election intervened. Preparing for that event, a team of the foremost personalities in the U.S. space program put together a policy study that structured the space policy efforts for the majority of the subsequent two Reagan Administrations.[6] Their "Report on National Space Policy Enhancing National Security and U.S. Leadership Through Space" structured language in Reagan's campaign speeches and the first Reagan National Space Policy. The document's contributors were the luminaries of the founding and evolution of

---

[4] Ibid.

[5] Ibid.

[6] *Report on National Space Policy Enhancing National Security and U.S. Leadership Through Space,* January 1981 (HQ SMC/HO Plans Folder, HO-020-03, 1981). With the exception of the revision of space policy following the destruction of the Challenger, this remarkable document captured all the key elements of the Reagan space agenda. One fascinating element of the document is its open reference to the function of reconnaissance at a time when this was still classified. It went so far as to ask, on page 8, "What are the future requirements for reconnaissance and its use for national, strategic, and tactical purposes? What is the optimum workable management concept to produce the highest effectiveness?"

national security and national defense space, including Bernard A. Schriever, David Bradburn, Edward David, Francis X. Kane, John McLucas, Simon Ramo, Charles H. Terhune, Albert (Bud) Wheelon, and Michael Yarymovych.[7]

Following the nascent theme struck by Quality Horizons, reflecting the tenor of the times, the study said:

> *The private sector should be provided greater opportunities and incentives to support an increasing share of our space effort and to maintain our competitive edge. Our leadership in space exploration should be supported and new opportunities exploited for science and national prestige.*[8]

Beyond that, the report expanded on the role of commercial activities in space. The writers wanted commercial and industrial activities in space to develop new technologies and "encourage economic advancement, to develop new skills and competence in space and engineering, and to assist the U.S. in maintaining a position of leadership."[9]

So, with clear economic, industrial, and national interests at stake, their solution was to encourage a stronger commercial sector. The downward trend in the government's launch rate (due to longer-lived spacecraft) was apparent. The nation should benefit if commercial enterprises picked up the slack. This eventually led to cost competition in commercial ELVs, where industry had to sharpen its pencils to deal with state-subsidized launch industries abroad. To do so without cutting costs would have been impossible. The U.S. advantage was its long string of steady improvement and constant successes. The report recommended "[t]he feasibility of creating a private entity for launching commercial satellites should be evaluated and possibly encouraged."[10]

The new Administration followed that agenda, and began broadening the commercial space enterprise. But in its admiration for successful commercial businesses and its pursuit of improved acquisition, the seeds were sown for serious unintended consequences.

*   *   *

Space Division's SD-STD-73-2C formed the core of two new military standards, the first of which was MIL-STD-1547, "Electronic Parts, Materials, and Processes

[7] Schriever had been the founder of WDD and the developer of the ICBM and the national defense space programs; Bradburn was a former ESD commander, David was a member of the President's Science Advisory Committee and Presidential special assistant, Kane was one of the originators of the idea for the Global Positioning System, McLucas was a former Under Secretary of the Air Force and Director of the NRO, Ramo was a space pioneer and co-founder of TRW, Terhune was Schriever's deputy at Western Development Division, Wheelon was a former CIA Director of Science and Technology and head of the CIA's part of the NRO, and Yarymovich was a former Air Force Chief Scientist.

[8] *Report on National Space Policy*, 1. Although I was unable to find the exact citation, the document appears to draw heavily on candidate Reagan's public announcements.

[9] Ibid., 4.

[10] Ibid., 7.

For Space and Launch Vehicles," released on 31 October 1980. A compendium from many sources describing technical lessons learned, MIL-STD-1547 established a coherent approach to qualification of parts for space. It prescribed the starting point of high reliability manufacturing. Qualification of parts meant subjecting a number of parts to various stressful conditions to determine if sufficient margins existed for the part's survival in space. The data supported estimates of the part's long-term behavior. If parts were able to handle the hazards and stresses of space qualification, then they could be de-rated, based on factors included in the standard. De-rating simply meant using less stressful levels of power, cycles, and other factors such as temperature extremes than the part was qualified for, in order to make the part last longer (lowering the problems in the middle part of the bathtub curve). By understanding the part thoroughly during qualification, then, designers could also find the design parameters useful for determining the likely long-term behavior. This allowed design of extremely high reliability systems and devices. So long as a manufacturer made no changes to the processes and materials in the part's production, subsequent lots of parts would inherit the qualification. Still, each lot of parts would be subject to sample testing at much reduced levels to verify they remained consistent with the original qualification lot.

On 12 February 1981, its companion standard, designated MIL-STD-1546, "Parts, Materials, and Processes Standardization, Control, and Management Program for Spacecraft and Launch Vehicles," laid out the means of managing a military space parts program. It grew out of the earlier Commander's Policy, "SPO Management Style Maximizing Effective Utilization Of Limited Technical Resources," which dealt mainly with general topics of program management.

An example of its content was its recommendation for membership roles and responsibilities in control boards for parts, materials, and processes. Membership in such boards might include government representatives from the program office, members from the Aerospace Corporation, and various parts of the prime contractor divisions. The new standard benefited from long-term observation of successful programs to provide guidance on what would work best. It answered questions about what relationships were most effective. Should government representatives have a voting membership on the board? If so, should the government's vote be counted as more important than the votes of the other members? The answers to such questions were not obvious. The new standard also underscores how every aspect of space system acquisitions was subject to scrutiny to wring out the highest possible reliability and chance for mission success.

When the new standards were released in the fall of 1981, selling them to industry and gaining support for them took a concerted effort. In November 1981, LtGen Henry sent letters to all the SD contractors to address their concerns. Specifically, Henry wrote:

> *For sometime [sic] I've been concerned about the continuing apparent discord between SD and industry on our space parts procurement specifications. This subject was discussed in detail at a Mission Assurance Executive Session last month. As a result of that session I've decided to form a team of contractor and government program managers to review MIL Standards 1546 and 1547. The purpose of this*

review will be to answer the question: "Are we on the right track in our parts specification for space systems from a program management viewpoint?"

The review was to be coordinated by Colonel (later BrigGen) Don Henderson.

Not everyone in government supported the newly developed standards. Recognition of the need for the standards depended on organizational perspective. Henry, his staff and managers, Aerospace, and RADC all recognized the need, value, and imperative for assuring quality. Within AFSC, from the headquarters down, the need was understood and supported. However, once inside the Pentagon, understanding evaporated quickly. Opposition came from unexpectedly high places. Under Secretary of Defense for Research and Engineering Dr. Richard D. DeLauer advocated removal of these standards as too costly.

To be sure, parts qualification and management controls raised initial parts costs, but saved many times their cost in later phases of the satellite development.[11] A small increase in upfront expenditure was a highly leveraged investment against future problems. DeLauer's criticisms did not stop with the apparent cost. His agenda included wider use of streamlined acquisition methods in use by "black" or compartmented programs. The streamlined programs came in two forms: space and aircraft.

Within the space community, SAFSP was a prime example of a compartmented organization, though other important examples existed. Streamlined acquisition owed much to Schriever's original methods in the first days of Western Development Division. These methods complied with the Federal Acquisition Regulations (FARs) but had not added the cumbersome bureaucratic overhead that had grown up in the national defense space programs. Streamlined acquisition took advantage of military standards the most sensible way by tailoring their application to the specific needs of each program. Tailoring military standards was fully acceptable and expected— not all elements of every standard applied to every program. Tailoring, though, required technical expertise with the standards themselves, which emphasized a difference between the level of expertise residing in the NRO programs versus their SD counterparts. While SAFSP complied with the FARs with a few special exemptions, some aircraft programs were setting new rules.

Aircraft programs were far larger and more expensive than satellite programs, so finding ways to economize in them was of great importance. Among the ongoing "black" aircraft developments were the stealth programs. These later were declassi-

---

[11] This dichotomy was neither new nor was it resolved then or later. The heart of the argument was that no program that had expended the effort to comply fully early could demonstrate what failures and additional cost it had avoided in the later, expensive, phases. That should be obvious; avoided costs had been avoided and therefore were unknown. In programs where less rigorous application had occurred and later costs had been increased, the source of the excessive later costs was rarely clear cut. Such costs could be argued to be related to high program risk, Congressional budget cuts, and other factors. Over time, the argument that spending slightly more early saved significantly larger expenses later eventually gained sufficient supporting program evidence to demonstrate its truth. However, spending more earlier would remain a contentious issue for reasons having little to do with the recognition that programs could pay a little now or pay more later.

fied as the Lockheed Martin F-117 stealth fighter and the Northrop B-2 stealth bomber. At the time, however, the F-117's approach to streamlined contracting was getting some notice.

DeLauer apparently believed these methods were more favored by industry and should be extended to the military space program. This was premature. Acquiring spacecraft was not like, or had ever been like, acquiring aircraft. Aside from the obvious fact that aircraft could be maintained and boxes replaced whereas spacecraft were out of reach once launched, a host of other dissimilarities were driven by the demand for high reliability. Many of these have already been discussed. Aircraft acquisitions were essentially different, and the stealth programs were different from other aircraft programs.

The F-117 was later noted for having broken all of the rules to demonstrate a streamlined method of acquiring systems faster, more efficiently, and at less cost. Conceptually, the new approach took McNamara's Total Package Procurement Concept to the extreme. The FARs contained a contract clause known as Total System Performance Responsibility (TSPR). The TSPR clause put full responsibility onto the prime contractor, which certainly streamlined things. Fewer government personnel were needed because TSPR virtually ended deliverable data and many progress measurement products. Government specifications and standards were eliminated in favor of the contractor's practices. TSPR seemed to succeed, though mainly because the contractor's practices were tempered by long experience and discipline created by compliance with military standards. When TSPR was exercised in the 1970s and early 1980s, the common approach to doing business was disciplined by the military standards, and all the people involved in the process, from designers and planners through the manufacturing floor technicians to the systems testers, were trained according to these standards, and all knew their jobs in terms of the standards. The best practices at the time were not commercial; they were defense-related. In addition, as with NRO space programs, the military importance of the programs in terms of capabilities, and as demonstrations of the new approach, demanded the best program office organization possible. The program offices paid great attention to the contractor's efforts, comprising hand-picked individuals with high engineering expertise.

Furthermore, the F-117 benefited extensively by a prototype aircraft, known as the Have Blue (or, more formally, the Experimental Stealth Testbed), which was essentially a scale model of what the eventual F-117 would be. This apparently allowed application of TSPR to only the final development and production.[12]

DeLauer did not appreciate the importance of the context within which TSPR was even possible, and was pushing to reform something about spacecraft acquisition that was actually working very well: military standards compliance.

---

[12] Ben R. Rich and Leo Janos, *Skunk Works: A Personal Memoir of My Years at Lockheed* (New York: Little, Brown and Company, 1994), 36–70 *passim*. Rich's description of his program leaves much out, of course. The extent of TPSR's application in these programs remains unclear. There is reason to believe, as indicated here, that TSPR was not intended at the time to acquire a full weapon system, but applied either only to segments, or to limited phases of development.

Henry verified that, and responded to DeLauer:

While at Kirtland AFB we discussed the application of Mil Spec 73-2C, since superseded by Mil Standards 1546 & 1547, to the IUS and other space programs. You indicated programs within Special Projects do not use this spec and they were getting the same performance at less cost. Since my return I have checked our policies with SAFSP's. We and SAFSP now apply Mil Spec 73-2C, or its equivalent, to all space programs in varying degrees depending largely on program maturity. It's important to note that the specification permits tailoring to accommodate individual program needs. We do the same on all our programs as each program has different radiation, mission duration and reliability requirements. Some special programs familiar to you, and which you may have had in mind, are mature programs and were initiated prior to 73-2C.[13]

Henry continued,

These apparently controversial documents were developed jointly with industry. Admittedly they were developed by specialists. I have had underway a review by management (Air Force and industry) to see if there are changes to be made from that perspective that can reduce cost and schedule impacts without reducing reliability.[14]

*   *   *

Elsewhere, Senator Sam Nunn and Representative David McCurdy began to act against developing the ALMV anti-satellite system. Rather than seeking agreement of other members of Congress (unlikely at the time), they developed an indirect approach, embodied in the language of their amendment to the Department Of Defense Authorization Act of 1982. The Conference Report contained several general provisions, a key one of which was titled "Reports on Unit Costs of Major Defense Systems," but has subsequently gone by its more popular name taken from its principal authors, "The Nunn-McCurdy Amendment."[15] The DoD argued that the United States should hold Soviet satellites hostage in the same way the Soviets held U.S. satellites at risk with their (long-standing) anti-satellite. Nunn and McCurdy disagreed, and developed a tactic to kill the program. Without citing the ALMV specifically, their amendment appeared intended to hold down cost growth in weapon

---

[13] HQ SD CC Letter to the Honorable Richard D. DeLauer, 23 Oct. 1981.

[14] Ibid.

[15] Department of Defense Authorization Act, 1982, Report No. 97-311, 3 Nov. 1981, Sec. 917. The actual language was simple enough, but contained one key giveaway: "If the Secretary concerned determines, on the basis of a report submitted to him pursuant to subsection that (A) the procurement unit cost of a major defense system for which procurement funds are authorized to be appropriated by this Act has increased by more than 15 percent over the procurement unit cost derived from the Selected Acquisition Report of March 31, 1981, or (B) the total program acquisition unit cost (including any increase for expected inflation) of such system has increased by more than 15 percent over the total program acquisition unit cost for such system as reflected in the Selected Acquisition Report of March 31, 1981, then . . . no additional funds may be obligated in connection with such system after the end of the 30-day period on the day which the Secretary makes such determination." The giveaway is the reference to the Selected Acquisition Report of 1981, which contained the Air Launched Miniature Vehicle Anti-Satellite's estimated production program at $1.4 billion.

systems. If the unit cost or total cost of a weapon system breached certain growth criteria, the program would have to be recertified as critical to national defense. Until such time as that certification, which was expected to be lengthy, could be made, no additional funds could be spent on the acquisition. In an accounting sense, Nunn-McCurdy simply recombined the phases separated by McNamara in TPPC—the cost of a unit or total system included *all phases* of the program. For highly complex, highly technological systems, their approach added a large amount of prior research and development costs into the equation. Prior to the Amendment, the cost of the ALMV program was $1.4 billion. That would change dramatically as the total cost approach was implemented, which would take some time.

*   *   *

Meanwhile, although MIL-STD-1546 and -1547 were significant steps forward for quality and reliability, they fell short in some of the key issues. To illustrate, both industry and the government had concerns about the quality and reliability of integrated circuits that would only be exacerbated as technology continued accelerating. The requirements for hybrid microcircuits, for instance, remained spread across four appendices of MIL-M-38510. Hybrid devices were lumped together with integrated circuits in a diffused way that did not improve the situation. Hybrid devices were odd ducks, not fitting with discrete devices, and not really integrated circuits either, so their qualification remained an elusive problem. The fact that they were spread across several appendices merely highlighted the problem of deciding if they were fish or fowl.

Early in the development of solid state devices, hybrids were considered a necessary aberration which time would eventually overcome with application specific integrated circuits or by some other means. That delayed addressing them in a coherent solution, since their acceptance was more a matter of tolerance than of understanding how they fit into the range of solid state devices.[16]

Thus, just after release of MIL-STD-1546 and -1547, JEDEC's JC13.5 committee took the initiative to develop a military standard to cover these increasingly important and problematic devices. JEDEC's original proposal was unsuccessful in gaining government interest, and was adopted as an industry standard, JEDEC Specification 1.[17] The topic lay fallow for the time being, but not for long.

*   *   *

Colonel Henderson's team reviewing the new military standards included the Aerospace Corporation's Allan Boardman. In nominating Boardman to the team, corporation president Dr. Eberhardt Rechtin told LtGen Henry, "He is already working the problem, has an interesting wealth of cost data." Rechtin added for emphasis, "parts cost, including management at the prime, is on the order of 10–13% of program cost

---

[16] William J. Spears, "MIL-STD-1772 and the Future of Military Hybrids," *IEEE Transactions on Components, Hybrids, and Manufacturing Technology* 12, no. 4 (Dec. 1989): 506.

[17] Richard K. Dinsmore and John P. Farrell, "MIL-H-38534: Military Specification for Hybrid Microcircuits," *IEEE Proceedings*, 0569-5503/89/0699, 699.

for a Class S or S equivalent parts program."[18] This 10–13% program cost was insurance against failures, and the price of the highest available reliability, or put another way, the cost of mission success. The fact Rechtin had to provide this (and Boardman had to collect and analyze it) reflects the differences between the early Minuteman days and the way satellites were being acquired at the time. The Minuteman program had gone to great lengths and expense to emphasize quality and reliability, but that approach had not completely translated into satellites. Partially, the reason was a lack of parts specifically qualified for use in space. Also, Boardman's figures only referred to the cost of the Class S parts and not for the overall quality control program, some portion of which would be in place regardless of the kind of parts being used.

The 10–13% figure withstood the test of time. The variability reflected spacecraft size and complexity, not parts price volatility. Historically, parts had cost about 5% of the total program cost, so Class S parts added between 5% and 8%.

Using cost avoidance as an excuse to buy lower reliability parts, then, was false economy. However, even were a program to try buying only Class S parts, industry had not expanded the range of available part types to support this. The JAN Class S industry grew slowly as more manufacturers were qualified to produce parts, and more parts came into the inventory. By 1982, about 50,000 Class S parts were delivered, while just two years later the deliveries were at about 100,000 parts.[19] These numbers pale to insignificance in the overall electronics marketplace, of course, but the Class S industry was never intended to challenge commercial numbers. At the time, a complex spacecraft had between 100,000 and 150,000 parts, and it was these quantities that drove demand in the Class S marketplace. Class S parts were clearly only a niche within the larger electronic parts marketplace, but one proving crucial to national defense and security. Spacecraft became more reliable and attained longer lifetimes, driving down the number of spacecraft needing to be built. At the same time as the number of spacecraft was declining, though, complexity was rising. The net effect preserved the level of demand for a while, but the balance was not stable.

The JAN Class S parts were an essential element of quality. AFSC Commander Gen Robert T. Marsh saw an essential difference between the predominant American approach to achieving the highest reliability systems versus the major foreign alternative approach. His responsibilities included national defense space programs acquisition, but he also acquired all the weapon systems (airplanes, missiles, and ground systems) for the Air Force. Addressing the Defense Logistics Agency (DLA, successor to DSA and the parent organization of DESC), the organization responsible for buying the military's parts, Marsh said he saw Japan, and other nations modeling themselves after Japan, with significant corporate commitments to quality. They were making inroads into the American defense market.

---

[18] Dr. Eberhardt Rechtin to LtGen Richard C. Henry, *Memo*, Subj: Your letter of 13 Nov. on nominees for a SD/industry team to review MIL Standards 1546 and 1547, 19 Nov. 1981.

[19] Don Seta, "JAN S Electronic Parts," *Institute of Environmental Sciences Proceedings*, 1985, p. 311.

Japanese firms have learned to design and build quality in, while American firms are still trying to inspect it in. Unfortunately, neither the Air Force nor your company can afford the price tag of inspecting quality in to make up for insufficient management and manufacturing processes.[20]

Trying to test-in or inspect-in quality remained a losing proposition. Test and inspection were important elements of quality and high reliability systems, but did not substitute for achieving the highest reliability, which began much earlier than the testing phase. Marsh recounted:

More than one half of our total quality cost is generated by the inspection process. Unfortunately, you can never catch everything no matter how much you inspect. A survey of twenty-one Air Force contractors revealed that they were processing over 370,000 material review actions per year.

Any potentially non-conforming part or material generated a material review action to check for its suitability. Marsh was citing the number of potential or actual problems, but only the ones that had been found. He continued,

One company was processing over 1800 material review actions per month with a scrap and rework cost of $275,000. This equates to over three million dollars per year. Is any company so big now that it can't be bothered with three million dollars?[21]

His message was clear: either put the quality in at the outset with the best available parts, or pay a heavy price when failures were detected in later testing. There would always be escapes—no inspection and testing system would catch everything. Tests and inspections were important, but they could not achieve high reliability by themselves. It was not an either-or proposition, though. High quality of input and a thorough inspection and testing program were essential to the highest quality output.

AFSC invested in the technologies that would support improved quality and reliability. As rapidly as the commercial electronics marketplace was moving, the military still had a hand in steering some areas. Just as the military invested in microelectronics technology at the industry's birth, so investment continued throughout the 1970s and into the 1980s. Beginning after World War II and throughout the 1970s, annual military investment in aerospace and electronics accounted for largest part of federal research and development expenditures.[22] In AFSC's Manufacturing Technology Quality Improvement Program, spanning the budget years from FY82 through FY87, AFSC had expenditures across eighteen different initiatives totaling some $38 million. The investments included some subassemblies such as traveling wave tube amplifiers. Importantly, investments also went into basic items: quartz

[20] General Robert T. Marsh, Address to *Bottom Line Conference*, Fort McNair, Washington, D.C., 13 May 1982, p. 5.

[21] Ibid., 6.

[22] David C. Mowery and Nathan Rosenberg, *Technology and the Pursuit of Economic Growth* (Cambridge, Massachusetts: MIT Press, 1989), 123–168; Daniel Holbrook, *Business and Economic History* 24, no. 2 (Winter 1995): 133.

crystal growth, gallium arsenide crystals, non-volatile memories, metal matrix composite sheet manufacturing, and nickel-hydride batteries.[23] The leverage of these investments against the huge electronics industry was infinitesimal, of course. Statistics about these investments typically point out how small a portion of the overall electronics industry the military had become by the 1980s. But such statistics overlook some fundamental and lasting truths.

The military investment in high reliability electronics was never an infinitesimal fraction of the high reliability market. In fact, the military investment in technology development and its purchasing of high reliability products dominated the high reliability electronics market. What appeared to some as a minor niche for the military was from a different perspective a marketplace dominated by critical national defense and national security needs. Some companies continued in the marketplace out of patriotism, others because the niche market remained very profitable. The manufacturers of high reliability parts continued to offer parts from domestic manufacturing lines. These domestic manufacturers benefited from the military investments in the kinds of technologies mentioned above which might not otherwise have received the same level of attention and progress. Though small in comparison to the overall marketplace, some of these technologies were able to affect the commercial processes in positive ways.

Misuse and misunderstanding of the gross level statistics and lack of insight into the reality of the high reliability industry were emerging as threats. Just as Henry's earlier exchange with DeLauer indicated, a serious dichotomy existed between those with "hands on" experience and those in top echelons of government management whose "hands on" was far enough in the past as to lose current relevance. The rate of change of electronics created a rift between current reality and past experience.

Some experiences can usefully inform future judgment while other experience gets quickly outdated. The management realities of a few years of rapid and accelerating technology change meant some powerful former industry managers who had moved to senior government positions were in jeopardy of incorrectly applying prior experience to a changed current reality. Experience can be relevant or irrelevant, and which is not always immediately clear. Learning lessons from experience requires expert thought and analysis because hard lessons are temporally fleeting. Added to that is the time span between performance and management. Senior managers, having learned to trust their experience, sometimes find it hard to recognize when lessons are obsolete in a matter of only a few years in a highly dynamic industry such as existed in the 1980s and beyond.

Very shortly, senior government officials, intent on reforming government acquisition, would lose sight of the distinctions between the military's infinitesimal market share in the larger market and its dominant position in the high reliability market. The consequences would jeopardize national defense and security.

* * *

---

[23] HQ SD Directorate of Productivity Business Management Review, 2 Feb. 1983.

The new Reagan Administration's space policies began addressing the challenges of making the Space Shuttle "operational." "Operational" meant too many things to too many people, each choosing to see the Shuttle in a different light. NASA's public relations pushed the rhetoric of routine access to space, spacecraft maintenance and repair in orbit, and using human presence to do such things as microgravity research and manufacture of important new medicines. These claims captured the imaginations of some who believed it was time to change the military status quo. Reliance on space systems no longer received the hand-wringing response of a decade earlier. Budget realities and critical enhancements to forces had won out to the point, by 1981, at which reliance was a done deal. However, most military space programs remained in an acquisition command—AFSC—not in an operational one.

Within AFSC, space was "a place, not a mission." Put another way, satellites were not an end unto themselves, but rather, they supported missions performed by the military. Space systems ownership by AFSC implied that they were still experimental. Changing that required creation of an operational command for space.[24]

General James V. Hartinger, commander of North American Aerospace Defense Command met with his West Point classmate, General Marsh, the AFSC Commander. As Marsh later put it:

> *Operating things wasn't our [AFSC's] expertise. It put us in a conflict of interest—to advocate these new capabilities and then turn around and operate them. Fox in the chicken coop. I was anxious to turn true operations capabilities over to some operator as soon as possible. I felt that was our fundamental responsibility.*[25]

Together with support from another of their classmates, LtGen Henry at SD, the three of them, and a select few from their staffs and the Pentagon, put together the plan to establish Air Force Space Command (AFSPACE). The new organization, announced on 21 June 1982, was activated on 1 September 1982.

AFSPACE was hardly a direct solution to the problem of high reliability and space qualified parts. It interposed itself as the advocate for satellite systems that could be considered "operational." Although this facilitated assignment of the ALMV anti-satellite to an operational command, establishing an operational space command implied satellites were not fragile research projects. Instead, satellites provided the kind of support for military operations that could be relied upon throughout the whole spectrum of military conflict. Achieving that tall order was now a military command's mission. And *that* contributed to the solution by raising the level of importance, oversight, and interest in high reliability.

---

[24] Temple, *Shades of Gray*, 520–521.

[25] Author interview with General Robert T. Marsh, 24 June 1991. It should be noted that Marsh's view really dealt only with his command and the military space program. The NRO and the national security space program continued to acquire and operate their own satellites, but as a separate agency, they did not have the same concerns as AFSC.

Chapter 13

Stocking the Shelves

Curiously, success in achieving high reliability created demands for even higher reliability. The longer a spacecraft lived, the lower the demand for more of the expensive spacecraft, and the less had to be spent on expensive space launches. High reliability was demonstrating its value in terms of returned savings. But that was just the microcosm—there were larger effects throughout the space industry that rippled through to the manufacturers of the highest reliability space parts affecting availability, and beyond.

LtGen Forrest S. McCartney took over from Henry upon the latter's retirement in 1983. McCartney had previously been the program manager for all the military communications satellites and had a long space career. When he assumed command of SD, he also became the second vice commander of AFSPACE. McCartney remarked that Henry "had assembled an efficient and smooth functioning organization, well staffed with competent people." He went on to set his priorities for his first year as "two major items: Mission Assurance and Cost Control."[1]

But because of his new operational responsibilities, he added:

Space systems are now recognized as having an indispensible role in combat capabilities. . . . In the early days of experimentation and exploration some degree of failures were expected, accepted and tolerated. However, this is no longer true, we have built such a good track record in space that more is expected of the space community. Risks are being taken now that would not have been considered years ago. Deployment of operational programs places a heavy burden on both Government and Industry Managers. In the 80s, there will not be the tolerance for failures that we experienced in the 1960s or even what we tolerated in the 1970s. This means that Mission Assurance, along with good cost control is a primary goal. . . . In the space business, the opportunity to fix a problem the second time around does not exist.[2]

Mission Assurance and cost control were complementary, challenging goals.

[1] Lt. General Forrest S. McCartney, Closing Remarks to the Mission Assurance Conference, Los Angeles, California, June 1983.

[2] Ibid.

Implosion: Lessons from National Security, High Reliability Spacecraft, Electronics, and the Forces Which Changed Them, First Edition. L. Parker Temple III.
© 2013 the Institute of Electrical and Electronics Engineers. Published 2013 by John Wiley & Sons, Inc.

Air Force space expenditures rose from 8% of Air Force Total Obligation Authority to 10% following the activation of AFSPACE. The increase was attributable in part to the new operational focus provided by AFSPACE, but also to a new initiative of 23 March 1983. On that day, President Reagan announced the nation's intent to become free of the threat of ballistic missile attack, realizing the Army's vision dating back to 1946. Known immediately as the Defense Against Ballistic Missiles, it quickly became formally called the Strategic Defense Initiative (SDI), and popularly "Star Wars." Several military space programs fell under the new SDI umbrella. The Boost Surveillance and Tracking System (BSTS) was the SDI follow-up to the DSP, which had been the source of EW since the early days of the space program. BSTS had far more stringent requirements to track multiple missile launches in order to hand off information to the Space Surveillance and Tracking System (SSTS) or to space-based interceptors. SSTS in turn would follow incoming warheads and their post-boost vehicles after the propulsive phase of flight. The information from these fed a battle management capability. Specific space- and ground-based interception capabilities would destroy the incoming missiles' post-boost vehicles or the warheads themselves. It was as though the long delayed move to an operational space command had caused the opening of some unsuspected flood gates. Space programs had not just become operational; they had become a national priority to turn back the fears of the Cold War and the doctrine of Mutually Assured Destruction. All of this meant a large infusion of advanced technology funding. And that infusion of funding went, in many cases, right to the microelectronics industry.

AFSPACE, joined by the other Services' obligatorily formed equivalent organizations, became the basis for a new unified command known as U.S. Space Command. The new operational space command's responsibilities included not only force enhancement programs such as DMSP and GPS, but it also now had two potential weapon systems in the ALMV anti-satellites and whatever came out of SDI.

*　*　*

After almost three years of activity, DoD investment in advanced VHSIC technologies had begun to have an impact. In February 1983, the program produced its first 1.25 micron chip, and was on the way to developing a pilot line to produce 0.5 micron, 100 MHz chips. VHSIC had expanded to include evaluations of manufacturing technology, design automation, and demonstration of VHSIC Phase 1 chips and technology by inserting them into a variety of military systems.[3] Although the investment was small in comparison to the broader commercial market, VHSIC was

---

[3] *Very High Speed Integrated Circuits—VHSIC—Final Program Report 1980–1990*, Office of the Under Secretary of Defense for Acquisition, Deputy Director, Defense Research and Engineering for Research and Advanced Technology, 30 September 1990, pp. 4, 11. An interesting sidelight is that VHSIC's use of gallium arsenide instead of silicon created problems of combining VHSIC chips with other conventional devices, and was solved by the use of hybrid manufacturing techniques so that both kinds of substrates could be used in a single device.

helping the industry with techniques to move feature sizes below the 1 micron level. Once again, military investment was small, highly leveraged, but crucial and influential.

By 1984, integrated circuits were proliferating in general with the rise of the personal computer, and in a widening variety of other uses, including toys. In the high reliability integrated circuit area, ASICs and hybrid devices were experiencing an equivalent explosive growth. While spacecraft might demand only a few devices of any specific design, other uses, such as automotive engine control systems, demanded production runs of hundreds of thousands to millions. In such large production runs, qualification considerations would be quite different, since some parts could readily be removed and replaced, and the price per device would be considerably lower. There seemed no single way to handle qualification of parts for such disparate use cases.

The first significant attempt at a standard for dealing with this topic addressed the specific issue of small production run ASICs. SD compiled its general guidance on qualifying such devices for space in MIL-HDBK-339, "Custom Large Scale Integrated Circuit Development and Acquisition for Space Vehicles," dated 31 July 1984. Because of its focus on Large Scale Integration (LSI), it immediately lagged behind available commercial technology. Industry leaders, such as Intel Corporation, had overtaken LSI's 5,000 circuit element upper limit more than eight years earlier. The combination of the Mead-Carter VLSI revolution, DARPA and VHSIC investment, and such private technology advancements contributed to VLSI's overtaking LSI. Although MIL-HDBK-339 was important and necessary, its shortfall in technology was a harbinger of problems with standards keeping pace with technology advancement. Furthermore, it did not solve the problem of hybrid device qualification, relegating hybrids to treatment under the general ASIC problem, which was a poor solution.

Hybrids were non-standard devices uniquely designed to a program's particular needs. To the frustration of some who thought hybrids too costly, hybrids were not going to "go away," or be replaced by ASICs. Histories of solid state electronics development often cite the cost of hybrids in particular, followed by the cost of ASICs, as the reason for changing how these devices were qualified. Cost was undoubtedly crucial in the commercial marketplace and in some smaller military applications. But more was involved. Neither JEDEC Standard 1 nor MIL-HDBK-339 was the right answer. Something else was needed.

They were a conundrum for the routine qualification of parts for spacecraft. Advanced tools allowed custom design of ICs, whose smaller feature sizes were becoming very difficult to test. Each hybrid's qualification was specified in a unique Source Control Drawing (SCD). Each SCD was developed by a company wishing to buy such a part, and was the means by which the company could detail in a somewhat similar way the custom manufacture, testing, and qualification of the end item. While some commonality existed across the SCDs for these devices, it was not absolute, and led to further complications in qualification testing and, ultimately, costs.

Until the semiconductor portion of a hybrid has been wire-bonded and packaged, it cannot be tested (it is not a complete device until that point). But once the semiconductor part has been bonded, it cannot be used again in a hybrid. "The only way to gain assurance that these bare chips would perform as required is to build the chips in a certified wafer lot, draw a statistical sample from that lot and to subject the packaged and wire-bonded sample to full testing."[4] Then, by statistical inference, once such tests were successfully passed, the lot of chips was satisfactory for further assembly into hybrids.

With a growing number of hybrids and ASICs, the number of SCDs grew. By their program-unique nature, SCDs defied any consistency in qualification for use in space. SCDs were understandably inherently more expensive than long production run parts. Their proliferation and cost started to attract management attention.

The cost of ASICs and hybrids, though, was really a classic example of the adage "where you stand is where you sit." For most sitting in medium to large military and national security spacecraft programs in particular, the cost of such devices was an irritant but not a major force. However, for the parts manager on a moderate to large spacecraft development, the cost of a hybrid (some approaching a hundred thousand dollars at the time) was important. To the parts manager's boss, the SPO director, the total cost of parts represented only about 10% of the total cost of a satellite. While that cost was important, the SPO director's interests were schedule, unexpected problems, budget cuts, and similar issues.[5] So, spacecraft hybrid costs were often lost in the noise by the time issues rose to the SPO director's level.[6] Certainly if parts costs could be reduced, any SPO director would be interested, but few would have taken significant time to correct the qualification problems of the industry.

However, device complexity drove the time to build and test individual devices, and their long lead times affected schedule. Schedule impacts *would* get the attention of a SPO director. Hybrid devices being qualified under the QPL rules required about six months (or more) of delivery lead time, and this hit the SPO directors right in the center of their concerns. Schedule impacts equated to large dollar impacts. So, schedule impacts tied directly to hybrid complexity and their qualification costs were getting noticed.

---

[4] William J. Spears, "MIL-STD-1772 and the Future of Military Hybrids," *IEEE Transactions on Components, Hybrids, and Manufacturing Technology*, 12, no. 4 (December 1989): 511. Much is owed in this section to Spears' fuller explanation of the relationship of hybrid standards, testing, procurement practices and costs.

[5] The SPO director's big four management attention items were typically cost, schedule, performance, and risk. Depending on the overall program's complexity and the director's ability to deal with this degree of attention to detail, some SPO directors were very interested in the costs of hybrids. This was certainly not universally true, and even when the cost of hybrids became a concern, it became a fleeting concern in the overall program.

[6] Michael J. Sampson personal communication with author, 9 January 2009. Of course, as this history shows, there are important exceptions to such sweeping generalities. Sampson, NASA EEE Parts Program Manager, recalled that from his perspective, QML was "driven by a need to cut qualification costs. My memory is that it was estimated in 1990 to cost $750,000 to qualify one microcircuit under QPL. I don't know what that is in 2009 dollars but it must be well over a million." Such a cost on a small- to medium-sized spacecraft would, of course, have attracted the interest of a SPO director.

Their complexity, qualification costs, and numerous problems demanded attention. The consequences of hybrid device complexity finally pushed the government over the edge to recognizing a standard was needed to qualify such devices.

Until 1984, manufacturing and testing of hybrid devices had been an appendix to MIL-M-38510 (for the microcircuit portion of the device) and a test method in MIL-STD-883. Test method 5008 was element evaluation, about which more will be said shortly.

RADC and DESC addressed the rising concerns about hybrids by adding a third hybrid-related specification, MIL-STD-1772, "Certification Requirements for Hybrid Microcircuit Facilities and Lines," released in March 1984. The key intellectual breakthrough was recognition that the one thing small lot sizes of unique devices had in common with each other was the *process* used in their manufacture. If the manufacturing processes could be controlled well enough to produce high reliability devices, then it would no longer be necessary to do a QPL-like individual part qualification of these low-volume, high-cost devices.

*   *   *

On 11 July 1984, the space shuttle *Discovery*, mission STS-41-D, aborted just prior to liftoff. One of the main engines had to be replaced. The failure was traced to an IC. Some form of contamination was found to be the cause, but the source of the contamination was a mystery. After considerable failure analysis, RADC's laboratories did a controlled study at various semiconductor clean rooms, and finally discovered what the source was. The contamination was from moisture droplets remaining after an operator in the room sneezed. The microscopic droplets contained sodium and other chemicals in sufficient quantity to cause varying degrees of degradation over the life of the electronic parts. The consequence of solving the mystery was an overall improvement of clean room practices industry-wide. The spittle failure mechanism was the root cause of one of the most famous non-life-loss events in the 1980s. This was but one of countless examples where the industry depended on RADC's high expertise.[7]

*   *   *

The initial control approach was a form of statistical process control. Statistical process controls found new favor as part of the 1980s management fad known as Total Quality Management (TQM). TQM itself was an amalgam of quality control methods adapted from the Japanese methods that had captured the attention of American manufacturers, as well as the works of W. Edwards Deming. Deming and his work had been influential a decade earlier. TQM made Deming's approach to process controls an essential part of recovering the perceived loss of American prominence in producing high quality manufactured items (about which General Marsh had lamented). Thus, a TQM approach to addressing the quality problems of

---

[7] Author interview with John Ingram-Cotton, 25 April 2009; George H. Ebel, "Failure Analysis Techniques Applied to Resolving Hybrid Microcircuit Reliability Problems," *Proc. Fifteenth Ann. Reliability Physics Symp.*, 1977, pp. 70–81. See Ebel's paper for the story of this investigation.

hybrid devices fit well with the need for some improved approach to process control in lieu of end item qualification.[8]

Procurement policies had to be amended to make all three hybrid-related standards mandatory on any program using such devices. Hybrid device proliferation meant these three standards would be placed on many military acquisition programs. Finally, on 29 November 1985, MIL-STD-1772 became mandatory when necessary changes to MIL-M-38510 and MIL-STD-883 were released.[9] One necessary change, for example, was MIL-STD-883's mandate on element evaluation, whereas before it had been only a strongly suggested good idea. Unlike individual integrated circuits or capacitors, hybrids could not be adequately tested at the end of the manufacturing process. Being a combination of integrated circuits and discrete devices, the end of the overall process was too late to find out that the integrated circuit or one of the added devices was defective. The individual elements, then, needed to be tested before assembly, and again as the final assembled device. The former tests examined the quality of the individual parts or elements, the latter the quality of the workmanship and assembly.

Not everyone in the hybrid industry agreed with element evaluation. In the commercial marketplace, element evaluation was simply expensive and unnecessary. Essentially, for commercial applications, if the device worked, that was good enough. For high reliability applications, however, proper functioning was only one consideration. Long lifetime and reliable operation depended on characterizing the individual parts; a complex device could be only as reliable as its least reliable element. As might be expected, the cost of such testing prior to the 1985 mandate was high, on the order of $35,000 per lot for medium complexity hybrids. (The use of cost as an argument against these devices proved wrong once market forces began to operate. By 1989 that cost had dropped to about $5,000 per lot through competition and other efficiencies.)[10] The changes did not drive down acquisition costs, as some expected, because they aimed at reducing total life cycle costs (about which more will be said shortly).

*   *   *

High reliability space-qualified electronic parts and their associated military standards began receiving an upbeat sales treatment. Part of the reason was to ensure continued support for AFSC space program budgets, but just as important was

---

[8] Spears, "MIL-STD-1772," 507. As an interesting sidelight, there is no definitive source for the origin of the term Total Quality Management. Both Japanese and American sources lay claim to the term, and differences in translations between the two only serve to further confuse who deserves credit. The statistical process controls and some related measures undoubtedly owed their refinement and expansion to W. Edwards Deming, as well as Philip B. Crosby, Joseph Juran, and Kaoru Ishikawa. The term, whatever its origin, was in vogue by the early to mid-1980s.

[9] Ibid. The specific revisions were MIL-M-38510 revision F, amendment 1, and MIL-STD-883 revision C, notice 4.

[10] Ibid., 511. This cost was for the element evaluation, an expensive portion of overall qualification of a device, but not the complete cost of the qualification. The cost cited, also, has to do with commercial grade parts, not the higher reliability parts, which were more expensive as previously noted.

changing the perception of satellites to be more robust and enduring capabilities critical to the American way of war.

When the DoD Inspector General (IG) began an investigation into parts control programs in 1984, SD fared very well. The IG asked about thirteen SD contracts "to determine compliance with existing Parts Control Program requirements." Only three of the thirteen were production contracts, and SD Vice Commander MajGen Bernard P. Randolph could explain, "The three production contracts identified exceed the PCP requirements per AFR 800-24 (which implements DODI 4120.29)."[11] But that should have been expected for space programs. Randolph added:

> *Since the mid-1970s, SD has had a Commander's policy requiring a more comprehensive PCP than those required by either DOD or the Air Force. This policy was established to allow our space systems to meet or exceed our unique mission reliability needs. The policy is implemented by SAMSO-STD-73-2C or the superseding MIL-STD-1546. On-going internal assessments indicate that SD programs are in compliance with the policy and our PCPs are increasing cost effectiveness and reliability of our systems.[12]*

Furthermore, the JAN Class S program appeared to be another important success story. According to Colonel David C. Bockelman, SD Deputy for Acquisition Logistics, "JAN Class 'S' high reliability space parts have been a real success at SD. Recent Space Systems data shows that built-in quality can decrease failure rate by a factor of 60 and that high-reliability parts are proving to be cost effective even before the spacecraft leaves the factory."[13] This addressed the issue of "inspecting in quality" versus "designing and building in quality" about which Marsh talked. Also, it demonstrated that even in the spacecraft assembly phase, reduced failure rates were saving money in spite of the higher cost of the parts themselves. This was exactly what earlier claims for mission assurance standards tried to get across: a small program cost increase in the earliest phases saved considerably more money overall.

The success of Class S parts in satellites led SD to consider their use in ground systems. Despite the premium initial cost, the benefit in lower maintenance and

---

[11] HQ SD, Memorandum for the Office of the Assistant Inspector General for Accounting, Subj: Audit of DoD Parts Control Program, 28 Feb. 1984. The three were the Defense Satellite Communications System III, the Defense Support Program, and the Atlas launch vehicle avionics program. As a sidelight, it was interesting that the DoD IG did not know more about the nature of the programs and contracts they were examining, since they asked about the parts control programs on ten contracts for which that was inapplicable. It seems further evidence that even in the early 1980s, familiarity with space programs was limited.

[12] HQ SD, Memorandum for the Office of the Assistant Inspector General for Accounting, Subj: Audit of DoD Parts Control Program, 28 Feb. 1984. Randolph's remarks also reveal the extent to which metrics about the SD contracts were being collected, which intervening events have precluded.

[13] HQ SD Message, Subj: Joint AFSC/AFLC Conference, Nov 1984; Don Seta, "JAN S Electronic Parts," *Institute of Environmental Sciences Proceedings*, 1985, 311. The latter reference expanded on the factor of 60 number, by explaining that JAN Class B parts in missile systems were experiencing failure rates of one failure in 200,000 device-hours, while the equivalent in JAN Class S programs was on failure in 12,000,000 device-hours. The comparison "was based on equipment and vehicle acceptance tests because field failure data was not available."

fewer system outages promised a positive payoff. Advocates for Class S claimed even entertaining such considerations illustrated the degree to which Class S parts had improved the reliability of space systems. This was not pursued in part because costs of commercial technology in ground systems continued to drop steadily. Also, the idea overlooked the inherent maintainability of ground systems. Satellites, of course, demanded the highest reliability because their maintainability was essentially nil. It was possible in ground systems, though, to trade acquisition versus maintenance costs. More to the point, considering Class S in ground systems indicated a marketplace problem.[14]

Class S parts availability did not meet space programs' needs, especially when such parts became mandatory. The causes might have been not enough suppliers, or not enough parts selection, or long lead times and low yields on the highest reliability parts. Higher qualification costs to the manufacturers may have created some hesitation in providing a full range of Class S parts. Whatever the reasons, Class S might have been a victim of its own success. Greater reliability became expected, and with demonstrated increased reliability came demands for even more. Longer lifetimes reduced the overall demand for parts because fewer spacecraft were being built. The downward market pressure was offset initially and briefly by increasing satellite complexity (parts counts), which increased the demand. By 1985, increased complexity driving a demand for more parts per spacecraft was completely swamped by the lower demand for spacecraft due to their higher reliability. The delicate balance had shifted in favor of increasing costs for Class S. The military and national security space programs were not the only ones being subjected to these offsetting pressures. The major space agencies were walking a tightrope between demands for greater reliability without letting costs go wild.

NASA's chief engineer, Milton A. Silveira, contacted McCartney about their mutual needs for high reliability parts.[15] In March, 1985, Silveira wrote:

> *Both the SD and NASA have been affected by the recent noncompliance by microcircuit manufacturers, as well as the continuing difficulty that our projects have had in procuring high reliability electronic parts. I believe that these problems could be reduced through the greater involvement and use of the Defense Electronic Supply Center (DESC). To this end, I would propose that we establish a working group with DESC management to address and develop solutions to the following and other topics:*
>
> * *Reduction of contractor control specification (use of DESC drawings and application reviews.)*
> * *DESC engineering support (. . . automated parts review).*
> * *DESC logistics support (stocking of Class/JAN S parts).*
> * *DESC technical support (new failure analysis lab).*
> * *Resolution of qualification procedural problems.[16]*

---

[14] Seta, "JAN S Electronic Parts," 310.

[15] Evidently, some of the prior arrangements for such things as joint inspections had fallen by the wayside.

[16] NASA Code DR Letter to LtGen Forrest S. McCartney, SD Commander, 12 March 1985.

The list included a number of possible ways to cooperatively improve parts quality, with the not-so-hidden agenda to offload some of the internal management of parts to DESC. McCartney asked his Deputy for Acquisition, Colonel Bobbie L. Jones, "What's in this for us?" From his very long experience in space, McCartney knew the importance of Class S parts, so he was not asking about the Class S program. What he meant was "Which of Silveira's ideas holds any benefit for the SD space programs?"

At the time, SD's prime spacecraft manufacturers had extensive laboratory facilities for inspection and testing, as well as ascertaining root cause of failures (for most parts). SD also had access to independent third part testing capabilities in the Aerospace Corporation's laboratories, which complemented and augmented the industry capabilities. In addition to these, the electrical and electronic engineering expertise at RADC provided the expertise to determine the qualification criteria for any specific kind of electronic device and its technology. Given these were not likely to change any time soon, McCartney was justified in asking why the Silveira suggestions had benefits for SD.

Jones saw merit in the idea of a Class S stockpile. He took Silveira's idea of having DESC stock Class S parts one step further, as he told McCartney:

We would like to work with DESC toward a DESC stockpile of JAN Class S parts. Individual programs are sometimes required to purchase only about 50 units of a particular part, but are forced to buy a full lot of as many as 250 units. We would like to have DESC buy and stockpile Class S parts for Air Force and NASA use, much the same as they now stockpile Class B parts. The stockpile would cut cost and ensure availability for our programs.

Drawbacks to a stockpile have always been (1) who is liable for a government-supplied part that fails, and (2) what do we do with parts that become obsolete? The first question, we think, can be handled by "seeding" and proper contractual language; the second may not be a drawback. Obsolete parts that are no longer manufactured can be a problem if one is required (such as an old satellite brought out of storage). A working group could address these questions, possibly reduce cost, and improve availability.[17]

This was similar to an idea LtGen Henry had floated several years earlier, but the concept had matured in the interim. One reason stockpiling had not taken hold was the use of SCDs drove parts to be other than the standard part offered by the manufacturer. Apparently, as long as the contractors were demanding SCD parts, which were essentially unique, stockpiling's benefits would be marginal. But was that really so?

Col Jones expanded on the thoughts of his memo.

Satellite and Launch Vehicle programs many times need only about 50 units of a particular electronic part type but are forced to buy as many as 250 units. They are forced to pay for these excess parts because of minimum buy requirements imposed by the parts manufacturers. In addition to this extra cost, space quality parts take about

---

[17] HQ SD Acquisition Logistics, Memorandum for General McCartney, Subj: NASA and SD Working Group with NASA.

one year to build, test and deliver. And many times, failures and schedule slips cause the deliver time to exceed one year. This extra cost and time involved in procuring parts often leads to contractors taking short cuts and making promises that affects [sic] the reliability of the system. Even though high reliability parts, especially JAN Class S parts, have been shown to be cost effective in the long run and reduce the chances of mission failures, the previously mentioned procurement issues need to be resolved.[18]

The idea of stockpiling was not a surprise to anyone. The idea had batted around in industry and government circles for some time as a solution to procurement problems. The new idea was "The parts could be stored at the Defense Electronic Supply Center."[19]

The issue was not as clear-cut as it might seem. One important concern was contractual liability. For instance, if a government-supplied part were to be found the cause of a failure in orbit, determination of liability would be a knotty question. While alternatives and variations on this theme existed, the apparent advantages of stockpiling slowly gained traction against the concerns.

The cost of acquiring military systems was a perennial budget topic, and in the mid-1980s most critics began to take aim at alleged abusive use of SCDs. SCD parts were more expensive than standard parts by their very nature. Starting with some standard part (not necessarily Class S), SCDs supplemented the treatment of that part with additional test cycles, environmental factors, or new and different types of tests to which the part would not otherwise have been subject. Hence, when examining the source of costs in military acquisition programs, the SCD parts were quickly identified as a target for economizing. There was a "DOD-Wide thrust to eliminate source control drawings."[20] Use of non-standard parts prevented some important economies in the manufacturing of parts (such as stockpiling). Eliminating SCDs would force use of standard parts in all military acquisitions.

That made superficial sense and thus became the target of budget cutters. In the specific case of hybrid devices, compliance with the triad of MIL-STD-1772, MIL-M-38510, and MIL-STD-883 forced up the cost of hybrids. These standards had meant to address total life cycle costs for hybrid devices going into military systems. Forcing more testing and design analysis early was expected to significantly lower costs in the field. Higher reliability translated to longer mean times between failures, and consequent lower repair and replacement rates. That was where the really large costs lay in the life cycle of terrestrial systems.[21] Hybrids for space, of course, had no such concern about repair and replacement, but benefited equally from longer times before failure. Longer times between failures equated to longer spacecraft lifetimes in orbit, reducing the number of new satellites needed, which was perceived as a good thing and an additionally significant source of cost savings.

---

[18] HQ SD Acquisition Logistics Point Paper, 7 May 1985.

[19] Ibid.

[20] HQ SD Acquisition Logistics Memo for LtGen McCartney, Subj: Monitored Line Program Briefing, 7 May 1985.

[21] Spears, "MIL-STD-1772," 510.

To some extent, the concern over SCDs was a tempest in a teapot where the concern was preservation of the JAN Class S parts manufacturing lines. Of the SCDs used for space programs, many, and perhaps most, began with specification of full JAN Class S parts, and then added requirements. Conceptually, JAN Class S parts were simply very high reliability, lowest common denominator parts that needed some specific adaptations to any given satellite's actual design, operating environment, and other considerations. Put another way, JAN Class S gave a high confidence, high quality, and high reliability starting point for customized space parts. Despite the higher cost of JAN Class S, had there been no Class S to start with, SCDs would have been even more expensive because their lot sizes were smaller and devoted to only one program at a time.

For those situations where a JAN Class S part did not exist, the JAN Class S specifications were used to obtain a higher reliability part by evoking the specifications in an SCD (and at much higher cost than starting with a JAN Class S part). Thus, the SCDs represented a greater cost than the offered JAN Class S parts, but they also preserved and expanded the JAN Class S parts industry.[22]

Just as electronics industry analysts overlooked the high reliability segment because it was too insignificant, the cost-cutters taking aim at SCDs were not seeing the Class S forest for the SCD trees.

* * *

This cost consciousness was taking place within the larger picture of the importance of the spacecraft for which the SCD parts were intended. Reagan intended to change the decades-long Cold War calculus, and the national security and national defense space programs were integral to achieving that.

The Soviets knew, from long experience, that when the United States set out to accomplish some goal, especially with the dedication of such a strong president, the goal would be attained. In the midst of the Reagan presidency, SDI took center stage, but much more was happening.

The Air Force was pushing ahead with plans for a revolutionary new kind of aircraft—one that could take off and land conventionally, but also reach orbit. GPS was about to redefine global navigation and timing accuracy. NASA's Space Shuttle was in its initial flights, promising tantalizing new possibilities. Each of these forced a response from the Soviets, stretching their ability to pay to the limit, and beyond.

Behind the scenes, though, near-real-time reconnaissance had advanced to the point where it was an important national capability. Space-based reconnaissance supported one of Reagan's most often repeated phrases: "Trust but verify." The phrase was not original to him, but it reveals much about a foundational part of his plan for dealing with the closed Soviet society. When Soviet Premier Mikhail Gorbachev pointed out his frequent use of the phrase at every one of their meetings, Reagan responded he used it because he liked the phrase. He did not have to explain why. Throughout the treaty negotiations for the Strategic Arms Limitation, Strategic Arms Reduction and the Intermediate Nuclear Force Treaties, such a capability had

---

[22] Seta, "JAN S Electronic Parts," 311.

become routinely accepted by both sides of the Cold War as an essential means of assuring themselves about what was happening in each others' heartlands.[23] As both nations reduced the number of nuclear weapon systems and warheads, both would have to show their good faith through actions the other could watch and verify. If the Cold War were ever to end, as Reagan hoped, both nations would have to gain mutual trust. Mutual trust meant dependably verifying such trust was warranted. National security space systems formed the fundamental core on which verification was built.

Verification, then, had to be extremely capable and the epitome of reliability. Technologies for capabilities and reliability began with piece parts. How much was the ending the Cold War worth?

*   *   *

While intent on economizing by eliminating SCDs, DoD once again lost the distinction of space programs as the exception to the rule. Obtaining the highest reliability parts for use in the nation's most important space programs was simply a difficult task which elimination of SCDs would have damaged. To obtain the quality needed for the highest reliability programs, SAFSP established an Order of Precedence that actually addressed some of the DoD concerns. They chose to use JAN Class S first, SPO-developed SCDs second, and then Monitored Line Program parts third.[24] Monitored lines were exactly what the name implies: SPO or other government personnel spent time in the plant ensuring adherence to procedures that, lot by lot, produced the highest quality attainable.

Within the military space program, an SD Commander's Policy put their selection precedence as:

- Microcircuits and semiconductors specified in MIL-M-38510 or MIL-S-19500 (respectively), which were listed in Part I or Part II of the JAN Class S Military Qualified Products Lists (QPLs).

- Passive components listed in respective Military QPLs that meet space quality requirements.

- Parts other than JAN CLASS S must be designed, constructed, qualified, screened, quality conformance checked, process controlled, and have failure analyses as called for in MIL-STDs 1546 and 1547.[25]

Although the SD Order of Precedence was more like that found in MIL-STD-1546, the order implied standard part types were to be sought first, followed by the MIL-STD-1546 and -1547 parts second. The latter covered SCDs. SD did not use monitored lines. The national security and national defense space programs were following

---

[23] Martin Anderson and Annelise Anderson, *Reagan's Secret War: The Untold Story of His Fight to Save the World from Nuclear Disaster* (New York: Crown Publishers, 2009), 347 *passim*.

[24] HQ SD Acquisition Logistics Memo for LtGen McCartney, Subj: Monitored Line Program Briefing, 7 May 1985.

[25] Commander's Policy, "Management, Selection, and Purchase of Electronic/Electrical Piece Parts," SD Regulation 540-12, 15 Feb. 1984.

a rigorous discipline to obtain the highest reliability parts in a manner consistent with the least cost, even though least cost was still expensive compared to commercial or avionics. They both still had problems with small quantity purchases below the economic order quantity of the manufacturers. That was where the stockpiling initiative was thought to play.

Pressure to eliminate SCDs aside, McCartney and his staff continued pursuing the stockpiling idea. To make the concept viable, DLA designated DESC, in Dayton, Ohio, to manage the JAN Class S stockpile. By early May, 1985, two key issues had arisen: (1) liability determination (who was liable if the government-supplied parts failed?) and (2) DESC's ability to support space-level requirements. Col Jones reported, "Although DESC seems willing to stockpile, we must verify that DESC can maintain (1) documentation control; (2) traceability; and (3) quality control on handling."[26]

Jones' concerns covered the necessity of knowing how, when, and where a part had been made and handled. Documentation was necessary for each stage of a high reliability part's manufacture because tracing a later problem to its origin sometimes required knowing if a process step had been compromised, skipped, or done out of order. That implied traceability, and included a given part's "lot date code," the date the lot was manufactured. One defective part might indicate that its lot date code siblings were also defective, and would require further investigation. Handling went beyond documentation and traceability, and included the temperature and humidity levels environments parts had experienced. Environments could lead to problems with static, more properly termed electro-static discharge. Solid state electronics were particularly sensitive to getting zapped by an ungrounded person. So, Jones was asking whether DESC's processes, people, and facilities could track, receive, store, and ship out sensitive space parts.

As part of the full court press to get the stockpiling initiative under way, the topic became a major discussion at the annual Space Parts Working Group (SPWG) held in June 1985. The SPWG hosted over 120 participants from prime contractors, subcontractors, parts manufacturers, DoD, AFSC, SD, NASA, RADC, DESC, Navy, Aerospace, other private companies, and government agencies.[27] The SPWG supported stockpiling JAN Class S and radiation-hardened parts.[28] The stockpiling

---

[26] HQ SD Acquisition Logistics Memo for LtGen McCartney, Subj: Class "S" Parts Stockpiling, 12 April 1985. A meeting between SD and DESC held on 7–9 May 1985 at DESC attempted to resolve the issues.

[27] HQ SD Acquisition Logistics Memo for LtGen McCartney, Subj: Space Parts Working Group, 12 June 1985. The SPWG that year started four working groups: (1) laboratory certification, (2) JAN Class S stockpiling, (3) JAN Class S streamlining (a revision of 1546/7), and (4) updating SD Regulation 540-12 (SD Commander's Policy on Parts Selection). The latter had to do with the order of precedence issue.

[28] Though it might seem incongruous, Class S parts are qualified for use in space, but are not necessarily innately radiation-hard. The latter term, along with radiation tolerant, are terms used to describe the part's ability to function in radiation environments. Resistance to radiation-induced degradation arises from design and other processing measures that are not routinely part of Class S part manufacturing. Radiation susceptibility testing and qualification are separate activities from Class S qualification. Conversely, radiation hard parts are usually considered usable in most space applications.

initiative's objective was to have 95% of all JAN Class S parts required by SD on ready inventory within two years. DLA obligated $50 million for the DESC-administered program.

After making the initiative known to the SPWG participants, McCartney followed up with a letter to all SD contractors in August 1985. His told them:

> *Although JAN Class S parts have proven to be cost effective because of their reliability, many procurement problems exist. The stockpile will solve most of these problems, since it will: (a) reduce procurement lead times, (b) allow for small quantity purchases, and (c) expand JAN Class S availability. The current plans call for stocking over 700 line items of JAN Class S semiconductors and microcircuits with inventory becoming available as early as next summer. This volume of stock, which is available off-the-shelf, will reduce your needs for nonstandard and costly Source Control Drawing parts.*[29]

McCartney solicited estimates of JAN Class S part types needed by each of the contractors so the work of determining the proper mix of the stockpiled parts could begin.

The September 1985 industry responses were not what McCartney expected (see Table 13.1). The preliminary response did not rise to the level of funding supported by DLA. The survey projections did not result in a sufficiently strong base of needed parts, which McCartney and his staff felt was not the case.

The responses reflected the slow contractual changeover to the standards mandating Class S parts. Ongoing contracts were not being changed to mandate Class S parts because such a mid-stream change affected not only the cost of parts, but also the actual basic designs of circuits. Many Class S parts had different electronic characteristics than equivalent parts of lower grades. A mid-stream change would drive a system redesign that would be more expensive than the cost of the higher reliability parts alone. But that was not the whole explanation.

Some of the responses revealed the reactive nature of parts buying at the time. Companies' ability to forecast what the government would want them to do was very limited. Spacecraft building was not, for the most part, a predictable assembly line business. Beyond that, budget vagaries made the year of purchase for any expected spacecraft uncertain as well. In effect, the government was asking the contractors for their projections of what the government was going to do.

Consequently, SD hired two consultants from Integrated Circuit Engineering to survey the industry and come up with the right parts and lot quantities for the operating stock program. The consultants did a considerable amount of study, but in the end, were not familiar enough with the space programs to make a valid forecast.[30]

The solution to most of these issues came from the members of the SPWG. In the end, much of the groundwork for the operating stock program was done without charge to the government as a patriotic duty because of the importance of the effort. The work was carried out largely by the SPWG Executive Committee, and included

---

[29] HQ SD letter, Subj: JAN Class S Electronic Parts Stockpile, 26 Aug. 1985.

[30] Author interview with David M. Peters, 19 Sept. 2008.

Table 13.1. Responses to SD Survey of Class S Need Projections

Corporation	Response Date	Summary of Response
RCA Astro	18 Sept. 1985	Supportive, but concerned about warranties of defective parts
McDonnell Douglas (MDAC)	17 Sept. 1985	Supportive, but Douglas Aircraft Division did not routinely procure semiconductors or microcircuits—these parts were in computers and avionic equipment they procured from Original Equipment Manufacturers (OEM)
Sanders	23 Sept. 1985	Commendable idea, and they had used small quantities of class S in the past, they could not project any utilization
AT&T Tech	24 Sept. 1985	None of AT&T's programs employed Class S, and they did not anticipate any, so they could not be of help
RCA	25 Sept. 1985	RCA Missile and Surface Radar Division did not have any programs requiring Class S. "We have been able to meet or exceed our contractual reliability requirements with the use of Class B microcircuits. On the Navy AEGIS program we are re-screening parts and the average cost per microcircuit, including re-screening, is substantially less than the average cost of Class S parts"
Aerojet	27 Sept. 1985	Fully supported effort and included a summary of their projected needs
JPL	27 Sept. 1985	Fully supportive, provided projections, though current programs were too far along to need any
Sandia	30 Sept. 1985	Had to use DOE quality standards, closely related to but not identical with MIL-STDs. The difference was largely in their unique radiation requirements. Did not believe they would use any such stockpiled parts
Hughes	30 Sept. 1985	Fully supportive, but did not currently use Class S. SDI might change that, but no projection was possible
Johns Hopkins University/ Applied Physics Lab	30 Sept. 1985	Supported, but added radiation requirements and their projection of needs
Boeing	Undated	Supportive, but only program was Inertial Upper Stage so they had no projected needs

Source: Original individual corporate responses in HQ SMC History Office files.

suppliers such as National Semiconductor, Signetics, Texas Instruments, MicroSemi, and International Rectifier, as well as contractors such as Honeywell Clearwater and Boeing Seattle. They traveled as a team with SD and other agency representatives to select the logistics center to stock the hardware. DLA's Ogden, Utah, facility was ultimately chosen for warehousing and distributing parts. They ran some sample cases of ordering parts, reviewing the data packages upon receipt, and then returning the parts, all in order to ensure the processes were working.[31]

The group also addressed the JAN Class S Operating Stock quantities to be ordered. Facilitated by the SPWG Executive Committee, the parts and quantities were based on spreadsheets provided directly from the parts suppliers, rolled together by the Integrated Circuit Engineering consultants.[32]

By the mid-1980s, high reliability spacecraft had achieved a high degree of success. Indispensible to the President and key elements of the Executive Branch, satellites had also achieved major advances in military command, control, and integration into the American way of war.

This happened because of longer lifetimes and robust capabilities. Working in opposition, though, was the lower number of satellites needed and the consequent shrinking of the Class S marketplace. By the mid-1980s, though, stockpiling addressed these issues as one of the many threads of activity necessary to keep high reliability spacecraft serving the nation's defense and security. Just as high reliability electronics were a critical but small niche segment of the overall electronics marketplace, high reliability satellites were a critical but small niche segment of the overall Department of Defense acquisition of weapon systems. That positioning would soon serve neither parts nor spacecraft well.

[31] Ibid.
[32] Ibid.

Chapter 14

Hammered

Every crusade needs some icon to capture the imagination. Reform of acquisition was about to receive one of the most enduring such icons, from the strangest of places, for the oddest of reasons. When Nunn and McCurdy took aim at the ALMV anti-satellite, they set in motion a mechanism that hit something they probably never intended.

The ALMV remained technically challenging even as it approached the production phase. Because of its risk, complexity, and cost, the ALMV was one of the programs selected for tracking in the Selected Acquisition Report of 1981. Not coincidentally, that 1981 report was specifically cited in the Nunn-McCurdy Amendment. At the time, the ALMV's cost estimate was $1.4 billion. Nunn-McCurdy's changed accounting, however, added back all the previous research and development costs, causing the program's unit and total costs to skyrocket, reaching $5.3 billion in 1986. Regardless of whether it was an accounting change or not, such an increase drew unwanted criticism of the program and the Air Force's acquisition process. In March 1987, the Air Force cut the number of anti-satellite systems to be acquired, dropping to 35 missiles from the 112 originally requested and 18 aircraft down from 40, but, in so doing, only whittled the Nunn-McCurdy calculated costs to $4.3 billion.[1]

Cutting the total cost by about a billion dollars would certainly seem to have saved money. But that is where Nunn-McCurdy's clever unit cost calculation kicked in.

[1] Paul B. Stares, *Space and National Security* (Washington, D.C.: Brookings Institution, 1987), 102.

Implosion: Lessons from National Security, High Reliability Spacecraft, Electronics, and the Forces Which Changed Them, First Edition. L. Parker Temple III.

It should be clear from the cost after the cuts that the actual costs of manufacturing the individual weapons and rockets were not a significant portion of the total program. Even considering the cost of the aircraft modifications and other factors, the production cost of the ALMV-booster combination was at most about $13 million apiece at the time of the reassessment.[2] However, using the Nunn-McCurdy formula, taking into account all of the research and development credited to the program, the pro-rated cost per ALMV and launcher would be about $47 million prior to the reassessment (from the 1980 cost) and $122 million afterward (based on the 1987 cost). A breach of Nunn-McCurdy was obvious—the cost had increased far more than the allowed limits per weapon.[3]

Curiously, the same accounting trick (including all research and development costs) led to one of the most infamous and enduring mistaken criticisms of defense procurement. When the total program costs are divided as in the above example, a missile and warhead that actually incrementally cost $13 million during production suddenly become $122 million, and the only thing that changed is the accounting. So, when the earlier research and development costs were used to track the cost per weapon, how exactly was that done? The most straightforward way was simply allocating a pro-rated amount across all the parts of the weapons and associated equipment. If there were 10,000 parts, for an arbitrary figure, from resistors and transistors up to rocket fins, rocket motors, guidance boxes, tool boxes, and warheads, then each and all 10,000 received an equal pro-rated spread of the research and development total. In the same way that a missile whose cost at the time of production goes from $13 to $122 million, a hammer actually bought at a government discount turns out to have $400 allocated to it when the toolbox becomes a program deliverable. The $400 hammer never existed in the sense that the government never paid anyone $400 for a hammer. That was the pro-rated allocation of the research and development that preceded the production of a complex, technologically sophisticated weapon system.

Hot on the heels of the $400 hammer came the notorious $600 toilet seat. Few paid attention when the DoD made clear that it did not exist, either. Later investigation showed that the toilet seat was not a toilet seat after all, but a "corrosion resistant shroud to cover the entire toilet system in a P-3 aircraft." The cost was actually justified, but by then, the damage was done.[4]

---

[2] Even that figure is high, since the total system was costed. The tracking, planning, coordination and targeting, and command and control systems included several new or highly upgraded ground systems. The figures here are not intended as precise details because the point is best served by simple calculations in which the order of magnitude is the significant change. The reality was that the exact production unit cost of the ALMV and its booster was less than the $13 million in the illustrative calculation used here.

[3] There is no flim-flam here—it is a straight forward calculation. In 1981, the $1.4 billion production cost covered 112 ALMVs and boosters (as well as aircraft modifications and extensive ground facilities). Straightforward division reveals the cost per unit to be about $13 million. In the identical calculation for 1987, the $4.3 billion was only split among 35 missiles, hence the increase in cost per unit.

[4] Stephen Howard Chadwick, *Defense Acquisition: Overview, Issues, and Options for Congress* (Washington, D.C.: Congressional Research Service of the Library of Congress, 2007), CRS-2.

In the mood of the times, though, with government waste and abuse as a cause célèbre, the $400 hammer became an icon and took on a life of its own. In fact, conflating the two errors, the hammer became most often reported as costing $600.[5]

The battle lines were starting to form up. The insurgents were going to fix weapon system acquisition, eliminating forever the $600 hammer. Or so they thought.

---

[5] The $600 hammer of later Congressional testimony was actually a confusion of an earlier confusion. It was a $400 hammer, and the $600 item was a toilet seat. Some reports also claim that the cost of the hammer was from "an incorrect invoice that the government never actually paid." It is not clear there ever was such a mistaken invoice, but within the Air Force Secretariat, Dr. Thomas Cooper, Assistant Secretary of the Air Force for Acquisition at the time, bore the brunt of the criticism for that hammer. The word he got was in line with the history presented here, and that the figure came from an accounting spreadsheet, not an actual invoice. See more at Chadwick, *Defense Acquisition*, CRS-2; Thomas L. McNaugher, "Weapons Procurement: The Futility of Reform," *International Security* 12, no. 2 (Autumn 1987): 64–65.

Chapter 15

Battlegrounds: Reorganization and Reform

Change was the norm in the histories of the electronic piece parts industry and its consumers. But the pressure to eliminate SCDs hinted at change among a larger climate of change bent on eliminating the problem of overpriced hammers.

The larger climate did not favor the space program's high reliability efforts. After all, each phase of a complex system optimized for high reliability cannot also be the lowest cost. Overall military acquisition efficiency was about to collide with the high reliability system, to the detriment of both.

What acquisition model was the most efficient by the mid-1980s? When DeLauer questioned Henry about streamlined acquisition, his interest was only part of a longer standing attention paid to eliminating waste in acquisition.

Inevitably, DoD-wide considerations resulted in some sense that commercial programs were the most efficient, since profit motives eliminated unnecessary expenses. The "Report on National Space Policy Enhancing National Security and U.S. Leadership Through Space" had staked out the need for greater opportunities in commercial space. The only commercial efforts that had been truly successful were communications satellites. By the mid-1980s, commercial space launch had been held at bay as NASA tried to maintain the field for its Space Shuttle. All was about to change.

* * *

A Defense Science Board (DSB) study in the summer of 1985 was an indicator of the changing tide. Ostensibly, the DSB examined how military systems' functional performance requirements were defined. The study was headed by Robert A. Fuhrman (chair, with Norm Augustine vice-chair), and the membership included former Under Secretary of Defense for Research and Engineering William J. Perry, future Secretary of Defense. They "noted significant difference in the development

Implosion: Lessons from National Security, High Reliability Spacecraft, Electronics, and the Forces Which Changed Them, First Edition. L. Parker Temple III.
© 2013 the Institute of Electrical and Electronics Engineers. Published 2013 by John Wiley & Sons, Inc.

process between commercial programs and military problem programs."[1] The report claimed, "some major lessons can be judiciously applied to military programs." They identified some perceived strengths of commercial programs from which the DoD could benefit.[2] The study group equated commercial practices with DoD streamlined practices, such as those used by the NRO and other classified programs where bureaucracy was severely limited and the program manager had the authority and accountability for success of the program. Among their ideas were: a single, powerful executive with authority to make unchallengeable decisions and settle disputes and a program manager with continuity, authority, flexibility, accountability, and direct access to the executive.[3]

The DSB's report held the germ of some important changes that were nearly ready to break upon the beach of military acquisition. The DSB identified some important ideas about management of systems acquisition. Their recommendations led directly to the President's 1986 Blue Ribbon Commission on Defense Management, with some of the major members of the DSB also participating in the latter.

Created by Executive Order 12526, dated 15 July 1985, and also known as the Packard Commission for its chairman, David Packard, the commission included William J. Perry as director of the Acquisition Task Force.[4] Packard had been a Deputy Secretary of Defense responsible for some forward thinking cost-controlling acquisition reforms in 1969–1970. Perry characterized the acquisition process as "hucksterism," which forced program managers to view cost, schedule, and performance optimistically, or be replaced. Clearly, such pressures resulted in understated cost and schedule, and overstated performance, for each new system.[5] Such an untenable situation had change.

Perry's task force was "to evaluate the defense acquisition system, to determine how it might be improved, and to recommend changes which can lead to the acquisition of military equipment with equal or greater performance but at lower cost and with less delay."[6] It is possible to see the role of the DSB, then, in a somewhat different light. It was a trial run for a number of its members' ideas, and resulting in a determination they were ready for the light of day. Thus, the President could charter a Blue Ribbon Commission with considerable confidence in the importance of the outcome.

---

[1] Final Report of the Defense Science Board 1985 Summer Study, "Practical Functional Performance Requirements," March 1986, DTIC accession number ADA170961.

[2] Ibid., 64. However, those streamlined programs were not devoid of the use of standards, and in some cases rigidly adhered to the parts standards.

[3] Ibid., 2.

[4] *A Formula for Action—A Report to the President on Defense Acquisition*, by the President's Blue Ribbon Commission on Defense Management, David Packard, Chairman, April 1986.

[5] Captain David A. Searle, *The Impact of the Packard Commission's Recommendations on Reducing Cost Overruns in Major Defense Acquisition Programs*, Master's thesis, Wright-Patterson Air Force Base, Ohio: Air Force Institute of Technology, Sept. 1997, pp. 28–29.

[6] *A Formula for Action*, 1.

The Blue Ribbon Panel report even mentioned that

[t]his leaves unanswered, however, what level of excellence can be achieved in defense programs. To answer this question, a landmark study was undertaken by the Defense Science Board (DSB) last summer. The DSB compared typical DoD development programs with successful programs from private industry.. . . These commercial programs clearly represent the models of excellence we are seeking, but it is not obvious that DoD, or any large bureaucratic organization, can follow successfully the management procedures used in industry.[7]

After all, it went on, "federal law governing acquisition has become steadily more complex, the acquisition system more bureaucratic, and acquisition management more encumbered and unproductive."[8] In essence, the acquisition culture and environment did not favor excellence. Open for question was the extent to which such a culture could change, and whether such a change could avoid irreparable harm.

While most of the report discussed streamlining reporting chains and authorities of the acquisition officials, this was not the exclusive topic.

First, the report recommended a new Level II Presidential appointee position in the Office of the Secretary of Defense as the Under Secretary of Defense (Acquisition). Each of the Services was to have a comparable senior official position filled by a Presidential appointee. These Service Acquisition Executives would appoint a number of Program Executive Officers (PEO) to act as general managers, overseeing specific programs whose program managers would report only to their specific PEO. These measures were to establish "short, unambiguous lines of authority."[9]

The Senior Procurement Executive position was created in 1983, when Congress amended the Office of Federal Procurement Policy Act to strengthen the federal government's procurement system. The amendments[10] specified the Senior Procurement Executive "shall be responsible for the management direction" of the agency's procurement system, "including implementation of the unique procurement policies, regulations, and standards of the executive agency."[11]

The lengthy qualification of high reliability parts, especially JAN Class S, drew the Commission's attention. Life testing requirements unique to military electronic parts and extensive environmental testing caused long delays from commercial release to acceptance. Such testing reduced the chance of early mortality and determined the expected lifetimes for the parts. Both were necessary considerations when designing long lifetime systems. As discussed here, these elements of high reliability had grown slowly, incrementally, for good reasons, and had proven their worth.

The Commission's laudable intent was to ensure the best and most capable parts were inserted into military systems as soon as possible to keep up with the

---

[7] Ibid., 11.

[8] Ibid., 16.

[9] Ibid., 17. The appointed positions and more were adopted as part of DoD Directive 5134.1 in 1987.

[10] 41 *U.S.C.* 414(3).

[11] Acquisition Reform: Military Departments' Response to the Reorganization Act, GAO/NSIAD89-70, U.S.GPO, June 1989, p. 13.

rapidly advancing high technology industry. Use of the most advanced technology in military systems was a hallmark of U.S. weapon systems acquisition. This had been so ever since the latter part of World War II and the Cold War only increased its importance. The Commission overlooked important differences between the military and commercial operating environments. From the start of the strategic missile and space programs, government acquisition agencies and electronics manufacturers had learned high reliability parts could not simply have reliability "tested in"; high reliability required a distinctly different manufacturing flow. This was overlooked, and sadly, the final report included false assertions, especially about the nature of commercial parts and their suitability in the extremes of space environments.

Another major thrust was to "Expand the Use of Commercial Products." On the face of it, this might sound reasonable. Instead of buying tissues or bandages to military specifications, their commercial equivalents would work just as well, and at a fraction of the cost. The problem was that the Commission did not stop at the point of doing a good thing. They went an extra step by infringing on elements of the overall acquisition system that were working well. "Products developed uniquely for military use and to military specifications generally cost substantially more than their commercial counterparts." The report held an important logic flaw especially relevant to the high reliability space qualified electronic parts industry. To make claims as it had done, the team assumed there were indeed commercial equivalents of the parts being developed for the military. For tissues and bandages, of course, there were. For satellites, of course, there were not. Lacking that distinction in the report meant the small niche occupied by high reliability electronics was going to get overlooked, or worse, lumped together with the larger commercial electronics industry. But not before the report went well beyond its own competence, and actually took on the subject of microelectronics.

"A case in point is the integrated circuit or microchip—an electronic device used pervasively in military equipment today. This year DoD will buy almost $2 billion worth of microchips, most of them manufactured to military specifications." The same way SCDs had become a target, when something had so much money put against it, it too became a target for economizing. "The unit cost of a military microchip typically is three to ten times that of its commercial counterpart. This is the result of the extensive testing and documentation DoD requires and of smaller production runs.[12] (DoD buys less than ten percent of the microchips made in the U.S.)" Here the report glossed over the important point: performance and reliability requirements for military systems were far different from their commercial counterparts.

---

[12] The assertion was only partially true. The stockpiling initiative was in part due to the problems of small production runs. Commercial parts, by the definitions in the standards, had to survive much more benign environments than military avionics grade, which itself was less stringent than the one for space parts. Commercial parts needed extensive testing because they were not necessarily adaptable to the environments in which the parts would be used, such as the extremes of heat and cold where military equipment must operate, which pale in the face of the greater extremes between satellites exposed to the full energy of the sun versus the times when they are in eclipse behind the Earth.

"Moreover, the process of procuring microchips made to military specifications involves substantial delay. As a consequence, military microchips typically lag a generation (three to five years) behind commercial microchips."

The Commission report made a great leap as it described the state of commercial electronics.

> *When military specifications for microchips were first established, they assured a high standard of quality and reliability that was worth a premium price. The need for quality and reliability in military equipment is as great as ever. In the last few years, however, industrial consumers of microchips have come to demand equivalent standards, and manufacturing processes and statistical methods of quality control have greatly improved. It is now possible for DoD program managers to buy the bulk of their microchips from commercial lines with adequate quality and reliability, and thus to get the latest technology at a substantially lower cost.*

There was a kernel of truth there. For *some* military avionics, commercial parts were adequate. But the Commission was not seeing the commercial trend. As the market grows, the unit price comes down, to be sure. But the next logical step beyond that is it becomes cheaper to dispose of and replace them than to repair. While that might work for consumer electronics, clearly it was not feasible in many military applications or in orbit. Furthermore, military aircraft systems routinely see the extremes of temperatures—within one flight—from −55°C to +125°C. Commercial electronics were not designed to do so and usually fare poorly under such conditions. Yet the Commission recommended "the DoD Supplement to the Federal Acquisition Regulations be changed to encourage streamlining military specifications themselves. Applying full military specifications, far from being ideal, can be wasteful." The Administration mindset was changing.

Packard also made broad recommendations on the use of military standards and specifications. They praised the Air Force's Electronic Systems Division (ESD) for revising procedures "to harmonize military specifications with the various commercially used specifications. For example, required military drawings for integrated circuits could incorporate a manufacturer's standard design specifications, test methods, and test programs." Military drawings were another name for Standard Military Drawings (SMDs), which were the basis of the "slash sheets" of MIL-M-38510. Parts made to SMDs were put on the Qualified Products List (QPL) as acceptable for use in military systems. The QPL contained the standard military parts, divided into the various parts classes, one of which was Class S. ESD and its subordinate RADC were unknowingly laying the foundation for an excuse to do away with the latter's existence, as we shall see.

Release of the Packard Commission report was a major event. It was accompanied by President Reagan's National Security Decision Directive 219, "Implementation of the Recommendations of the Blue Ribbon Commission on Defense Management," directing immediate implementation where no further legislation was required.[13]

---

[13] "Implementation of the Recommendations of the Blue Ribbon Commission on Defense Management," NSDD-219, White House, 1 April 1986; Searle, *The Impact of Packard*, 30.

*   *   *

The recommendations that did require legislation took an interesting, if circuitous, route, and some background is necessary.

Shortly after establishing AFSPACE, the Joint Chiefs of Staff established a unified command called United States Space Command. Both were signs of the times and part of a long-standing movement toward more joint (that is, multi-Service) actions. To take the steam out of inter-Service rivalry, Eisenhower had attempted two defense reorganizations in the 1950s. The second of these, in 1958, strengthened the office of the Defense Secretary, but had not completely solved his problems with the organization of the Joint Chiefs of Staff. U.S. law vested warf-ighting ability only in the unified and specified commands—the former comprising components of all the services' elements concerned with some region or function. Under U.S. Code Title 10, the Services were not to fight wars. The Services' com-ponents in the unified and specified commands would go from being minimal staffs with paper forces to being in control of actual forces in time of conflict. The Ser-vices' responsibilities were organizing, training, and equipping the forces to supply to the unified and specified commands. Because of these responsibilities, the Ser-vices would be the ones most concerned with the Packard recommendations, but this left out the actual operational users. Since the 1970s, it had been clear that future military actions demanded the closest possible cooperation among the Ser-vices, affecting their equipment and the way they trained. It affected their equipment because of the need to interoperate among the Services. It affected training because they had to train as they would fight—and fighting was going to be joint. Unified and specified commands' inputs on their equipping needs had to become more effective.

Few in Congress understood this better than Arizona Senator Barry M. Gold-water. Together with Representative Bill Nichols, he created the Defense Reorgani-zation Act of 1986, finally completing what Eisenhower could not. As part of the larger effort to improve "jointness" and affect organizing, training, and equipping, the Defense Reorganization Act also folded in some of the elements of the Packard recommendations that required legislation to be implemented.

Thus the 1986 Defense Reorganization Act packed a huge wallop. From two different but very powerful directions, it affected the key Service responsibility of equipping forces. The Services' equipping responsibility was implemented through buying activities, dealing with everything from tissues and bandages through equip-ment and weapons—in short, the acquisition activities.

Along with the beginning of significant changes in military standards, other activities revised the way in which military systems stated their requirements for the acquisition processes. The unified and specified combatant commanders' staffs tended to express their needs in terms of capabilities, not in the arcane language of systems acquisition requirements. But that fed the gradual change of the way requirements were described, and then addressed, in more qualitative rather than discrete quantitative terms. The reorganization also intended to strengthen civilian control over the acquisition process and reduce layering and duplication within the headquarters. Thus, the Air Force merged the Chief of Staff's acquisition office with their civilian Secretariat's counterpart and designated an assistant secretary as the head of this structure. For the time being, no problems were noted, but much more was to come.

More reform was on the way. On 8 April 1986, the Under Secretary of Defense for Research and Engineering, Donald A. Hicks, chartered the DSB to conduct another study on "The Use of Commercial Components in Military Equipment." Hicks asked specifically for "an examination of some past programs where commercial components could have been safely used but MILSPEC items were used instead. Include an estimate of cost savings that could have been realized." In so doing, he implicitly set his sights on reducing the use of military standards. To provide balance, however, he asked for an "estimate of 'down side' risk if commercial components had been used including evaluation of logistics issues such as proprietary rights." Finally, the DSB was asked for "identification of the impediments to the use of commercial components and recommendations for making it easier to use commercial components in military equipment if this is a wise course to pursue."[14]

Ignoring for the moment the biased questions aiming for a preconceived outcome, there was also the issue of his having overlooked national defense space systems. His only explicit example of "down side" risk had to do with logistics tails. In terrestrial systems, that translated to lower availability rates, or more down time for scheduled and unscheduled maintenance. For space systems, the equivalent was premature loss of a valuable national capability caused by unsuitable components. That was not in the scope of Hicks' direction.

Aside from standard contractual language, of course, stood the military standards themselves, which made use of commercial parts unlikely. Hicks' intent was clear—he wanted ratification of military standards costing too much so commercial parts could be substituted. There was no intent to undermine the high reliability parts industry, but to aim where most of the DoD parts expenditures were—in tanks, ships, trucks, and so on. These all came from the same acquisition system which procured spacecraft.

The 1986 DSB study, co-chaired by Dr. James R. Burnett and Dr. William J. Perry, included key players in subsequent reforms: Robert A. Fuhrman and Jacques Gansler. The goals of the study seemed reasonable. They were to develop an integrated set of military and industrial specifications, standards, and procedures to:

- "Lower cost and procurement cycle time of existing military semiconductor products.

- "Develop an effective process for selective use of commercial industrial grade semiconductor products in DoD systems to take advantage of lower cost, shorter design-to-production time, and new technology.

---

[14] Under Secretary of Defense for Research and Engineering Memorandum for Chairman, Defense Science Board, Subj: Defense Science Board Summer Study on the Use of Commercial Components in Military Equipment, 8 April 1986.

- "Ensure that modified specifications and procedures do not result in sacrifice of required quality and reliability."[15]

The problem here was the high reliability electronics standards and specifications had been honed for a very long time, based on hard-won lessons from failures and anomalies, deep insight into industry processes, and the failure modes of the different devices and technologies. The standards were very directive as a result, with as little tolerance for variation as possible in order to ensure the best chances of success. Modifying the standards and specifications along the lines of commercial alternatives simply made no real sense. Nevertheless, that was the tenor of the times.

The study soon created serious distress within the microelectronics industry by its preliminary recommendation eliminating electronic parts orders of precedence. In effect, military systems could use any kind of part in a system; none was preferred over another. The order of precedence for the highest quality parts first, followed by successively lower quality, was found in MIL-STD-454, "Standard General Requirements for Electronic Equipment." The DSB proposed "to abolish the order of precedence for electronic parts used in military applications as currently specified in Requirement 64 of MIL-STD-454 (JAN Class B parts first, then parts tested to MIL-STD-883C requirements, then parts described by military drawings)." One of the major manufacturers, National Semiconductor Corporation, found out about the upcoming DSB recommendation. They objected, along with Texas Instruments, Signetics, RCA, Intel, Raytheon, Sprague, NASA, and others. Instead of focusing on eliminating SCDs, an initiative that was then popular, they were on the verge of throwing the baby out with the bath water. At SD, McCartney received his staff's assessment of National Semiconductor's concern. His staff told him:

> *National's summation of the situation is fairly accurate—if MIL-STD-454 is modified to eliminate the current order of precedence, and specifies that all three types of parts are equal, there will be no incentive for a manufacturer to continue to produce JAN Class B parts. Our sources for JAN Class S parts will disappear at the same time.*[16]

The furor resulted in a letter from co-chairs Burnett and Perry to DSB Chairman Charles A. Fowler, on 15 December 1986. They told Fowler, "There have been numerous questions about the semi-conductor parts recommendations of the Summer Study on Commercialization. I will attempt . . . to state what the Panel had in mind when making their recommendations."[17] The letter includes some valuable (and rare) insights into what was in the minds of the key figures in a DSB study whose recommendations, while not immediately implemented, formed the basis for William J. Perry's eventual initiatives known as Acquisition Reform in the 1990s (about which more will be said).

---

[15] Ibid.

[16] HQ SD Acquisition Logistics Memo for LtGen McCartney, Subj: Your Note Regarding National Semiconductor's Letter Concerning Class S Parts, 7 Aug. 1986.

[17] S.J. Burnett and W.J. Perry, Letter to Charles A. Fowler, Chairman, Defense Science Board, 15 Dec. 1986, cited in Final Report of the Defense Science Board 1986 Summer Study, "Use of Commercial Components In Military Equipment," Jan. 1987, DTIC accession number ADA180338, p. 69.

They recounted for Fowler, "The DoD process for ensuring the reliability of semiconductor parts for use in DoD systems was started many years ago, long before the required disciplines in process control of microelectronic design and production were well understood. Also, at that time, the DoD used a large fraction of the world output and virtually controlled the market."[18] This overstated the case significantly, because not since the early 1960s had military procurement held a dominant position in the microelectronics marketplace, and it had for only a couple of years. The exception was high reliability parts, which were unmentioned. The DSB, like Packard before it, erroneously made no distinction between commercial electronics and the highest reliability parts.

> *The situation today [1986] is quite different. The DoD usage of the world market is approximately 10 percent; the commercial use of semiconductors has exploded. The tremendous commercial market forces of demand, competitive prices and quality have forced the semiconductor supplier to develop highly effective design/manufacturing processes and related process control systems which produce the high production yields required for business survival. These disciplined processes have resulted in dramatic improvements in the quality and reliability of commercial, especially industrial-grade, semiconductors. The industry recognized the higher production yields (which resulted in dramatic device cost reductions over the typical JAN part) tightly correlated with dramatic improvements in quality.*[19]

However, the typical JAN part was not a JAN Class S part, and not keeping that distinction in mind skewed the DSB arguments to cause outcomes they did not intend.

The co-chairs claimed they had achieved agreement with their team and industry on a number of fronts. The agreements included streamlining SMD and SCD processes and preparing an industrial IC specification and a quality/reliability reporting system. They said they had gotten partial agreement on the contentious issue of using plastic encapsulated integrated circuits.[20]

What was contentious about plastic encapsulated parts? The issue of plastic encapsulated parts deserves more explanation, because it was a microcosm of the issues separating military avionics parts and Class S parts.

Plastic encapsulated parts had little or no reliability data behind them at the time. Absence of data was the limiting factor in reaching any consensus on the suitability of plastic parts in military avionics. Furthermore, the use of plastic encapsulation in space had other problems. Two such problems were tin whisker growth leading to short circuits and outgassing of chemicals, which would degrade the devices in which they were embedded.[21] Tin finishes grow dendrites, referred to as

---

[18] Ibid.

[19] Ibid.

[20] Ibid., 71.

[21] To be clear, plastic encapsulation itself did not create the microscopic whiskers. Plastic was then, and has continued to be, used as a passivation technique to suppress the tendency of some underlying material (usually pure tin) to grow whiskers. Some plastic encapsulation techniques have been found to be so ineffective that whiskers will grow though the plastic. Without data, there was no ability to support exaggerated claims about plastic's effectiveness.

whiskers. These thin filaments could grow long enough to bridge the gap between different parts of an active circuit—creating a short circuit and causing device failure. One means of suppressing whisker growth was encapsulation with a material the whiskers would not penetrate. Encapsulating with plastic was unproven, and evidence pointed to plastics being ineffective in suppressing such whisker growth. Plastics also contained residual materials from their processing, and could, under high temperature or as a result of other environmental effects, produce damaging gases. RADC was investigating the reliability and maintainability implications of the devices Perry was pushing because of their promise of cost savings. The use of plastic encapsulated parts—a deeply technical issue—had become politicized in the sweeping drive to save costs, overlooking significant technical concerns about their use in space. Even eight years later, use of plastic encapsulated devices in high reliability systems remained unsupported due to insufficient data.[22] Though the extent of the claimed partial agreements was unstated, it is hard to imagine how a politically driven agreement could overcome problems of physics and metallurgy.

Their claim about the acceptability of plastic encapsulated parts indicates they were not really paying attention to space applications, though their discussion did not explicitly exclude space use. They had absolutely no concept of the impact of such parts in space applications. Instead, the DSB focused on whether plastic encapsulated parts could be ruggedized and used in "*selected* military applications."[23] In one of its less fine hours, the DSB was doing cost cutting without regard to the science involved. Even without a full appreciation of the impact, they at least understood there would be some applications, though unspecified, where plastic encapsulated parts were inappropriate.

Despite claims of agreement, full or partial, Perry and Burnett could not get manufacturers to agree to removal of the MIL-STD-454 Precedence of IC Selections. McCartney and his staff agreed with the manufacturers this would be very damaging to both his satellites and to the high reliability industry. When Burnett and Perry came to that, they paid lip service to McCartney's complaint.

> *The final area of disagreement centers on the recommendations to remove the MIL-STD-454 precedence for specifying JAN, MIL DRAWING and MIL-IC-883C parts. This concern is primarily based on the fear that this action would virtually eliminate U.S. manufacturing of JAN products creating (1) a foreign source dependency particularly of critical JAN Class "S" space application components, (2) a reduction in quality and reliability and (3) increased JAN part costs because of lower volume production. It is our belief that these issues have been adequately addressed.*[24]

Only the study team agreed the area had been adequately addressed. SD and the industry suppliers were not part of such an agreement.

---

[22] Rome Laboratory Command and Control Systems Fax, Subj: Back-up Data, Programmatic Impacts of SECDEF Recommendations, 21 April 1995, p. 3.

[23] Burnett and Perry, Letter to Charles A. Fowler, 71.

[24] Ibid., 72.

Furthermore, Burnett and Perry dismissed foreign dependency as a significant issue. They saw only the strength of the U.S. industry and were not considering foreign dependency as a rising threat:

> *With regard to the foreign dependency issue, we do not believe that the revision of MIL-STD-454 will "kill" the JAN program. It is obvious that JAN "B" certified facilities in the U.S. will decline over time as adoption of the DESC MIL DRAWING program takes effect. If no other steps were taken, this would put at risk an "off shore" dependency of another 20 percent of DoD semiconductor procurement (or another 2 percent of total U.S. procurement). Please recognize that 80 percent of DoD semiconductor procurement* currently *falls in the offshore dependency class. Another DSB study under the chairmanship of Norm Augustine has been addressing this issue in total. Our recommendations are consistent with the Augustine study preliminary recommendations. If these (Augustine) recommendations are adopted, we feel the dependency issues could be resolved and the savings and the program management prerogatives attendant to our recommendations could be realized. The "business issue" concern of certain semiconductor companies (e.g., more* business *will go off-shore) is a fundamental issue of U.S. competitiveness, and must be addressed in a much broader commercial/military context.*[25]

Somewhat defensively, they finally addressed the Class S issues McCartney had raised. But instead of recognizing the legitimacy and importance of these issues, their defense was dismissive:

> *Please note that our recommendations did not suggest any change in JAN Class "S," or space qualified, device procurement. Concerns that JAN Class "S" production facilities will not survive if Class "B" capacity declines do not take into consideration our total recommendation package. Our recommendations do provide for MIL device facilities that are specifically certified (under the same process disciplines as respective JAN lines), thus there should be appropriate capacity and facilities for quality, space qualified devices. The real issue is one of off-shore dependence for JAN Class "S" type parts and, as noted before, this would be resolved by the Augustine recommendations.*[26]

While it was true they did not specifically recommend a change to Class S production, they should have seen their recommendations affecting Class S, mainly because no exceptions to their recommendations were taken for Class S. They seemed to view the complaints as a special interest group's whining.

They next turned to an argument that would prove to have a much longer life than it deserved. They asserted that commercial (or non-JAN) parts were essentially equivalent in quality to JAN parts. To achieve this claim, they had to use some fascinating logic about quality and reliability across different classes of parts:

> *There is good quality data available for decision making which shows the quality of non-JAN military semiconductors from reputable suppliers is comparable to the*

---

[25] Ibid., 72.

[26] Ibid., 73.

quality of JAN parts. Gross data show the incoming quality of non-JAN ICs as measured by OEMs had higher failure rates than JAN devices. However, controlled studies by several OEMs showed that this was due to test correlation problems and handling damage. . . . The panel believes that there is a high correlation between excellent quality data and excellent reliability; however, we should emphasize that this belief requires further substantiation. There is enough quality/reliability data on JAN parts in the field to indicate that of the relatively old MIL part population, JAN parts show the highest reliability. The implied statement that MIL DRAWING parts and MIL-STD-883C parts from properly certified lines are, on average, less reliable than JAN parts is not known to be true.[27]

By making this argument, they fueled what became a long-standing controversy. The question was whether all the additional high reliability process controls and testing made any difference. Saying the commercial parts had the same quality as JAN parts had three real flaws.

The first was the use of statistics, because the larger production runs of commercial parts swamp the statistics of the smaller JAN lots. They were incorrectly using the statistics of large numbers, which serve insurance companies well, but in this case led to invalid apples-to-oranges comparisons incorrectly asserting the parts were equally reliable. A 90% yield rate in a huge production run was not the same thing as 90% yield in a single lot of JAN Class S parts.[28] In the larger case, whole lots might be useless and discarded without affecting the percentage yield. In the single lot case, the yield makes a huge difference—the 90% that are good have to be right on the first shot. The larger run can be dealt with using minimal testing from one lot to the next, if field returns and warranty exchanges are also considered. The latter needs a statistically valid sample or a part-by-part test to determine what the yield actually is. Comparisons of the large production run statistics to individual lot statistics were not mathematically valid, giving the false impression that quality of commercial parts was as good as that of the parts for which extra measures were taken to ensure the highest possible quality. The DSB did, to its credit, recognize that such comparisons required additional study, but that was a sop to Cerberus. The study results and executive summaries were the grist of the policy mill—not the technical details buried in the report.

The second flaw said quality implied reliability. As this history has shown, they are separate considerations. Quality cannot be tested in, but is built in because it derives from the materials and processes used in manufacturing (for the most part). Reliability is affected by design and other factors, but it depends on test or other statistics for its determination. Reliability can be estimated, as MIL-HDBK-217 showed. Actual reliability determination, though, depends on testing results. There

---

[27] Ibid., 73–74.

[28] The numerate reader familiar with sampling and population statistics will also recognize that large production runs approach a population's normal (Gaussian) distribution, but the small lot sampling is modeled with a Poisson distribution whose shape depends on the size of the sample, slowly approaching a normal distribution as sample sizes increase. The technical aspects of this important distinction are beyond the scope of this present history.

should be no question that quality and reliability are related. The likelihood that a highly reliable part will emerge from a base quality set of materials and processes is low, and the reverse is also true.[29] Blurring the distinction of quality and reliability, though, led to a foolish set of conclusions, which is the third flaw.

The third flaw allows nonspecialists to claim a commercial part will survive and operate within specifications under the same conditions expected of military parts. To understand this, consider the temperature ranges expected of commercial parts and military parts. The performance of solid state electronics piece parts depends on temperature—this is the basis for accelerated life testing (about which more will be said later). Occasionally, some commercial part types might be able to operate out of their design parameters (the commercial temperature range), and still perform close to specification limits across the military temperature range. The latter is a very wide range, from −55° C to +125° C, and is very challenging. Few consumers go from the depths of the Antarctic winter to environments more extreme than boiling water, but military avionics and some spacecraft do, sometimes several times in a day. Thus, commercial parts could be designed to lesser performance requirements and still satisfy the need of the vast majority of consumers. Equating reliability in consumer electronics with suitability to military environments was wrong-headed.

The DSB also attacked the argument that supported maintaining the order of precedence of parts. Their language seemed to compliment the JAN parts' reliability, but was actually doing the opposite. If they were of equal reliability (which was the case so far as the DSB said it knew), then JAN parts were simply expensive parts doing the same things as MIL Drawing and Mil-STD-883C parts. Their argument would rapidly have fallen apart if they had examined the equivalent to JAN Class S parts, but the DSB focus was elsewhere. They wanted reform.

Since JAN process lines and other MIL process lines must be certified to the same stringent standards (only the certifier *is different—DESC for JAN, the manufacturers themselves for other MIL) any data from suppliers which shows a significant JAN quality difference indicates they have not upheld their obligation to DOD for self certification and process control.*[30]

Using this argument, the DSB confused line certification with part quality. Line certification meant that a manufacturer had been approved to produce certain classes of parts. This was about the processes involved. These processes surely supported the quality argument, but in the end, the parts themselves were the objects that did or did not possess the desired quality. MIL parts and JAN parts were, therefore, different, despite the similarities of their process certification.

---

[29] When JAN Class S devices are not available, the usual practice is to buy the next lower quality parts and, though individual testing, "up-screen" the parts. This does not make the lower quality parts equivalent to the higher—it simply screens out the lower performing parts, resulting in parts that may work nearly as well as the desired Class S parts. Up-screening does not make the parts Class S—it simply identifies those parts that will perform acceptably in lieu of the Class S.

[30] Burnett and Perry, Letter to Charles A. Fowler, 74.

With disregard for the impact on the small high reliability market, as they had been warned by SD and the industry's manufacturers, Burnett and Perry said:

Compared to the overall cost savings achievable by removing the order of precedence in MIL-STD-454, any increased cost associated with a reduction in the number of JAN parts produced is not an overriding issue in our opinion. Furthermore, as the concept of process certification gains acceptance for all semiconductor devices including ASICs, the production of high quality—reliable parts at competitive prices will continue.[31]

They believed that if a manufacturer's processes were certified, then as soon as they developed a new part or new technology while using these processes, the new device would be satisfactory for inclusion into military systems.

Our original recommendation to remove MIL-STD-454 precedence remains valid; however, we believe the slight modifications are indicated:

JAN devices and MIL DRAWING devices should be made equal in precedence with MIL-STD-883C devices a slightly lower precedence. This will permit use of qualified devices (MIL-STD-883C) prior to their becoming registered in the MIL DRAWING program but would not permit proliferations of these non-standard devices.[32]

Thus, any qualified manufacturer could produce a part and never go so far as to register it, claiming it would be obsolete by the time that lengthy process was followed. With the move toward process certification only, McCartney's concerns became more likely. All the MIL STD requirements would not be of interest if the bottom line predominated and cheaper parts were bought. Their assertion quality would be maintained was a strained piece of logic unsupported by the reality of the system they were undermining.[33]

They also did not see that the changes they were promoting would undermine some important features of the military electronics marketplace. A move to commercial parts undermined RADC as the focal point of expertise for the entire industry on reliability. However, it was RADC's Reliability and Diagnostics Branch that worked "at the microcircuit level as the lead for the SECDEF's acquisition reform initiative with the goal of replacing most MIL-SPECs with best commercial practices. This group lead [*sic*] the DoD Qualified Manufacturers List efforts under this initiative."[34]

The DSB released its results at a time when the political atmosphere was right and change was the order of the day. Accordingly, the study proved to be very influential, both directly because of its immediate impacts, and also as the members went on to assume greater levels of responsibility within the DoD. The tide of weapon systems acquisition was turning.

---

[31] Ibid., 75.

[32] Ibid., 75–76.

[33] Clearly, though, Perry was convinced of the logic, so when it became his turn at the tiller, he would go forward with the changes that were not accepted from the DSB study.

[34] Rome Laboratory Command and Control Systems Fax, Subj: Back-up Data, Programmatic Impacts of SECDEF Recommendations, 21 April 1995, p. 2.

Whether the reformers considered concerns about high reliability electronics as special pleading (when they considered them at all) or whether they assumed that by not specifically including them in the changes they were exempting high reliability electronics is debatable. Less debatable, based on the foregoing, is they had only superficial appreciation for how the industry and markets were tied together. They assumed things about high reliability parts remaining untouched that later proved damaging and untrue.

The 1986 study revealed that the DSB considered that the commercial line processes were well understood, so the military standards were excessive and unneeded (an argument that worked for tissues and toothpaste, but not satellites). The standards brought not only process controls, but lot-to-lot quality control, interchangeability, and discipline, plus RADC's key role in the standards and in failure modes, reliability improvement, and advisors to industry and the government. The DSB was dealing with a problem out of its technical depth. Like the blind men of fable, its members knew only part of the elephant.

The real damage they did was to get a number of balls rolling that would snowball, reaching maximum impact seven to eight years later. Nor was the 1986 study the end of their influence.

Commercial electronic part costs, like water, were going to seek the lowest level. The initially strong position of the United States in microelectronics manufacturing had been eroded by Japan's entry, which had slowly whittled away at the United States since the mid-1970s. This presented a quandary for high reliability electronics users, since these users were essentially military and national security organizations who generally abhor overseas connections. One thing about commercial electronic piece parts, however, was their abundance. As the costs of military systems continued to rise (with a lingering focus on SCDs as a cost driver), the DSB was again asked to examine part of the maze of issues. The formed the Task Force on Defense Semiconductor Dependency to examine the dependency of military systems on domestic sources of supply.

The DSB 1986 Use of Commercial Components In Military Equipment summer study dismissed the threat of overseas manufacturers, believing instead in the robustness and technological lead of the United States. Many of their conclusions depended on the U.S. ability to ignore the threat of manufacturing going off-shore.

Yet, conducted at about the same time as that study but not published until February 1987, the new DSB Semiconductor Dependency study challenged this confidence. What it found was not comforting:

> The superiority of U.S. defense systems of all types is directly dependent upon superior electronics, a force multiplier which not only enhances the performance of the weapons themselves, but also maximizes the efficiency of their application upon which much of our defense strategy and capabilities are built. . . . [S]uperiority in the application of innovation no longer exists and the relative stature of our technology base in this area is steadily deteriorating. As evidenced by market share and the perception of the technical and financial communities, the United States' semiconductor device and related upstream industries, such as those which supply silicon materials or processing equipment, are losing the commercial and technical

leadership they have historically held in important aspects of process technology and manufacturing, as well as product design and innovation.[35]

An essential part of the Cold War calculus was U.S. technological superiority in the face of a numerically superior enemy. Anything that dulled the technological edge of U.S. forces, shifting the balance of power, was a serious concern.

The range of options for dealing with erosion of this domestic capability was limited. Senior defense and other agency officials struggled with the direction of the commercial electronics industry and how to provide incentives for its rejuvenation or retention.

While they struggled with that problem, the space acquisition community continued to deal with their need for access to the small volume, high reliability parts. Despite the national importance of these parts, whatever was decided for the high reliability, space qualified parts would hardly affect the larger forces at work determining the underlying economics of the commercial electronics parts industry.

The SPWG effort on stockpiling resulted in a forecast for the key national defense space programs, including the principal launch vehicles and upper stages: the ALMV, DMSP, Defense Satellite Communication System (DSCS), Global Positioning System, Inertial Upper Stage, Milstar, FLTSATCOM, Titan IV, and Centaur, and some NASA requirements.[36] With these projections in hand, SD was ready, by November 1986, to lay out their initial Supply Support Requests for the JAN Class S Operating Stock program.

The program felt short of initial expectations. The forecast included only 283 different kinds of parts (National Stock Numbers or NSNs), which was down from McCartney's anticipated 700+, but a significant number nonetheless. The highest use categories were fast/low power microcircuits, linear devices, diodes, and transistors. Microcircuits were almost twice as numerous (at 84 NSNs) as the next highest, which was diodes at 48. The value of the initial program came in at an estimated $34.7 million, well within DLA's $50 million budget.[37]

Although larger military production programs were accustomed to using parts stockpiling, the space programs had long since lost that familiarity. Almost immediately, billing problems arose due to the novelty of the new procurement procedures, but were quickly solved.[38]

They no sooner fixed that bump in the road when they hit another pothole in December 1986. Gen Slay's legacy was a few fixed price contracts still in use at SD. LtGen Aloysius G. Casey, the SD Commander, was told a quirk in the Federal

---

[35] Report of the Defense Science Board Task Force on Defense Semiconductor Dependency, Feb. 1987, DTIC accession number ADA178284. The study does not appear to have been correlated with the other 1986 DSB study.

[36] HQ SD Acquisition Logistics Memo for LtGen Casey, Subj: JAN Class S Operating Stock Forecast, 4 Nov. 1986. Eliminated from stockpiling were parts with fewer than three users or an estimated usage less than 200 parts, and forecasts from low probability users.

[37] Ibid.

[38] HQ SD Acquisition Logistics Memo for LtGen Casey, Subj: Request for Presentation of JAN Class S operating Stock Briefing to Deputy Assistant Secretary of Defense (Production Support) Mittano, 17 Oct. 1986.

Acquisition Regulations (FARs) did not permit operating stock use on fixed price contracts. To use the stockpile, those SD contracts required a deviation to the FARs. This was a concern because deviations to FARs were neither automatic nor quickly given and could cause delay in the programs. Moreover, the reason for the FAR clause was obscure. Casey found no one on his staff or at AFSC headquarters could explain it. So, they requested a class deviation to the FARs.[39]

Luckily, the billing and deviation problems proved relatively minor startup problems that simply had to be worked while the stockpiling effort moved ahead. Shortly, the list of total JAN Class S part types for stockpiling grew from the 283 to 384 due to new requests. Again, the program ran into an unanticipated problem, which proved to be a bit more serious. The number of requested part types was one thing, and the sources of supply quite another.

The first solicitation for participation ran in the Commerce Business Daily from 10 November 1986 to 19 December 1986. The solicitation covered only 164 of the 384 NSNs. They expected the first proposal being received on 2 January 1987, and the final one for that group by 23 February 1987. A second group of 165 NSNs was due out in February 1987. But 67 NSNs were found to have no qualified sources, so no procurement actions were in process.[40] The consequence, to the surprise of some, was that lower quality parts would have to be bought as individual needs for them arose, and these then would have to be handled through SCDs to compensate for their lack of space qualification. Curiously, the stockpiling effort, justified in part by a desire to do away with SCDs, provided an independent demonstration that SCDs were necessary for producing satellites of the highest attainable reliability.

*   *   *

Before the first delivery could be made to the JAN Class S operating stock, faint rumblings of other changes began from the most unexpected places. Soviet Premier Mikhail Gorbachev's first book, *Perestroika: New Thinking for Our Country and the World*, raised Western eyebrows. The book began with the obligatory homage to Lenin, and otherwise might have been met with a collective yawn, but Gorbachev then did something most unusual. He openly questioned how a country that had set so many records in space could not produce quality home appliances. Not since the "kitchen debate" between Khrushchev and Nixon in the 1950s had the world taken so much notice of Soviet home appliances. The fact of the contradiction was not news in the West; the news was that Gorbachev had put it in writing for the whole world to see.[41]

The world was about to change radically, bringing with it unanticipated consequences. Those consequences would have profound implications for national security and defense, and thereby on the high reliability electronics industry. Reforming acquisition was only one force affecting the electronics marketplace.

---

[39] HQ SD Acquisition Logistics Memo for LtGen Casey, Subj: JAN Class S Operating Stock, 22 Dec. 1986.

[40] Ibid.

[41] Brian Crozier, *The Rise and Fall of the Soviet Empire* (Rocklin, California: Forum, 1999), 405.

* * *

Consequences of the Packard Commission and the DSBs continued to roll on. All had made claims about the adequacy of commercial parts that had held kernels of truth, but not complete truth. Instead, they were actually seriously misleading. Military avionics systems were not representative of the highest reliability uses, such as satellites. Packard and the DSBs were looking at what *was*, not what *was becoming* true. So, they were not only looking at only one part of the overall electronics issues without regard to national priority, incorrectly using that part to stand for the whole, they were also not taking into account the ongoing changes rendering their conclusions invalid.

Qualification of highly complex devices such as hybrids was changing, and not cosmetically. The situation with hybrids had simply become untenable. The changes meant assertions that commercial devices met most military needs were completely wrong for the highest reliability applications.

The government hybrid qualification specifications crossed four appendices of MIL-M-38510G. JEDEC Standard 1 never gained acceptance as a military standard, so government and industry were somewhat at odds on how to proceed. This state of affairs ended when RADC began a new effort to develop a hybrid microcircuit standard. RADC and NASA collaborated with JEDEC JC13.5's Task Group 75, created specifically to work the hybrid issues, in 1987. Starting with JEDEC Standard 1, they embarked on what became a two-year effort to produce a specification that would deal with the practical realities of the rate of technological change and the needs for qualification.

As that work got underway, the first parts were destined to arrive for the Ogden stockpile in early April.[42] The routine began at DESC. DESC ensured all shipping containers included prominent JAN Class S marking, and all parts were in electrostatic discharge-protected packages. They took randomly selected test samples from every lot, and performed destructive physical analysis tests or contracted for such tests at appropriate testing facilities. Once DESC was satisfied, the parts were shipped to DLA's Ogden facility for storage in a suitable environmentally controlled facility. Anyone wishing to draw from the stockpile had to deal with the Air Force's controlling Air Logistics Center. For most space programs, that would turn out to be in Sacramento.

A memorandum of understanding (MOU) covered each organization's responsibilities. The MOU was signed by the end of March, prior to receipt of the first parts. DLA was to "purchase, stock, issue, and replenish NSNs based on projected program needs." SD was to "estimate the needs of its programs, as well as contractually authorize and encourage its contractors to use the parts centrally managed under NSNs when they are available." The parts were not considered government-furnished material because contractors were to be directly billed for the parts.[43] DESC was

---

[42] HQ SD Acquisition Logistics Memo for LtGen Casey, Subj: First Delivery to the JAN Class S Operating Stock, 6 March 1987.

[43] This distinction was important, having to do with liability in case of problems. For Government Furnished Equipment, the contracting government agency (SD for national defense space) was responsible. Since SD was never involved in purchasing or handling, they did not want to assume that liability.

"responsible for item management and quality assurance. This includes the responsibility to develop sources; compute requirements; process original equipment manufacturer (OEM) requisitions; and procure, test, store, and issue JAN Class S parts." DLA Depot Ogden (DDOU) at Logan AFB, Utah, was the place where the parts were to be stored, and from which they would be shipped to the contractors. DDOU had good controls in place, and was not an especially busy depot.[44]

DDOU was to store everything in a separate storage area, and ensure electrostatic discharge handling procedures were in effect at all times. They were to remove one part from each date code pack (lot number) and visually inspect it, matching its particulars to the bill of lading. Although they were to avoid handling parts whenever possible, the visual inspection requirement left a part from each lot date code without any further disposition, so these were the first parts selected for testing.

Aside from what these handling procedures tell about the care needed in handling space qualified parts, they also illustrate a source of the cost of dealing with such parts. These handling procedures had to be in place to protect the parts. The comparatively small number of parts had to be free from handling accidents because there were so few replacements, and in many cases the time to obtain replacements would be lengthy. Commercial parts needed to be handled similarly, but with abundant replacements and lower cost, the consequences of handling problems would be less severe. If all parties to the stockpiling arrangement complied, SD's Acquisition Logistics division expected government savings of "34 percent over current parts procurement procedures."[45]

Although the parts had already arrived, an event was planned for 6 May 1987 to bring together all the officials of the various involved agencies. Such events take longer to plan than it took for the parts to arrive. The parts were expected only on 29 April, and actually got there 24 April, but the celebration date was padded to make sure it was not premature. Present at DESC for the event was Assistant Secretary of Defense for Acquisition and Logistics, Dr. Robert B. Costello. In addition, Lieutenant General Vincent M. Russo, Defense Logistics Agency (DLA) Commander, Rear Admiral James E. Eckelberger, DLA Director of Supply, and Brigadier General Richard H. Huckaby, DESC Commander, attended.

In May 1987, the operating stock program hit its spending limit. As of 7 May 1987, DESC processed purchase requests for $35.7 million, with more on the way.[46] Combined buying power for all programs showed the manufacturers for once that space programs made a bottom-line difference. For instance, contracts had been let with Texas Instruments and Signetics for 37 part types and 26,920 parts, worth $1,208,153.04.[47]

---

[44] Memorandum of Understanding Between SD of the Air Force Systems Command and Sacramento Air Logistics Center of the Air Force Logistics Command and the Defense Logistics Agency, 31 March 1987; author interview with David M. Peters, 19 Sept. 2008.

[45] HQ SD, Progress and Problems Report, 16 June 1987.

[46] HQ SD Acquisition Logistics letter, Subj: Staff Historical Reports, 1 April–30 Sept., 15 Oct. 1987.

[47] HQ SD Acquisition Logistics, Staff Summary Sheet, Subj: JAN Class S Operating Stock Status, 30 June 1987.

On 17 June 1987, SD received the deviation to the FARs allowing use of operating stock parts on fixed price contracts.

Almost immediately, though, another startup problem occurred, and this was related to the required procedures. In June 1987, the DCAS inspector, without authority, waived the pre-seal visual inspection of 67 lots of JAN Class S microcircuits produced by National Semiconductor Corporation at Tucson. All devices sealed inside some kind of package required a visual inspection for quality-related concerns such as ensuring the integrity of any connections and freedom from debris. Once assured everything was as it should be, the device would be hermetically sealed, after which any attempt to get back inside would destroy the device. Lack of the pre-seal inspection created serious questions about whether the 67 lots could still be considered JAN Class S. The DCAS inspector was removed with a disciplinary action, and other DCAS inspectors had to be trained on MIL-M-38510, with an additional DCAS inspector being added in Phoenix. Among the many other resolution actions, numerous additional inspections of the lots had to be performed to convince everyone the parts were suitable as JAN Class S. This underscores just how seriously the program was taken. It also shows the importance of in-plant representatives, who were key elements of the JAN Class S parts program. They supported the Government Source Inspection occurring in their plant before the parts were shipped. The parts were then checked again upon receipt to make sure nothing had happened in transit, as covered by the MOU.[48]

* * *

By this time, space systems finally gained acceptance by the operational military, leading to reliance on their advanced capabilities, while weapon systems acquisition came under increasingly pointed scrutiny by reformers bent on fixing acquisition. The problem with the reformers was that they painted with a very broad brush. The momentum for change accelerated, such that anything being done before reform was clearly wrong-headed and needed to be made more efficient and effective and less costly.

Change was in the wind. Unrelated winds of change blew at all levels, from the manufacturing and qualification of parts, to acquisition of parts, to top level policy studies, all intent on some kind of change. While all participants would have agreed the status quo was inadequate, the kinds of changes they espoused were hardly compatible. Normally, changes were simply evolutions of prior efforts, intended to refine and improve some process, such as the need for a new hybrid qualification standard. Other changes responded to the accelerating electronics marketplace and the manufacturing sector. Some reformers called for fundamental, revolutionary changes to military systems acquisition, requiring revolution at the highest levels of policy and legislation. Thus, in a sense, the battle lines were forming up for the inevitable clash of these different forces for change.

---

[48] HQ SD Acquisition Logistics, Staff Summary Sheet, Subj: Failure to Perform Government Source Inspection on JAN Class S Parts, 29 June 1987.

Chapter 16

Implementing Change in a Changing World

In 1987, nonlegislative implementation of the Packard Commission and the President's Directive proceeded slowly. Each of the Services adopted a slightly different arrangement of organization and reporting for its Service Acquisition Executive, with subordinate Program Executive Officers and Program Directors. This three-tier chain of command was how streamlined acquisition had been finally understood and implemented, excluding intermediate management levels to prevent their getting a bite at the apple before the real issues reached the Service Acquisition Executive.

The Acquisition Executive was "designated by the Secretary of Defense or a Service secretary to oversee major acquisition programs . . . and those otherwise designated by the Service secretary as systems or programs of high national and departmental priority and interest." Reporting to the one and only Acquisition Executive was the Program Executive Officer who was "responsible for implementing the guidance and direction of the Air Force Acquisition Executive on Executive Programs." Everything then fell to one of the Program Directors as "the single individual responsible for executing the program." Such was the theory. The net effect of implementing Packard was to vastly reduce the reporting chain for program managers, linking them as quickly and directly as possible to a senior Air Force or, occasionally, DoD official.

Like the other Services, the Air Force's implementation, called Acquisition Streamlining, was an imperfect reflection of what the Commission had wanted, due to practical considerations. Acquisition Streamlining was the first major reform for AFSC since Slay's earlier emphasis on early planning and fixed-price contracting. Gen Randolph, by then AFSC Commander, passed down direction to his center commanders to develop Acquisition Streamlining Plans.

That was the easy part. Things were more difficult in the Air Force Secretariat. Within the Air Force, Dr. Thomas C. Cooper, Assistant Secretary of the Air Force for Acquisition, became the Acquisition Executive—a clear, logical, and obvious choice. But that meant he was also the Acquisition Executive for *all* Air Force

Implosion: Lessons from National Security, High Reliability Spacecraft, Electronics, and the Forces Which Changed Them, First Edition. L. Parker Temple III.
© 2013 the Institute of Electrical and Electronics Engineers. Published 2013 by John Wiley & Sons, Inc.

acquisition programs, including the military space programs. However, since the start of the Reagan Administration, one person was in charge of both military space and national security space, though the latter's existence was publicly unacknowledged and highly classified. That person was Edward C. Aldridge, the Undersecretary of the Air Force. Within the NRO, the acquisition chain was already very streamlined and required no changes. The same could not be said for the military space program.

Events from 1986 through 1987 greatly complicated the management of the two major space programs. The 1986 loss of Challenger completely revised a decade and a half of national policies predicated on the Space Shuttle program's success and eventual replacement of ELVs. The nation was fortunate to have both national security and national defense space reporting to Aldridge. Had national security and national defense space acquisition and operations not been united under him, Aldridge could not have been as effective in directing and accelerating the recovery from Challenger's loss. Aldridge, in part due to his steadfast determination to gain some kind of ELV backup for the Shuttle in case of a failure, came right down to the wire to achieve this. As later revealed by several ELV manufacturers, the United States was within weeks or months of losing the ability to rapidly reconstitute its launch vehicle production lines.[1] His actions averted a delay of four or more years before ELVs would return to operation. Aldridge was justifiably acclaimed for his efforts, and upon the retirement of Air Force Secretary Verne Orr, Aldridge was promoted to fill that position. As others before him, he chose to retain his responsibilities for both space programs. So, while the Secretariat had Cooper in charge of all acquisitions, he was not really in charge. Cooper was Aldridge's subordinate, but had no involvement in space.

The in-depth changes to military reporting chains began in September 1987. Final agreement on how things would work was reached in November. The problem was that the single chain of reporting would bypass the AFSC headquarters staff, whose responsibilities included putting together the annual Air Force acquisition budget. Cutting the staff out of the reporting chain reduced the staff's insights into the programs and needed budget details. Such contentious issues were not readily or easily solved.[2]

As can be imagined, there was no universal happiness at the changes. LtGen Casey levied the implementation on his program offices, keeping faith with the goals of reform. "Acquisition streamlining can reduce the costs of our systems and compress our acquisition schedules. It behooves us all to get into the spirit of the effort."[3]

Streamlining ended up with a compromise. For LtGen Casey and his staff, the challenge was to maintain a consistent acquisition approach for implementation of military standards and use of the operating stock. Not all space programs fell into

---

[1] The reconstitution would have been delayed because the manufacturers were not able to sustain the manufacturing workforce, losing talent and skills, but also they were about to scrap production tools and equipment and allocate factory floor space to other activities.

[2] HQ SD, Command Plans Staff Summary, Subj: Acquisition Streamlining Plan, 18 Nov. 1987.

[3] HQ SD Commander letter, Subj: Acquisition Streamlining Plan, 24 Nov. 1987.

the categories to report to the Acquisition Executives. For those that did, abbreviated direct reporting applied to six Defense Acquisition Executive Programs: ALMV, DSP, DMSP, DSCS, GPS, and Inertial Upper Stage (IUS). For two Air Force Acquisition Executive Programs (Milstar and boosters), the SD Commander was the Program Executive Officer. The chain of command had been split and abridged, which would lead to inevitable confusion without considerable management effort.

This was the first crack in an edifice whose integrity depended on consistent, enterprise-wide efforts in such things as applying military standards. Top level reforms made it harder to be consistent at the lowest levels where the parts came together. While trying to solve a high level policy issue, the lowest levels were simply not considered because they did not seem to be affected.

*   *   *

Reforming acquisition and realigning reporting chains had little to do with the imperatives driving the national security and national defense space programs. Having emerged from the effort to convince the military of space programs' dependability, a new challenge emerged. As command of the Air Force shifted from former strategic bomber pilots to former tactical fighter pilots in the early 1980s, space again needed to prove its relevance. The new Air Force leadership had been mid-career fighter pilots in Southeast Asia, where most pilots knew little or nothing about satellites. While some knew communications satellites had a military role and most were aware of the importance of the DMSP's weather data, the national defense and national security space programs seemed generally foreign, incoherent, lacking unifying themes or relevance to the tactical warfighter.

Along with the organizational changes creating the Air Force Space Command was an important planning document, known as the Air Force Space Plan. Despite its name, there had been very few such plans, and none for a very long time before the Air Staff's new version in 1984. Though lacking agreement with the rest of the Air Staff, the Space Plan included four topics that it considered space missions. To the tactically oriented Air Staff members, space was still a place, not a mission. Nevertheless, these proposed space missions brought a vital military coherence to the national use of space.

Setting aside the question of their terminology as Air Force missions, the four descriptors replaced the disparate and unrevealing program names with a readily understood set of categories. Air Force leaders got the big picture quickly. The four descriptors were space support, force enhancement, space control, and force application. Together, these were a clear and understandable set of bins into which all current and projected national security and national defense space programs fit.

Space support grouped together all activities necessary to provide ground stations, command and control, space launch, and global tracking for satellites of all types, as well as the rest of the elements of infrastructure that made everything work.

Force enhancement, as the name implies, included the set of space capabilities that improve the performance of U.S. and allied forces. Thus, weather, early warning, reconnaissance, communications, and GPS navigation satellites fell into this category. The majority of the program names known to outsiders were contained in force enhancement. Since force enhancement was only one of four, part of the message was that satellites were not the only things relevant to military operations.

Space control activities ensured the U.S. space programs could do their assigned tasks in space without interference. It also contained the means for the U.S. to deny space capabilities to an adversary—namely, the ALMV anti-satellite. Although the ALMV was distinctly unpopular with Congress, it was an essential element of overall military planning, countering the long-standing Soviet anti-satellite. It made another strong point about the relevance of space activities to the average warfighter. Any adversary was likely to have capabilities to enhance their own forces, so possession of the means to deny that adversary use of their force enhancements appealed to the warfighters.

Force application was typified by the President's SDI, and comprised the means to influence the outcome of conflicts on the Earth. SDI was also the most complex arrangement ever of interlinked space systems dedicated to accomplishing a single objective. In the parlance of later decades, it was the first true "system of systems." Without the satellite robustness gains of the 1970s and the advances in high reliability electronics, even the conceptual arrangement of SDI would have been difficult.

The recovery from Challenger's destruction afforded a unique opportunity to make these topics something more than an interesting set of discussion points. The 1982 DoD Space Policy, built on the premise of the Space Shuttle's capabilities becoming operational, had to be rewritten to address the radically changed future of the nation's space programs. Aldridge, recognizing the opportunity for the DoD to structure the future, asked Secretary of Defense Casper Weinberger for the charter to bring the old space policy up to date. The new version, as a DoD document, used the four topics as an organizing principle, and referred to them as the four space mission areas. Though resisted by the Services, the Secretary of Defense was in a position to impose his views. Weinberger agreed the four former topics were indeed the best way to think of national security and national defense space missions. Publication of the 1987 DoD Space Policy was delayed because Weinberger did not want to beat the President to publication of a revised policy. However, the draft (though in reality final) DoD policy was given to the National Security Council and the President's National Security Advisor for Space. When Reagan's next space policy was published, announcing the future path for the major U.S. space sectors and programs, the four missions became national policy, missions, and organizing principles.[4]

It took about four years to overcome the pause in top Air Force leadership support. The new national policy gave the greenest light possible to going beyond space reliance concerns. Space capabilities were to be incorporated into the American way of war, and there was no obvious way to pull back from that. It was national policy. And national policy's *sine qua non* was high reliability electronics.

---

[4] L. Parker Temple, *Shades of Gray: National Security and the Evolution of Space Reconnaissance* (Reston, Virginia: American Institute of Aeronautics and Astronautics, 2005), 549; "Department of Defense Space Policy," Office of the Secretary of Defense, Washington, D.C., 10 March 1987; "United States Space Launch Strategy," National Security Decision Directive 24, White House, Washington, D.C., 16 Jan. 1987.

*   *   *

Despite the efforts to help him manage his space programs, Casey's concerns inside SD did not change much. When presented with a draft policy for implementing the operating stock program in December 1987, his main concern remained the use of the operating stock. Despite anticipation of having over 300 part types and over 600,000 piece parts within the year, its success would come only when contractors used it. All known obstacles had been cleared away, yet there had been no rush to the depot. Approving the policy, Casey pointedly told his Acquisition Logistics director, "Remember, I'd like to hear when we get a hit!"[5]

Casey directed all his program managers to use the operating stock:

> *All future contracts shall include the appropriate provisions for use of the JAN Class S Operating Stock inventory. . . . All RFPs currently released, including proposal under negotiation, shall be amended to include the necessary provisions into the contract. . . . All existing contracts shall be assessed to determine if it is advantageous to modify these contracts. . . . Each program office shall prepare a list of all active spacecraft and launch vehicle contracts and RFPs and identify those which will be modified.*[6]

Surely that would be enough to make the operating stock a success.

Casey did not fully realize buying Class S parts from the operating stock was a very different way of doing business. The glitch of needing a deviation to the FARs indicated some amount of incomplete anticipation of how things were changing. The ripples of change did not appear to be anything more than minor disturbances requiring a slight change to the status quo. The operating stock required a new (actually a very old) delivery system known as the Military Standard Requisitioning and Issue Procedures (MILSTRIP). MILSTRIP was slow, and had been out of use in the space programs for so long, that none of the participants knew their DoD Activity Code (DODAC) number. Their DODAC number had been unused by many of them for a couple of decades. SD found a retired person who could identify the DODAC numbers over the telephone, solving that part of the problem rather quickly.[7] But the problem is revealing about the underlying difficulties of changing acquisition approaches, even in something as simple as the operating stockpile. There were many unintended and unanticipated consequences, and factors impinging on its acceptance.

Some less-than-obvious reasons slowed the movement to the operating stock. From the standpoint of the government acquirers, MILSTRIP was old and few in the military space program wanted to use it. Some would later claim, with justification, that use of the stockpile reduced or eliminated a few of the nicer perquisites of the parts business, such as the needs for fact finding trips, inspections, and so on. Yet, the Commander's Policy mandating use of JAN Class S was having an effect.

---

[5] HQ SD Acquisition Logistics, Staff Summary Sheet, Subj: Policy for Implementing the JAN Class S Operating Stock, 16 Dec. 1987.

[6] HQ SD Letter., Subj: Policy for Implementing the JAN Class S Operating Stock Program, 21 Dec. 1987.

[7] Author interview with David M. Peters, 19 Sept. 2008.

As evidence, the unit price of the parts was coming down and approaching the volume price of the JAN Class S Operating Stock. Also, buying JAN Class S from other than the operating stock still included all the fact finding trips, inspections, and so on.[8] But there was more to it than that.

* * *

As changes to technology, policy, acquisition, and organization continued ever more rapidly, the world situation also emerged with far greater complexity and uncertainty. In February 1988, anti-government demonstrations took place in Yerevan, capital of the Soviet Republic of Armenia. Similar demonstrations had occurred in Soviet Bloc countries before, but this time the characteristic brutal Soviet suppression was absent.[9] Some form of Soviet response was anticipated, so nerves were frayed as each day went by and no response was evident. If this kind of instability became widespread, what would that mean for global security? The world was changing, indeed, and as the momentum of change increased, eventually the effects would be felt in the national security and defense areas. If the Soviet Union was becoming a less dangerous adversary, then the need for national security and defense built up since the end of World War II would have an unpredictable alteration. What would happen in the electronics industry if the need for high reliability electronic parts radically altered?

* * *

Casey's requested first "hit" remained unsatisfied for months. Then, on 11 May 1988, "forty-five Signetics 54LS153 electronic parts were delivered from the JAN Class S Operating Stock inventory to the Jet Propulsion Laboratory for installation into the Magellan spacecraft. . . . [T]he Defense Depot Ogden Utah is currently processing requisitions for two other part types which will be used on the NASA Topological Experiment Program (TOPEX)."[10] That was *not* what Casey wanted. Unsatisfied, he complained, "I'm looking for a 'Hit' by *Space Div Contractors*! For our H/W."[11]

Nevertheless, this was spun into good news by the Commander's Action Group, which hastened to point out "using the operating stock saved the Magellan program approximately $6850 by lowering per-unit costs and avoiding a minimum buy requirement of 100 pieces. . . . Finally, if purchased directly from a manufacturer, JPL would have had an additional six months wait until delivery."[12] Avoiding delay was actually the more significant saving, eliminating a six month period of paying

---

[8] Ibid. It is not entirely clear what caused the lower prices of JAN Class S parts. The manufacturers may have been influenced by the increased buys for the operating stock, leading to a temporary decrease in prices. The decrease might have been a pricing strategy to influence direct buys from the manufacturers.

[9] Brian Crozier, *The Rise and Fall of the Soviet Empire* (Rocklin, California: Forum, 1999), 413.

[10] HQ SD Acquisition Logistics, Staff Summary Sheet, Subj: First "Hit"—Initial Usage of JAN Class S Operating Stock Parts, 19 May 1988.

[11] Ibid. Emphasis in the original.

[12] Ibid.

Table 16.1.   SD Program Parts Usage, May 1988. Cost effectiveness of high-reliability parts briefing, 12 May 1988. Prepared by Aerospace for the Ad Hoc Committee on Systems Costs

	Semiconductors (Micro Circuits / Diodes / Transistors / Hybrids)			Non-Semiconductors (Resistors / Capacitors / Magnetics / Relays)		
	MIL-SPEC %	SCD %	CLASS S %	MIL-SPEC %	SCD %	CLASS S %
DSP	5	85	10	65	32	3
DMSP	35	35	30	0	60	40
DSCS III	9	90	1	30	70	0
IUS	0	100	0	0	100	0
FLTSAT(last build)	2	93	5	20	80	0
GOS	2	95	3	40	59	1
MILSTA	3	80	17	3	92	5
TITAN IV	10	85	5	10	85	5
COMMERCIAL	1	94	5	10	85	5
MEAN	7.4	84.0	8.6	19.8	73.7	6.5

Source: HQ SD Acquisition Logistics, Staff Summary Sheet, Subj: Usage of Class S Parts on SD Programs, 23 June 1988.

for the standing army of engineers and technicians to be idle or at a very low productivity level.

Casey was unsatisfied. He wanted to find out what kinds of parts the SD programs were using if they were not "hitting" the Operating Stock. In May 1988, he ordered a survey of the Aerospace parts managers working on SD programs (that is, only military space programs, not national security ones). He got his answer a month later.

The Aerospace parts managers' data came from the manufacturers' actual parts lists. In Table 16.1, the "Mil-Spec" column represents the purchase of military grade, B-Level parts. The "SCD" column includes parts purchased to SCD or MIL-STD-1547 requirements. Colonel Jon F. Reynolds emphasized the SCD column included some Class S parts to which additional requirements were added, so, in his words:

> Class S parts usage is really greater than the chart indicates. . . . For example, the MILSTAR contractors buy Class S microcircuits, but have additional radiation requirements. [NOTE: Thus, they would have been counted as SCDs, not Class S parts.] Also, fewer Class S manufacturers and part types were available when the programs listed were purchasing parts. Today however, contractors are using more Class S parts as a result of increased availability. For example, the IUS program office recently determined approximately 25 percent of their semiconductor and integrated circuit device requirements are currently listed on the JAN Class S Operating Stock status and availability list.

Which was to say, when IUS bought its parts they would either have to buy a few parts as SCDs or they would have to buy a full lot of Class S parts, and it was more expensive to go the latter route. Had the operating stock been available, they could have met their needs through the new stockpiling program.[13]

The information is more subtle even than Col Reynolds' explanation. Each of the programs was at a different stage of development or production, which meant that as a snapshot in time, it represented only amounts actually on hand. A snapshot in time provides no appreciable trending information to determine if the amount of JAN Class S was increasing. Furthermore, the amount of Class S included in the SCD column was unknown and suspected of being variable across the programs, but thought to be a significant number of parts. Parts suppliers' data, though, indicated the volume of JAN Class S was actually increasing. That increase was obscured by the chart's lack of trending, and burial of many class S parts in the SCD column. This presented another problem for space programs, because of the earlier push from the Office of the Secretary of Defense to decrease the use of costly SCDs. Here was data showing some of the most complex (and expensive) space program parts (microcircuits and hybrids) were mainly SCD parts. The data, if taken without understanding, seemed to undermine the usefulness of military standards, and emphasized the opportunity to save a significant amount on military space programs by forcing use of commercial parts.

* * *

Another problem of concern was making sure American military systems had the very latest technology available to them. Technology was accelerating, which made keeping up a greater problem.[14] Countering numerically superior Soviet forces should the Cold War ever heat up depended on advanced technology. The Reagan Administration brought an even greater drive for advanced weaponry. One major technological factor the politicians began to recognize was the increasing scale of integration. Put another way, the feature size of integrated circuit devices was decreasing, allowing more functions on a single chip. Moore's Law was on track. MSI in the early 1980s rapidly became LSI. LSI enabled devices popularly known as single chip microprocessors or microchips. By 1988, Very Large Scale Integration (VLSI) was pushing the hypothetical upper boundaries of scale. Even laypeople were routinely discussing the clock speed and scale of integration in their personal

---

[13] HQ SD Acquisition Logistics, Staff Summary Sheet, Subj: Usage of Class S Parts on SD Programs, 23 June 1988.

[14] Author interview with Thomas Turflinger, former parts engineer with Navy CRANE, July–Aug. 2010. A few years earlier, semiconductor manufacturing equipment was relatively inexpensive, and many government laboratories and universities had their own fabrication facilities. As the industry got seriously into CMOS and Moore's Law held true, the cost of fabrication facilities skyrocketed. How to recover the cost was a major issue of the day as the cost of fabrication facilities reached into the billions of dollars each. This was the key time for the military desire to incorporate the latest technology and its desire to keep to the tired and true military parts system. The funding was no longer available even within an order of magnitude of what was required to do both, and industry found it harder to serve the military needs.

computers. The rapidity of change soon outstripped adjectival designators based on the scale of integration. Shortly, physical feature size (measured in microns or millionths of a meter) became the scale of integration measure.[15]

In the early to mid-1980s, ASIC designers had several choices for implementing their designs, but usually defaulted to a specific manufacturer's tools. This situation created an early problem for designers, of course, because lack of standardization either locked them into a specific manufacturer or reduced their efficiency by having to learn new design tools.

The real breakthrough in custom integrated circuits depended on the explosion of VLSI device production. Continuing on the path of Moore's Law, with ever increasing circuit density and consequently reduced cost per transistor, the VLSI revolution included two paths to greater capability. Instead of using larger general purpose standard integrated circuits, custom applications could produce smaller, more specific, integrated circuits. This required the development of computerized IC design tools and a new generation of circuit designers who worked independently of the fabrication facilities. Alternatively, standard integrated circuits could be combined with specific custom logic functions, often referred to as "glue logic" for their role in putting together the whole assemblage into a custom integrated circuit. Fusing the standard integrated circuits and custom design elements provided a reliability boost by eliminating external connections. The logic synthesis tools and other advances helped make components of integrated circuits more standard. VLSI devices began exceeding the million transistors per device threshold, coupled with the onset of a broad range of ASIC design tools, complexity, and the rapid pace of the industry's change, all of which created a serious problem for the military standards. The QPL specifications simply could not keep pace with the design, fabrication, and testing requirements changes.[16]

Despite MIL-HDBK-339 guidelines, qualification of IC devices defied the routine approaches for qualification bound up in the other space-related military standards.

As the commercial industry moved forward with VLSI, the tri-service VHSIC program continued. After nearly a decade, VHSIC devices had become quite complex and large, so the VHSIC program took on the additional challenge to develop advanced design tools. In December 1988, the VHSIC Hardware Description Language was adopted as an IEEE international commercial standard.[17] VHSIC successfully found its way into some communications satellites, such as FLTSATCOM and

---

[15] *The NASA ASIC Guide: Assuring ASICS for Space, Draft 0.6* (Pasadena, California: Jet Propulsion Laboratory under grant from Office of Safety and Mission Assurance, National Aeronautics and Space Administration, 1993). After VLSI came short-lived terms such as ULSI (ultra LSI), WSI (wafer-SI), and SySI (system-SI), before the more obvious movement to simply describing the relevant feature size in microns and (more recently) nanometers.

[16] *Ibid.* Also of note, the advancement to one million transistors fit nicely into Gordon Moore's 1965 projections known as Moore's Law.

[17] *Very High Speed Integrated Circuits—VHSIC—Final Program Report 1980–1990*, Office of the Under Secretary of Defense for Acquisition, Deputy Director, Defense Research and Engineering for Research and Advanced Technology, 30 Sept. 1990, p. 9.

Milstar, and other space applications such as the Generic VHSIC Spaceborne Computer and the Advanced Spacecraft Computer Module. Another important effort, run by RADC, was the Radiation Hard 32-Bit Processor begun in January 1988 for SDI. Its advantages were improvements in reliability and capability with reductions in size, weight, and power.[18]

Altogether, as VHSIC was approaching its end, it was a positive experience for the electronics industry. VHSIC had led to the use of multiple layers of metal in advanced semiconductors, which had become routine by the end of the 1980s. The periodical *Electronics* claimed "most experts agree that without VHSIC, semiconductor development in the U.S. wouldn't have progressed so quickly toward submicron geometries, even in the commercial world."[19] Once again, this underscored the military's investments' influence on advancing electronics technologies. Though military and national security space applications still took an insignificant share of the total electronics market, their advanced high reliability niche requirements made them most influential, because many of their advances were needed for future commercial advances.

These advances, reflected in the VLSI technologies, brought some new problems for SD programs. First, suppliers of VLSI integrated circuits were retooling rapidly at enormous costs, and the older (larger feature) integrated circuits were becoming scarcer. These older parts, however, were the ones that either had proven space flight heritage or had had the necessary testing done as qualification for space flight. Such qualification was a difficult and expensive thing to do, involving increasingly sophisticated test equipment and time. Manufacturing and testing equipment unique to military lines could not amortize its cost over millions of parts, and thus drove up the costs of space qualified parts, especially as most test equipment became obsolete as the feature size decreased in a couple of years. Thus, the second point was becoming a serious consideration: when and how should new technologies be qualified for space?

MajGen Robert Rankine, SD Vice Commander, asked in June 1988 about qualification of the new VLSI devices. Rankine had had extensive experience working with the Air Force laboratories in new technology work. He wanted to keep up with the newest device technologies, which held important features for use in space *if* they would be qualified. VLSI represented the knee in the curve of high reliability standards keeping up with space technology qualification in technology-specific detail. The approach that evolved in the 1960s and had worked so well since the 1970s needed to change. This "traditional" approach was no longer able to keep pace with the rate of advances in electronic devices. No longer guiding the industry, it was seeing its last gasp in VHSIC. The electronics industry was innovating fully on its own. The flea was about to awaken to the realization the territory it commanded was really only the tip of the dog's tail. That realization did not obviate the military standards process responsibility to establish space qualification rules for new VLSI devices which would themselves be rapidly surpassed.

---

[18] Ibid., 162, 164, 165, 168, 169.

[19] "What Did We Get from VHSIC," *Electronics*, June 1989, p. 97.

Col Reynolds responded to this challenge by confessing MIL-HDBK-339 was out of date and too general in nature.[20] He knew this was due to the state of knowledge at the time of the handbook's writing. As soon as they set pen to paper, they were out of date. He went on to say there was another problem:

> *In addition, we realize that JAN Class S specifications are not adequate to cover many of the VLSI requirements of our newer programs, and that an alternative approach must be developed. Rather than modify the existing military specification for microcircuits (MIL-M-38510) to accommodate VLSI, we feel that it would be more appropriate to develop new specifications for this purpose. MIL-M-38510 should be maintained essentially in its present form, since, in addition to VLSI, we project space programs will continue to use these less complex devices for some years to come.[21]*

Once a part was qualified for space use, a strong cost motivation existed to continue using that kind of part. But that strategy only worked as long as manufacturers kept such lines open or held stockpiles. The rate of change, and limited cleanroom floor space, did not favor keeping obsolete lines open for very long. Producers faced recurring consideration of obsolescence issues.

Clearly, the electronics technology rate of change created the obsolescence problem. Programs had to confront the use of legacy equipment in ongoing production lines versus the drive of new spacecraft programs to convert to a new, but not yet space qualified, technology.

Maintenance and update of the underlying military standards, at least as they had been developed to that point, were having a hard time keeping up. Reynolds continued:

> *It will require significant development work over the next few years, along with a revised approach to qualification, to develop Class S specifications for VLSI, and to keep them current. Further, to fully serve our program offices, this new approach must facilitate the qualification of foundry processes and custom VLSI devices, in addition to standard monolithic VLSI.[22]*

Reynolds was, in effect, supporting some of the 1986 DSB conclusions regarding use of commercial components. That controversial study had also espoused the idea of doing away with the MIL-STD-454 order of precedence. The DSB contended military standards were not able to keep up with the rapid rate of change in the electronics industry. Thus, standards kept the very latest technology out of U.S. weapon systems, which they considered bad. But they did not really examine the alternative. Instead of forcing the most advanced (and therefore generally unqualified) technologies into space programs, capability could have been used as the deciding criterion. That is, if an older qualified part would work, the program should keep using it so long as supplies existed. This idea actually fit very well with the stockpiling effort, but apparently was never seriously considered.

---

[20] In fact, MIL-HDBK-339 was out of date when written due to the rate of change.

[21] HQ SD Acquisition Logistics, Staff Summary Sheet, Subj: "Class S" Specifications for Very Large Scale Integrated (VLSI) Circuits, 24 June 1988.

[22] Ibid.

The DSB's solution was to do away with the preference for military standard parts and, in the case of integrated circuits in particular, to shift the procurement process to "certifying design and process versus parts." Reynolds was leaning in the same direction as far as process qualification was concerned. In essence, both were saying the one element of production that remained constant in making these unique or short run parts was the underlying manufacturing process itself. Rather than increase the cost of the part through mandatory, possibly inappropriate or out-of-date testing procedures, the manufacturing process was the key. Once satisfied the manufacturing process would yield the kind of quality desired, the process would be qualified, and all parts made under that process would be accepted as qualified parts (subject to subsequent checks, screens, and reliability testing, of course).

Much of this was no different from the previous changes driven by new technologies. Reynolds was saying, in effect, RADC, the Aerospace Corporation, and the manufacturers had their work cut out for them. That was business as normal. However, in addition to some parts qualifications, the system of building high reliability spacecraft would have to incorporate a means of qualifying manufacturing processes.

Reynolds also saw the acceleration of decreasing feature sizes, and this impacted space qualification in new ways.

> *This new approach would require a significant change in emphasis in the way we qualify devices. First, some traditional Class S tests are not technically feasible at higher levels of integration, and alternatives must be found. Second, because of the increased complexity, more emphasis must be placed in areas, such as design methodology verification, test structures, testability analysis and simulation. . . . We have begun to develop short and long range objectives in these areas. However, the era of putting entire subsystems on single chips has greatly expanded the overall systems responsibilities of our parts engineers; current engineers are not sufficient to pursue the identified objectives.*[23]

That was as far as Reynolds and his staff could carry their thinking. They needed to get the experts involved.

> *We need to form a VLSI Qualification Working Group with program office, contractor, lab and staff participation, to develop a comprehensive VLSI Qualification Plan, with adequate funding provided to support this plan. We feel that such a working group would efficiently use limited resources by allowing staff and SPO personnel to resolve program problems and policy problems in parallel, essentially "killing two birds with one stone."*[24]

Clearly some new approach was needed to address the rate of change issues with VLSI's acceleration. Despite having only a partial suggestion on the table for the path ahead, Rankine was not one to admire a problem for too long, and responded, "Get on with it," immediately approving additional resources to form the group and get moving with the plan's development.

---

[23] Ibid.
[24] Ibid.

VLSI qualification accentuated some important ideas. The most important, perhaps, was that current processes were no longer viable. Trying to keep the slow, bureaucratic process of highly prescriptive military standards apace with the rapid acceleration described in Moore's law simply was past its time. It was the space age equivalent of what Thomas Malthus observed about food production, and the inevitable consequence that the world's population would starve. He had not seen that linear growth in food production would be overtaken by new methods that multiplied output. High reliability spacecraft were starving, so parts qualification had to be innovated to multiply output. But Malthus was not really wrong; he simply did not see far enough ahead. A new qualification approach was critically important, but could not solve the problems for all time. For the time being, though, the new approach had to be found.

SD sought alternative ways to keep up, using surrogate measures to control or assess quality and reliability. They were confronting the physics of failure of materials with feature sizes that determine a number of failure modes.

If tantalum, for example, is used in a capacitor, then materials science dictates what kinds of things will happen to it. Tantalum is vulnerable to moisture, so any such device must be kept dry and isolated from humidity. As another example, device sizes are important for several reasons, but two of the more prominent ones are dendrite growth and radiation effects. Most metals used in electronics, when present in pure form, will grow dendrites. As mentioned earlier, tin is especially prone to whiskers that can reach across voids between parts of a circuit, creating a short circuit. The result may be a dead device. Another possibility, with sufficient energy in a confined space, is that the vaporized whisker then forms a conductive plasma that can create other forms of damage, sometimes explosively. Smaller feature sizes are more susceptible to whiskers because short circuits can be created by shorter whiskers. Radiation effects, beyond the scope of this history, have similar issues at smaller feature sizes. Voids and displacement damage due to energetic particles hitting the materials can create a variety of different failure mechanisms, some temporary and some permanent. As device feature sizes decrease, the amount of energy needed to cause these undesirable effects decreases. Devices with smaller features can therefore be harder to qualify for space use.

These feature size issues were only then becoming clear. Lessons being learned with VLSI devices pointed the way to qualification of the very much smaller device sizes produced in the early twenty-first century. In twenty-first century terms, when feature sizes have moved below 45 nanometers, VLSI's 1 micrometer dimensions were immense, but VLSI highlighted some trends that were already under way that would change aspects of design and testing for future smaller feature sizes.

VLSI was the level of integration at which the first true challenge to the military standards system occurred. Changing the qualification approach in military standards was no small challenge, but there was more to it. The test equipment at prime contractors, the government laboratories, and the Aerospace Corporation needed to change. Feature sizes were getting smaller than the detection levels of the contemporary test equipment. Defects at the smaller feature sizes were just as devastating to an integrated circuit as at the larger feature sizes, but at the time were no longer detectable. The ability to test was thus another key element of keeping up with technology. The ability to test was going to have to undergo at least as fast a pace of change to stay ahead of the reduction of feature size and materials changes. That meant further cost increases in systems using advanced technologies, because such advanced testing equipment became exponentially more expensive.

SD's VLSI working group relied on the Aerospace Corporation's experts and laboratories, industry's (then) extensive in-house expertise, as well as RADC to address the issues. But despite this wealth of resources, the tide was changing in the electronics industry. An analogy might serve to explain better. Anyone who has ever been on a beach where there is an undertow will recognize there is a time when the wave that most recently crashed upon the beach has portions that are still surging up the beach while the underlying receding part of the wave is already dragging away from the beach. In 1988, the high reliability, space qualified parts industry still had momentum going up the beach—the high water mark had not yet been reached. But there was an undertow slowing momentum and threatening to reverse it. That undertow was caused by the acceleration of the electronics industry, but not that alone. The forces at work were far more complex. The extent of the complexity was not yet appreciated.

As mentioned earlier, to keep pace, the government and the spacecraft industry had to continue investing in ever more expensive and elaborate testing facilities. They also had to attract and retain a high level of expertise independent from the electronics manufacturers. Military standards were beginning to fall behind because the bureaucracy needed to promulgate updates and additions. A burgeoning electronics industry meant the cadre keeping track of the industry also had to expand its expertise and levels of effort. There would be an inevitable Malthusian point of diminishing returns, unless something changed.

*   *   *

When, in December 1988, LtGen Donald L. Cromer replaced LtGen Casey, the Class S stockpile usage concerns remained unresolved. By then, SD was no longer the only concerned agency. At a management review in early December, Colonel Cecil R. Hopping, who had replaced Reynolds as the Acquisition Logistics director, was hopeful regarding the Operating Stock. He told Cromer he had discussed "the growing concern within the Defense Logistics Agency, the funding source for the JAN Class S Operating Stock Program, over the lack of inventory sales to Space Division (SD) Contractors. We presented information showing the status of activities on individual SD programs, and expect within the next few months SD contractors will be ordering stock."[25]

Hopping suggested Cromer sign a letter to DLA reaffirming SD commitment to the Operating Stock program, and another to the SD program offices reiterating SD's implementation policy and the importance of the program. Cromer signed out

---

[25] HQ SD Acquisition Logistics, Staff Summary Sheet, Subj: JAN Class S Operating Stock Program, 12 Dec. 1988.

only the latter. The policy, "JAN Class S Operating Stock Program," 16 December 1988, said in part:

> It continues to be SD policy that all satellite and launch vehicle systems acquisition contracts, awarded as of Jan 88, shall include the appropriate provisions for use of the JAN Class S inventory. Additionally, existing contracts which have not completed the parts acquisition phase and require use of JAN Class S parts shall be modified to include the provisions.[26]

Only time would tell.

*   *   *

Attempts to shave the rough edges off hybrids to make them conform to MIL-M-38510 could not continue. In the twelve years since the release of Appendix G and test method 5008 in 1977, MIL-M-38510 had been changed 35 times, and MIL-STD-883 13 times.[27] Just within the previous two years, the changes to last member of the hybrid triad, MIL-STD-1772, had been so extensive its revision B did not include marginal change bars.[28] To be clear, not all the changes related to hybrids, but many did, and the need for a more appropriate approach had been recognized and addressed. The JEDEC JC13.5 Task Group 75's work finally resulted in the release of the new MIL-H-38534, "General Specification for Hybrid Microcircuits," on 31 March 1989.[29] MIL-H-38534 established the general requirements for hybrid microcircuits and for them.[30]

Hybrids had not been qualified under QPL for some time since they tended to be unique, low production volume parts. Hybrid devices were inherently custom devices, and generally of very small lot sizes, especially for space programs. Treating such devices as either integrated circuits or diodes or capacitors drove up individual part costs for the wrong reasons. Now, with the new approach for

---

[26] Ibid.

[27] William J. Spears, "MIL-STD-1772 and the Future of Military Hybrids," *IEEE Transactions on Components, Hybrids, and Manufacturing Technology* 12, no. 4 (Dec. 1989): 507.

[28] *Certification Requirements for Hybrid Microcircuit Facilities and Lines*, MIL-STD-1772B, 22 August 1990, 6. Section 6.2, "Changes from Previous Issue," said, "Marginal notations are not used in this revision to identify changes with respect to the previous issue due to the extensiveness of the changes." The previous issue, MIL-STD-1772A, had only been issued 15 May 1987.

[29] Richard K. Dinsmore and John P. Farrell, "MIL-H-38534: Military Specification for Hybrid Microcircuits," *IEEE Proceedings*, 0569-5503/89/0699, p. 699. Dinsmore and Farrell actually reported that the standard was produced in Jan. 1989, and while it is possible that was when editorial changes stopped, the actual release date was 31 March 1989. Also, the sequence of MIL-H-38534's publication before MIL-I-38535's release in Dec. 1989 allows insight into the approval process. While hybrids are more complex devices than integrated circuits (combining as they do integrated circuits with discrete devices), one might be tempted to think this would take more time to gain approval. However, the approval process requires considerable industry input and adjudication of their comments, and there were many more integrated circuit manufacturers than hybrid manufacturers. The coordination of the final product was simply easier because there were fewer manufacturers involved.

[30] *Military Specification: Hybrid Microcircuits, General Specification for MIL-H-38534* (Rome Air Development Center, Griffiss AFB, New York, 31 March 1989).

hybrids, *a manufacturing line itself* would be qualified. MIL-H-38534's emphasis was on the manufacturing processes necessary to produce qualified devices without getting into the specifics of the devices themselves. Approval of manufacturing processes placed the producer on the Qualified Manufacturers List (QML). QML represented a clever and necessary means to address the issues stemming from low volume, high quality, high reliability, advancing technology, and affordability. QML was revolutionary compared to the QPL process, and a partial recognition of the maturity of processes in the semiconductor industry.

A deeper look into the QPL and QML processes should help to clarify exactly how qualification was changing. The differences were more significant than either a listing of qualified parts or manufacturers. Understanding the steps to gain entry to either list shows that the differences were not cosmetic.

QPL qualification comprised four steps before a part could be listed. The process began with a government agency developing a specification. Next, the manufacturer then made some parts complying with the slash sheet to demonstrate the part could be made as specified. This manufacturer had to use the specific manufacturing line from which any future approved parts were to come. The third step involved extensive characterization testing of the parts to demonstrate they were indeed made as specified, but also to gain the necessary characterization data to support designers' needs. This was when the device's actual behavior became understood, including the ways and circumstances in which it might fail. The final step was end-of-line testing to detect manufacturing deviations that could have produced bad or off-nominal parts. Depending on the grade and type of the part, different tests might be done. The tests came from MIL-STD-883, "Test Methods and Procedures for Microelectronics," or for microcircuits, MIL-STD-977, "Test Methods and Procedures for Microcircuit Line Certification." For space qualified parts, the most stringent testing requirements were imposed, including testing each production lot for conformance to the specifications.[31]

Upon satisfactory completion of all this, the specific part type and its grade were entered on the QPL. Adding another grade for that part type or adding a different part type required a separate qualification. This assured very high quality and deep insight into the different parts listed on the QPL, but compliance was expensive for manufacturers, especially for a very low production volume individual part.

By contrast, the new QML listed no parts, cataloging instead those manufacturers that had gone through a two-part procedure governed by MIL-STD-1772. First came a certification step, followed by qualification.

The first step required DESC certification of a manufacturer's facilities and assembly line. In the case of hybrids, which were the first devices used to define the process, line certification focused on a manufacturer's satisfaction of MIL-H-38534 and MIL-STD-883. Certification against these standards ensured all documentation, training, and quality controls conformed. The changes allowed for element

---

[31] *The NASA ASIC Guide: Assuring ASICS for Space, Draft 0.6* (Pasadena, California: Jet Propulsion Laboratory under grant from Office of Safety and Mission Assurance, National Aeronautics and Space Administration, 1993).

evaluation. The combination of integrated circuits and unique design elements such as magnetics, coils, discrete devices, and so on had forced hybrids to have a mix of different kinds of tests for these individual elements. Testing needed be done in line (as opposed to the end of the production line). The in-line element evaluation testing served as a "go–no go" point for the completion of the hybrid.

After subsequent DESC review and approval of the manufacturer's purchasing, assembly, and testing documentation, DESC performed an extensive audit to examine machine calibration, appropriate levels of cleanliness, proper training of operators, and documentation of all these. If any major discrepancy were noted, the manufacturer (if it still wanted to proceed) had to correct these, and another audit was scheduled. Line certification was hardly for the faint of heart, because this was only an interim step, known as "Certification to MIL-STD-1772."

Upon achieving line certification, DESC authorized proceeding with the next step. The manufacturer had six months in which to assemble and test a production lot of hybrids according to the procedures and processes in their documentation. These hybrid devices had to pass detailed inspection (involving non-destructive and destructive testing). Successful production of one or more lots of qualification samples was completion of the second step. Upon DESC approval, the manufacturer's processes had achieved "Qualification to MIL-STD-1772," and were able to advertise that considerable accomplishment. They had become a QML manufacturer.[32]

This clever approach to qualification recognized the realities of hybrid production and was a radical departure from the QPL end-of-the-line inspection of parts. The production lot requirement made it expensive, but this was simply an entry fee for the QML. Once qualified, *any* parts made on the line were qualified, so long as the manufacturer maintained the processes, documentation, machinery, and staff according to what DESC had approved.[33] The manufacturer would generate a "slash sheet," so called because of the numbering system used, for each part type manufactured under the process. Only by following the requirements of the new MIL-H-38534 could a device be marked with a CH as a Compliant Hybrid microcircuit.[34]

MIL-H-38534 covered manufacturing options for a variety of high reliability customers, only one of which was for space use.[35] By 1989, of the twenty-one qualified manufacturers, only two had gone so far as to receive the most technically stringent space qualification.[36] Considerably refined statistical process controls

[32] Spears, "MIL-STD-1772,"509.

[33] Dinsmore and Farrell, "MIL-H-38534," 700.

[34] Ibid., 699. In the words of MIL-H-38534, "Only hybrid microcircuits which are inspected for and meet all the requirements of this specification and the associated acquisition documents shall be marked as compliant and delivered."

[35] Ibid., 700. Dinsmore and Farrell actually reported that the standard was produced in Jan. 1989, and while it is possible that was when editorial changes stopped, the actual release date was 31 March 1989.

[36] Spears, "MIL-STD-1772," 509; *The NASA ASIC Guide*. To avoid being misleading, the hybrid devices were classified as Class V, which was the hybrids' equivalent to JAN Class S for discrete parts. For the purposes of this history, I have not made such a distinction that would further muddy the waters for some. Technically astute readers will understand the distinction.

(compared to the original release of MIL-STD-1772), meant MIL-H-38534 took a huge step toward a process-oriented specification to redress the complexities and cost of producing such devices under the same terms as QPL devices.[37] Its release stands out as the final realization by the government and industry that hybrid devices were distinctly different from ASICs and integrated circuits.

Thus, MIL-H-38534 stood as the gateway to the QML. Then, in December 1989, a similar approach for integrated circuits was included when the companion specification, MIL-I-38535, "General Specification for Integrated Circuits," was released.[38] The most challenging military standards had evolved in a valiant attempt to keep pace with technology.

*   *   *

At the opposite end of the spectrum of activity, the Packard Commission's changes were slow to be implemented, but that was going to change. In February 1989, President George H.W. Bush issued National Security Review (NSR)-11, directing Secretary of Defense Richard Cheney to fully implement the Packard Commission's recommendations and the Goldwater-Nichols reforms. In addition, NSR-11 asked to "improve substantially the performance of the defense acquisition system; and manage more effectively the Department of Defense and our defense resources." Addressing a joint session of Congress, Bush explained Cheney would develop "a plan to improve the defense procurement process and management of the Pentagon." Cheney's Defense Management Review would examine acquisition practices and procedures, and recommend "actions the Congress could take which would contribute to the more effective operation and management" of DoD.[39]

The Defense Management Review reported its results to the President in June 1989. Cheney's Review emphasized the need for a single, clearly defined chain of responsibility for systems acquisition, a central tenet since the Packard Commission. The Review did not rubber stamp Packard, however. Cheney's group found implementing the Packard Commission recommendations made some things better and others worse (not an uncommon problem when implementing any complex policy). Bush approved Cheney's report, and issued it in July.[40]

The result was a significant reorganization of the acquisition functions within the Services, with few as important as those affecting national security and national defense space programs. The Air Force Secretariat created a new office accentuating its newfound prominence in space in the wake of Challenger. Internal discussions suggested the Under Secretary of the Air Force should head the officially designated space office. This would formalize the precedent for the combination of the classified

[37] Dinsmore and Farrell, "MIL-H-38534," 700–701; Spears, "MIL-STD-1772," 511.

[38] Joe Barrows, "QML Specification To Become Performance-Based," *EEE Links* 1, no. 2 (April 1995).

[39] Richard Cheney, *Defense Management, A Report to the President by Secretary of Defense Dick Cheney,* July 1989, pp. 1–2.

[40] Major Daniel C. Brink, "Acquisition Reform Why? What? Is It Working?" Research Paper, Air Command and Staff College, Montgomery, Alabama, March 1997, p. 13.

NRO and the military space programs that had been set in the 1960s. Air Force Secretary Donald B. Rice rejected that idea. His Under Secretary would have no divided loyalties reporting to the Secretary of Defense and the Director of Central Intelligence for the NRO, and to him for military space responsibilities. Whatever happened below the level of the Under Secretary did not concern him. He insisted any space official would be no more than an Assistant Secretary, which became a new Assistant Secretary of the Air Force for Space.

Such a change was not a problem in itself. Although the Air Force Under Secretary had usually lead both the national security and national defense space programs, this was not always the case. When Alexander Flax was the Assistant Secretary (Research and Development), he had both responsibilities. Also, when both Hans Mark and Edward Aldridge had been promoted to become Secretaries of the Air Force, both had retained their dual space responsibilities. All who had served in these dual capacities recognized their increased effectiveness by being able to control both.

However, Cheney's recommendations led to a separate Assistant Secretary of the Air Force for Acquisition. That created a scope and responsibility problem for the new Assistant Secretary (Space). Acquisition streamlining stripped the office of some of its most important responsibilities. Although as the NRO Director the Assistant Secretary (Space) retained his original national security acquisition authorities, the office no longer had authority over the military space program. Aldridge's broad-ranging effectiveness in the Challenger recovery demonstrated the importance of having these positions combined, but that was lost in the reform.

Although space was officially prominent in the role of the new assistant secretary, the reality was a demotion of its influence, effectiveness, and its importance. Splitting space responsibilities meant a less effective rebuttal to changes adversely affecting space programs. This impact would be more obvious in the face of more severe impacts within a few years as Acquisition Reform got underway in the next Administration.

In a curious way, though, creation of the Assistant Secretary (Space) was a backhanded compliment to those who had worked so long to make space "operational." Although the national reconnaissance programs remained highly secret, much had been done in the past decade to make their information available to warfighters. The military space program remained critical to every phase of warfare from early warning through precision weapons delivery. A largely unanticipated consequence of making space operational was that the use of space had become routine.

Thinking of space use as routine could only have happened through great effort aimed at bolstering satellite robustness, reliability, and removing objections about their fragility as "science projects." Space was no longer special from the warfighters' perspective. Loss of awe about satellites, though, led to a betrayal of the special nature of space systems: they fundamentally relied on the maturity of high reliability parts and processes developed to a high art over the prior 30 years. Demotion of their influential position made space programs vulnerable to changes on the horizon. The leader of national defense and national security space programs would be less able and effective in resisting the adverse effects coming in the next administration.

Efficiency was becoming the watchword of the day, as the Air Force restructured its laboratory system. Thirteen Air Force laboratories were reorganized into four "super laboratories," one of which was Rome Laboratories, emerging out of RADC. Taking effect in December 1990, the change was more than a cosmetic shuffling of reporting chains. The AFSC reorganization, including the laboratories, and subsequent inclusion of Air Force Logistics Command as part of the new Air Force Materiel Command, achieved its goals by an estimated reduction of 20,000 jobs. Also, the goals of improving efficiency and reducing duplication may seem, at first blush, unrelated to the importance of Rome Laboratories in the electronics industry.[41]

The laboratory reorganization took place while the Air Force reorganized its acquisition system. AFSC would no longer be the Air Force's advocate for new weapon systems. Reflecting the shift toward operational advocacy (evidenced by Air Force Space Command's formation several years earlier), the new approach had a subtle and ultimately near fatal effect on support for the laboratories. Goldwater-Nichols emphasized the importance of warfighters getting operational capabilities to meet their needs. For missiles and aircraft, the linkage was clear and unequivocal. The laboratories and their products, though, were something else altogether. No operational command took advocacy responsibility for the work done in the laboratories. Support for the laboratories' work, unless explicitly tied to a recognized weapon system development, became a tougher "sell" in the annual budgets. After all, which warfighting commander would recognize an operational need for VHSIC? Even less obvious to many of them was their need for a military specification for hybrid microcircuits. Whereas long-time acquisition officials might readily recognize the importance of the laboratories' work in development and maintenance of standards, such work now moved one step further from perceived relevance. The further from relevance, the less warfighter advocacy, and without such advocacy, the more vulnerable laboratories and their critical expertise became.

Deep bureaucracies usually move slowly, so the Cheney Review streamlining recommendations took a while to get folded into top level DoD directives and instructions. Additionally, a General Accounting Office (GAO) report concluded "streamlined" acquisition reporting might not be so streamlined. The report revealed the authors of acquisition streamlining had not checked deeply enough into other sources of reporting requirements. The GAO "found that nearly all of its regulations, policies and directives are connected to some law. Consequently, a program manager's regulatory burden may not be significantly eased under this program."[42] Thus, considerable work had to go into the DoD's 5000 series directives and instructions,

---

[41] *Shortchanging the Future: Air Force Research and Development Demands Investment*, Air Force Association Special Report of the Science and Technology Committee, (Arlington, Virginia: Air Force Association, Jan. 2000), 9.

[42] "Acquisition Reform, Implementing Defense Management Review Initiatives," General Accounting Office, NSIAD-91-269, Aug. 1991, p. 11.

which were changed and finally reissued in February 1991. This demonstration of the reformers' superficial understanding of the consequences of their reforms should have served notice to subsequent generations of reformers.

*   *   *

At the lowest levels of acquisition, high reliability piece parts processes were evolving and reoptimizing to address the realities of rapid technology change. High reliability spacecraft acquisition programs faced the internal issues of cost control, delivery of systems on time, and responsiveness to warfighter needs while external demands changed acquisition relationships.

Technology change accelerating for nearly two decades finally outstripped the ability of tightly controlled processes to keep up. Had that been the only kind of change, the situation would have been manageable. However, acquisition streamlining made program managers and intermediate acquisition management respond to rapid changes of a very different kind. Instead of arriving at a stable end condition, though, additional, even more extensive and intrusive changes were on the horizon.

For the moment, that part of the story has to wait, because in the interim, the whole world changed.

Part IV

Shorting to Ground (1989–2002)

Change became politicized in the late 1980s to such an extent that in the early 1990s, change was the objective. The mandate to change overlooked and ignored critical, factual, contrary evidence.

Going through the motions and appearing to perform due diligence, reformers made fundamental changes undermining the quality and reliability advances that began in the 1940s. Changes being driven by forces not controlled by the government, unappreciated by the reformers, served to intensify the consequences of changes made without the in-depth knowledge necessary to assure the desired outcome.

Momentous changes were about to occur. The use of space systems was routine, but untried in conflict. The crumbling Soviet Union took with it a reliable enemy and left an uncertain void in world power politics. The winds of change were whipping into a storm that was about to burst, pouncing on illusory "peace dividends" before the world again proved to be as dangerous, or more. High reliability electronics reached their apex, with nowhere "up" to go. The question was how to maintain this pinnacle of engineering achievement.

Implosion: Lessons from National Security, High Reliability Spacecraft, Electronics, and the Forces Which Changed Them, First Edition. L. Parker Temple III.
© 2013 the Institute of Electrical and Electronics Engineers. Published 2013 by John Wiley & Sons, Inc.

Chapter 17

Leap First, Look Later

On 9 November 1989, the Berlin Wall came down. German reunification followed quickly. A month after the fall of the Berlin Wall, the Lithuanian Communist Party declared itself independent of the Communist Party of the Soviet Union (CPSU). The CPSU's Central Committee assertion such a claim had "no standing" turned out to have little standing of its own. Romania crumbled on 25 December 1989 with a revolt during which President Ceausescu was executed.[1] The Cold War appeared to be coming to an abrupt halt. That did not mean the space systems that had played such a vital part in keeping the Cold War cold had been overtaken by events like the communist countries over which they had flown. The world remained a hostile place, as only the adversaries changed.

Responding to the Iraqi invasion of Kuwait in 1990, the U.S.-led coalition of Operation Desert Storm forces entered Iraq on 16 January 1991. By 6 April 1991, the coalition forces had defeated the world's fifth largest military force, the Iraqi Army. General Donald J. Kutyna, commander of U.S. Space Command (USSPACE-COM), presented his thoughts to Joint Chiefs of Staff Chairman General Colin Powell on 8 March 1991. Kutyna considered Desert Storm the first "space war." As he explained to Powell, USSPACECOM was no longer

> merely a provider of strategic nuclear intelligence. Desert Storm illustrated how USSPACECOM had become a major supplier of operational and tactical intelligence for the Army and Navy. Space . . . had spent 30 years emerging from the umbrella of the intelligence community, then of research and development experimenters, and finally of an offensive-minded SAC headquarters. Today, an Army cook used a pocket-sized global positioning system to deliver meals to soldiers scattered across the sands of Saudi Arabia. Within two minutes of an Iraqi SCUD missile launch, Patriot missiles were alerted and primed while troops had time to don their protective gear and F-15s flew toward mobile SCUD launchers—all using target coordinates provided from space.[2]

[1] Brian Crozier, *The Rise and Fall of the Soviet Empire* (Rocklin, California: Forum, 1999), 423.

[2] W.S. Poole, J.F. Schnabel, R.J. Watson, W.J. Webb, and R.H. Cole, *The History of the Unified Command Plan, 1946–1993* (Washington, D.C.: Joint History Office, Office of the Chairman of the Joint Chiefs of Staff, 1995), 109–110.

Implosion: Lessons from National Security, High Reliability Spacecraft, Electronics, and the Forces Which Changed Them, First Edition. L. Parker Temple III.
© 2013 the Institute of Electrical and Electronics Engineers. Published 2013 by John Wiley & Sons, Inc.

Space had surpassed operational expectations by anyone's standards, and was integral to the American way of war to such an extent the rest of the world took notice.

Boris Yeltsin became the chairman of the Russian Supreme Soviet on 29 May 1990. He was going to separate Russia from socialism. The Soviet Union collapsed and Yeltsin chose to control the process of its collapse rather than be its victim.[3] A bloodless coup on 19 August 1991 ousted Gorbachev. On 8 December 1991, Yeltsin and his counterparts Leonid Kravchuk of Ukraine and Stanislav Shushkevich of Belarus announced, "The U.S.S.R., as a subject of international law and a geopolitical reality, ceases to exist."[4] The Cold War was over!

In such a reversal of everything that had gone before for seven decades, only one thing seemed certain. If the collapse of the Soviet Union was an irreversible reality, the U.S. military would face a serious economizing effort. But of what form?

*   *   *

The space systems that had served the Free World so well for decades had become increasingly complex. Enabled by the rapid microelectronics advances, complex spacecraft demanded new ways of dealing with complexity and assuring they would work as required. The same rapid advances that forced the most complex standards to move from QPL to QML also affected how spacecraft were built and tested. The crossover concept and the waterfall model addressed requirement decomposition, but were simply no longer adequate. Integrated systems testing had evolved alongside component, device, and system complexity. The extremes of complexity demanded an end-to-end approach to engineering entire systems, including the piece parts.

Just as the 1970 waterfall approach had come from the software development community, once again a means to deal with extensive complexity arose there. This time, the NASA Software Management and Assurance Program was the genesis. The approach finally explicitly tied together requirement decomposition, electronic parts, assembly, and testing. In addition, the new approach provided important new feedback from testing to requirements. Until 1970, every requirement had to be testable, and considerable effort was expended to ensure testing proved each requirement had been met. The level of complexity of the systems even in 1970 made this a daunting task until the V diagram was developed in 1992.[5]

Acquiring space systems, from broad requirements to piece parts through integrated system testing into operations, had become highly disciplined, logical, and adaptable to the level of complexity. The V diagram process fed on data, whether from the contractors and vendors or from FFRDC insights, analysis, and assessments. It ran successfully because of the government officials' deep involvement

---

[3] Crozier, *The Rise and Fall*, 435.

[4] Ibid., 449, 462.

[5] Kevin Forsberg and Hal Mooz, "The Relationship of Systems Engineering to the Project Cycle," *Engineering Management Journal* 4, no. 3: 36–43; Hal Mooz, Kevin Forsberg, and Howard Cotterman, *Communicating Project Management: The Integrated Vocabulary of Project Management and Systems Engineering*, (Hoboken, New Jersey: John Wiley & Sons, 2003), 334.

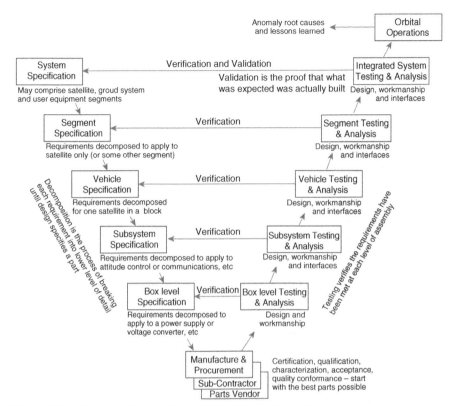

Figure 17.1. The Space Systems Engineering V-Diagram. This follow up to the crossover and waterfall diagrams illustrates how testing starts at the lowest level (at parts) and then ties to increasingly higher levels of requirements as parts become boards, which become boxes, then subsystems, and so on, until a whole spacecraft is assembled for testing. The kinds of testing, and the purpose of the tests, are different from the bottom of the V up the right hand side until the spacecraft's responsiveness to the original requirements is validated.

and commitment. These were the constituent parts of a well oiled, optimized machine. This assemblage combined high art with systems engineering to produce the most complex and sophisticated machines. Most marvelous of all, these machines ran remotely for years in the far reaches of space. But, as with any optimized machine, change or remove some part, no matter how seemingly insignificant, and the result will usually not be good.

* * *

The 1992 change of administrations led to economizing by further acquisition process reform. Spearheading the reform movement was Deputy Secretary of Defense William Perry. Perry's long DoD career made him intimately familiar with the workings of acquisition at the highest levels. Called upon to testify before the Military Acquisition Subcommittee of the House Armed Services Committee, he explained

Acquisition Reform right at its start in June 1993. Citing the time since the Packard Commission, he noted progress had been made, but basically, weapons systems cost too much and took too long to field. His view was that management and control of acquisitions involved as much as 40% of the total cost.[6]

He illustrated his concern in light of the Clinton administration's defense drawdown. If overhead costs were 30% of a program before,

> *as the procurement budget goes down, the temptation is to reduce the variable cost of the program, leaving the fixed costs, overhead costs behind. If that happens, what was at 30 percent overhead before will go up to 50, 60 percent cost as procurement base goes down. What was a serious problem before now becomes a problem that is urgent to address ourselves to.*[7]

Who knows better about cutting unnecessary overhead costs than a company that competes commercially? Covering the standing manufacturing army was a huge problem, because the army was not just a number. The number also represented a large body of experience and competence. More importantly for his audience, the number was the people and jobs being affected in the congresspeoples' districts. The number was people who would have to go elsewhere or do something else, and therefore affected voting constituencies.

Perry insisted the way to approach this difficult problem was to understand how the acquisition system was constructed. He believed the system was built gradually over time to "protect the Defense Department and the country against various abuses that might occur . . . [and] military users of the equipment against poor quality components or workmanship and that led to the military specifications around which we design systems today."[8] However, as this history has shown, his assertion was wrong; more was needed than understanding how the system was constructed. However, he felt he had identified a substantial culprit:

> *The first of those are changes which we can make within the Defense Department just by changing our regulations. I already have described to you that our military specifications are a substantial part of the problem. Our military specs come from regulations, and, therefore, we can change them. We are in the process now of conducting an intensive study to determine what we should change them to.*[9]

So military standards were on their way out. If a program manager wanted to use a military specification, that manager "will have to get a waiver to do that. . . . That is, they will have to justify the use of military specifications."[10] But the waiver option would be good for only a brief time, since the system maintaining the standards was

---

[6] "Acquisition Reform: Fact or Fiction?" Hearing Before the Military Acquisition Subcommittee of the Committee on Armed Services House of Representatives, HASC No. 103-26, 103 Congress, 1 Session, 15 June 1993 (US GPO: 1994), 3.

[7] Ibid., 4.

[8] Ibid.

[9] Ibid., 5.

[10] Ibid.

going to be largely wiped away, along with the standards themselves. The standards would quickly become out of date or only partially usable, making any request for waiver moot.

Perry missed the point that microcircuits were not like bacon. Commercially obtained bacon would work just as well as military spec bacon. Without referring to a specific issue, he claimed that some components cost as much as ten times their commercial alternative, which performed the same function. For some commodities, that was excessive, of course. For Class S parts, that was about right. The danger was that space quality parts were going to be swept together with bandages, tissues, and toothpaste. Unlike tissues or toothpaste, however, space quality parts were more than commercial parts masquerading as something special—they actually were special. They were made to withstand the rigors of the space temperature and radiation environments, and last for a very long time. Commercial alternatives were not. Nevertheless, to use a military specification space part, Perry said, "Therefore, I think it is reasonable and logical that if there is a need for that sort of a system the burden of proof will be on the program manager to justify it."[11]

He expanded his rationale for reforming acquisition, citing three challenges as the basis for their proposed changes and his commitment to seeing them through. First was the "radically changed threat" with the end of the Cold War, followed consequently by declining defense budgets and the need to change the DoD's approach to technology development. In this last category, he specifically addressed the rate of change:

> *In addition, technology and the way DoD needs to approach maintaining our technological superiority is [sic] changing. Technology is advancing at a pace that requires dramatically reduced acquisition leadtimes. Key underlying technologies for both commercial prosperity and military superiority are common, e.g., information technologies, simulation technologies, advanced materials, and agile manufacturing.*[12]

In the questions and answers that followed, Representative Gene Taylor's questions illustrated the mood regarding perceived past acquisition abuses:

> *Number one is going back to the $600 hammer [sic]. Obviously the person who ordered that hammer had to have some idea of the cost of what he was ordering and yet, in his mind, it wasn't a big deal. . . . In taking this back to the private sector, if someone had done this sort of thing for a private sector company, he would have gotten a pink slip, a severe reprimand, a reduction in pay. Something would have happened to him.*[13]

Despite evidence to the contrary, the $600 hammer remained a real thing to him and others determined to end such absent abuses.

Not all in congress were similarly in favor of major reform. Representative Herbert H. Bateman was worried about unintended consequences:

[11] Ibid.

[12] Ibid., 8.

[13] Ibid., 14–15. As noted earlier, neither the $400 nor the exaggerated $600 hammer ever existed, but by 1993 the latter was firmly established in the public's mind as a real thing.

I have been here now for 11-plus years. My experience seems to have been that the cumulative effect of congressional efforts to reform the Department of Defense acquisition process have [sic] been substantially counterproductive. There has been an enormous amount of unintended consequences that came into play.[14]

He recognized the danger of changing a massively complex system with what amounted to radical surgery. Well intended reforms of the past, far less extensive than those being proposed, had unintended consequences that he intimated had called into question the wisdom of the reform. However, he recognized that some military standards had gone too far:

Some years ago, I used an example on a procurement committee of one of those nightmares, in my opinion. This was the example of my wife who finds it very easy to go out and buy a premium package of bacon without elaborate specifications. She is very cost conscious. . . . Yet there is a distinct and very elaborate military specification in order to end up with a package of bacon that is cooked and served in the mess halls in the Armed Services, even to the extent of requiring a separate and distinct inspection program for the producers of bacon which adds substantially to their costs, which they pass on to the Government.[15]

So, how much reform was the right amount? No one really asked because concerted reform was well under way. On 15 July 1993, Under Secretary of Defense (Acquisition) John M. Deutch chartered another defense acquisition review. This one was more than a summer study. Calling itself the DSB Task Force On Defense Acquisition Reform, the effort subsequently lent its name to the resulting wave of changes known collectively as Acquisition Reform. Dr. Robert J. Hermann, a former director of the NRO, chaired the Task Force which included two strong acquisition reformers in Jacques Gansler and Robert Fuhrman. Colleen Preston, one of the advisors, was soon to become the Deputy Under Secretary of Defense for Acquisition Reform.[16]

This review was different. Previous DSB studies, the Packard Commission, and Cheney's Defense Management Review had "all addressed acquisition reform in a context of a stable or growing defense market." That was no longer the case. The Cold War was over and its end reduced the apparent need for a large military force. "[R]esulting defense spending cuts will reduce by at least half the amount spent for modernization compared to the last decade and further shrink the dedicated defense industrial base." The defense industry juggernaut, given its size, could not be stopped overnight, and certainly not without serious economic consequences. So, a way had to be found to realize "a higher political priority for the national requirement for economic growth and competitiveness."[17]

---

[14] Ibid., 16, 17.

[15] Ibid., 16, 17.

[16] Report of the Defense Science Board Task Force on Defense Acquisition Reform, Office of the Under Secretary of Defense for Acquisition, Washington, D.C., July 1993, Letter of Transmittal.

[17] Ibid., 1.

Paul G. Kaminski, Chair of the DSB and an acquisition reformer, sent the DSB's newest report off with a letter of transmittal. He wrote, "[T]he Task Force determined that its primary thrust needed to be the identification of those measures which would reconnect and integrate defense acquisition with the commercial workplace from which it has been drifting apart at a steady rate." Such a problem statement implied that the thing that was moving was defense acquisition, known for rather slow change, when in fact the fast moving entity was the commercial electronics industry. However, with that as the primary thrust, the task force had to identify such things as the "major barriers to use of commercial practices, facilities, and equipment," and "primary sources of excessive costs in the current acquisition process."[18]

Cutting right to the bottom line, then, Kaminski offered the task force's major recommendations, including "Adoption of commercial practices to the maximum extent possible, while assuring the mixture of tools available to the DOD and the commercial marketplace to protect public trust."[19]

A lot rode on their recommendations, in a number of ways. The United States was at a crossroads. The means had to be found to "give DoD access to those technologies, products, and processes which are dominated by the commercial marketplace," and to "broaden the industrial base upon which the department depends. The current, essentially dedicated and thus isolated, base is eroding, is not attracting capital, is losing technology leadership, is not using the most advanced industrial practices, nor is it capable of the required surge capability for crisis response."[20]

Emphasizing commercial practices, the task force believed, would enable DoD to "stretch its available resources significantly." Just how large the savings would be was not a subject determined easily by calculation. Nevertheless, the members judged something on the order of "tens of billions per year could be achieved after four or five years of determined reform." Finally, if the task force's recommendations were followed, they felt what they called "greater integration of the industrial base" would make the large R&D and production resources of the DoD more readily available to the U.S. economy overall, "to foster economic growth and industrial competitiveness."[21]

Anyone with any sense of public responsibility would have wanted all of those good things to happen. In their hope of achieving these, the task force believed "Commercial functional specifications must be applied. DOD unique product and process specifications must not be imposed which inhibit the delivery of defense products and services by commercial sources." That was the crux of the upcoming issue. JAN Class S were not *commercial* parts, in the sense that was meant. No commercial sources of JAN Class S existed, nor were any commercial electronics equivalent to space-qualified electronic parts. The same might not be said of other

[18] Ibid., Letter of Transmittal.

[19] Ibid., i.

[20] Ibid.

[21] Ibid.

defense applications such as computers, but for satellites, commercial sources were essentially nil.[22]

This distinction was quickly lost by those unfamiliar with the difference between the high reliability electronics market and the mainstream commercial consumer electronics market. This same problem had affected high reliability electronics many times over the decades when reformers lost sight of the real differences between the narrow niche occupied by high reliability electronics and the vast commercial electronics landscape. In times past, such oversights were merely annoying. Previously, global statements about the electronics marketplace, reliability of electronics, or usage of electronics by the military had mentioned military quality or commercial grade parts, overlooking the high reliability, space qualified electronics exception to the rule. Times had changed. Driving for efficiency and minimizing cost while stimulating the commercial marketplace were about to seriously impact the high reliability electronics industry.

The DSB task force tried to make the proper allowances. By being unspecific, they lost key distinctions when they said DoD should "[s]ubstitute commercial item descriptions for milspecs in every procurement of a commercial item. The use of a DOD specification or process standard should be prohibited unless it is the only practical alternative."[23] The intent might seem obvious, but the implementation would be less so. Who would determine when use of DoD standards should be prohibited, or when the only practical alternative was a unique DoD item? After all, who was to say a JAN Class S was not just an expensive version of a standard military or commercial part? Silly as it might seem at this point, this exact argument continues nearly two decades later.

Exploring the sources of costs, they concluded the most important source was cost-based contracting, reminiscent of Slay's move to fixed-price contracting in the 1970s. Right behind that was "the systematic imposition of a host of systematically applied product and process specifications, which are different from those used in the commercial market place, and which prohibit the integration of design, manufacturing, and support between the defense and commercial businesses."[24] In short, perhaps without so intending, they leveled their sights on the high reliability and space qualified electronics enterprises. Hermann may not have intended this, as he and others on his team such as Gen Randolph were very familiar with the importance of high reliability electronics to satellites.

Having said that, however, the DSB report could not have been clearer when it cited "a number of primary elements which impose inefficiencies in the current acquisition system." Two of these elements were "unique government specifications, processes, and practices [that] compound the problem and often require the creation of separate development and production activities," and the "large number of government people who monitor and audit industrial activities for compliance to government-unique requirements with at least as large a set of private sector people

[22] Ibid.

[23] Ibid., iii.

[24] Ibid., 4.

needed to deal with this force of government overseers."[25] The unique specifications included standards as well as SCDs. Auditing was a fundamental part of quality assurance and qualification of facilities and manufacturing lines. All were essential for highly reliable satellite systems.

The DSB said efficiency could be improved by using commercial products equivalent to military specification items. The usual examples were bandages, facial tissues, and toothpaste. The same process of doing a sensible thing, though, empowered those who wanted to cut the cost of high reliability systems. The DSB, explaining the inefficiency of unique government specifications, said such things included "prescribed use of milspec parts, special materials handling procedures, quality control and testing mandates, and associated record keeping." This fit precisely with the way high reliability satellites were made. Although the DSB probably did not intend to target national security and national defense satellites, they would have been hard pressed to give a more succinct description of satellite acquisition.

The DSB did not stop there, pointing out such specifications could "add from 20% to more than 50% to the cost of the product compared to best commercial practices." High reliability and space qualified parts did not exist apart from the military specifications that caused them to be in the first place. As a result, there were no best commercial practices which could be applied. Best commercial practices applied only to commercial parts, and no equivalence existed between commercial and high reliability parts. High reliability space parts were different from their commercial counterparts from the very outset. The history of the U.S. space programs had demonstrated, and those involved in supplying the parts knew, that no amount of testing, inspection, screening, or wishing could make commercial parts into high reliability space qualified parts.

Seeking any source of proof about their assertions of cost savings by eliminating military specifications and standards, the DSB glommed onto any examples of savings. But were they applicable to the reforms they were trying to support? More importantly, were they sufficiently general to make the case for all of defense acquisition? They claimed the following examples proved commercial practices were beneficial:

- Clothing: Military specification requirements forced a commercial glove manufacturer to set up a separate production line to produce gloves for the government. The price of the military version was $32 per pair; the nearly identical commercial version sold for $20.
- Construction: A survey by the American Consulting Engineers Council showed that social action clauses, business protective clauses, labor statute clauses, and other unique government purchasing requirements caused construction prices to be 5 to 17% higher for DOD projects than for comparable commercial construction projects.
- Jet Engines: The Air Force agreed to use commercial jet engine casings. This reduced labor by 30 to 50%, production time by 70%, and prices by 50%.

---

[25] Ibid., 4–5.

- Computers: Digital Equipment Corporation developed a military version of their VAX computer. DOD waived 300 process specifications and permitted the use of Commercial-Off-The-Shelf (COTS)[26] equipment. The result was a 20 to 30% price reduction.

- Simulators: DOD procured a Hewlett Packard commercial Frequency Agile Signal Simulator in lieu of new development. The capability exceeded the military alternative at one-fifth the price.

- F-22 Electronics: Analysis of electronic component costs of the F-22 fighter proposed wide use of "best commercial practice" components in lieu of military specified components. Savings of 30 to 50% in components were expected to be realized. These actions would cut the cost of the electronics suite of the F-22 ($1 billion) by 15% or $150 million.

- Westinghouse Modular Avionics Radar: DOD bought a variation on Westinghouse's commercial avionics radar for use on the C-130 and A-300. This equipment was comparable to the avionics radar on the F-16, but took five months to develop instead of twelve months. Its price was 60% less than the comparable DOD unique equipment. Even though military specifications were waived, the DOD was pleased with the performance of the radar.

- Allison 250-C30R Helicopter Engine: The Army bought this FAA-certified engine in lieu of the new development. Off-the-shelf procurement saved five to eight years and an estimated $150 to $200 million in development costs. The commercial warranties proved to be as good as, if not better than, military equivalents.[27]

All of these were, indeed, good things; the taxpayer had benefited. But did these examples actually prove *all* government acquisitions *would* benefit? Or did they show *some* acquisitions *could* benefit? Were these the only examples that could be found?

Further evidence the DSB had conflated high reliability electronics and routine commercial electronics comes from the list of claimed successful uses of commercial standards in lieu of military standards. Acquirers of national defense and national security space systems had traveled a very long road to prove space systems required different parts, processes, and procedures, finally culminating in their own military standards. Thus, being lumped together indiscriminately with clothing, ground

---

[26] Specific definitions became important concerning what exactly was commercial-off-the-shelf. An interchange in *Military and Aerospace Electronics* magazine in Nov. and Dec. 2001 highlighted two key differing perspectives. One held that COTS was equivalent to the consumer electronics used in cell phone technology, and did not include higher reliability military quality and space quality parts. The rebuttal was that any military or space quality parts held on manufacturers' shelves were immediately available, and were therefore commercial-off-the-shelf. This latter argument also (in the mind of some) held that the same production lines manufactured both commercial and military-grade parts, making them equivalent. As done throughout this history, we distinguish between commercial quality parts and those expected to meet the higher quality standards of military avionics or satellites.

[27] Report of the Defense Science Board, C-4–6.

systems, and airborne systems must have been particularly frustrating. The DSB repeated the errors of earlier electronics industry analysts, drawing conclusions on the larger market while ignoring the niche, though extremely important, high quality, high reliability space market. In so doing, the DSB put another devastating nail in the coffin of high reliability, space qualified parts, despite having members who had every reason to know better.

The DSB mentioned some items bearing peripherally on high reliability and space qualified applications. One such "success" was integrated circuits. The evidence they cited was that for many military applications, commercial ICs worked well, with much lower cost and shorter lead times. "The producer (Rockwell/Collins) states that commercial ICs meet up to 50 percent of all DOD needs."[28] The statistic, turned around, says 50 percent or more of military needs for integrated circuits (which included satellites) could *not* be met with commercial parts. The report was mute on the latter 50+%.

This report's appendices expanded on the examples. It was clear in every case the cost savings would occur in non-critical ground systems or some critical aircraft avionics packages where the systems could be easily maintained. The examples were isolated and narrow, as though chosen to prove a larger point.

One case, entitled "Use of Plastics in components," appearing only in the appendix, serves to illustrate the disingenuous and inaccurate reporting. The language in the appendix was more candid and less politically massaged. The DSB latched onto plastic encapsulated parts, saying:

> *A long standing DOD prohibition against plastic parts has been largely eliminated because of industry efforts to improve plastics reliability (reduced phosphorous content in glass passivation and reduced extractable halide content of molding epoxies). Use of improved plastics allows for a 60–70% savings over ceramic or metal packages.*

The technical-sounding discussion seemed to indicate great advances had overcome military objections to such parts. That was not the case. Passivation materials calm negative effects such as whisker growth. Coating some surfaces and materials with glass or plastic could reduce or eliminate negative effects. However, not all such materials are compatible with the space environment, which was another DSB oversight. For terrestrial uses in commercial electronics or military avionics, plastics might have had a positive effect, but their suitability for use in space was overlooked.[29]

Plastics used for encapsulation had indeed evolved since their initial commercial releases. However, all such parts were restricted in their temperature range because

---

[28] Ibid. "In the integrated circuits arena, when DOD uses commercially developed and produced printed wiring assemblies the quality, reliability and performance are comparable, but the price is 30 percent lower. An additional case involved the Navy use of commercial integrated circuits for its new inertial navigation system (CAROUSEL). Costs of ICs ranged from 1/2 to 1/8 the price of military specification versions. The procurement cycle was much shorter and the capability and reliability at least equivalent."

[29] Ibid., D-3.

the manufacturers knew such materials could not withstand the full military range (−55°C to +125°C) over which military avionics must be able to operate to support military operations all over the globe, in any weather. The DSB's conclusion was limited, then, to asserting similar functionality over a restricted temperature range. The DSB glossed over this, but it goes to the heart of whether such parts were truly suitable replacements for military parts.

The language further implied DoD intransigence and lack of knowledge of the state of the electronics industry, reflecting poorly on the expertise of the military. Rome Laboratories was at that time the world-class, leading-edge-of-technology center of excellence, well aware of the advances in various passivation techniques. Multiple government laboratories were studying the plastics issue, which remains, to this day, understudied and unresolved. What the DSB did not examine was why the DoD was so intransigent, making it appear the DoD was deliberately dragging its heels. The truth was, despite advances in passivation, considerable study remained to prove plastic or other new passivation techniques were effective. For instance, the DSB, had they wished to examine the reasons for the apparent intransigence, would have found the advances in passivation had still not addressed the space qualification issues of (1) differing coefficients of thermal expansion (leading to crazing and cracking of the passivation over time in the space environment), (2) outgassing of damaging chemicals creating plasmas allowing electric arcing, which caused damage to other components within sealed boxes, and (3) outgassing of chemicals, which would corrosively damage other materials within the sealed boxes. So intransigence in the face of inappropriate or unproven materials was a good thing, contrary to what the DSB wanted to find in their search for the "primary sources of excessive costs in the current acquisition process."[30]

Subsequent analyses allowed the actual facts to become clearer, but by that time, the damage had been done. The DSB report's claimed cost savings could not be generalized. The reform attempts built on the DSB report would run afoul of various laws and other directives that precluded the savings from being realized, most of which had nothing to do with military standards and specifications.

A good case could be made, in fact, that military standards and specifications were not the cause of increased costs so much as they were a convenient target for reformers. As momentum gained and such reports fed on one another, military standards and specifications became scapegoats for high cost in all military systems acquisition.

A later review of 197 defense programs between 1960 and 1990 demonstrated an average of 20% cost growth. During that time, as this history has shown, military standards grew substantially in number and in specific detail, while the average cost growth of acquisitions remained roughly constant. Another way to say that is, for a large cross-section of military acquisitions, the initial cost estimates had understated final costs by 20%. The average had not changed significantly in those thirty years, despite "implementation of more than two dozen regulatory and administrative

---

[30] Ibid., Letter of Transmittal.

initiatives."[31] Some mechanisms other than military standards were causing cost growth.

The DSB overlooked one major element of defense contracting: the process of determining contract awards through competition. Competitively awarded contracts were usually seen as a good thing, though even good things can be taken to excess.

> *A factor contributing to defense cost growth is estimation error, possibly due to excessive competition. Contractors have an incentive to understate initial costs to win new contracts. Likewise, because programs compete for limited resources, there is an incentive to accept low estimates as reasonable.*[32]

In the end, the contractors making the systems determine the price, and the price they choose has to cover such things as their facilities' maintenance, personnel, and the return on stockholders' investment. Reforming acquisition in any way that ignores or undermines that or does not effectively improve it simply cannot control cost.

Whatever the actual causes of cost growth were, the pendulum was swinging in a dangerous direction for the high reliability space qualified electronics industry.

*   *   *

Revision C of MIL-M-38510 (for microcircuits), released on 30 April 1993, was a portent of the changing standards environment. Slash sheet 608, for a new gate array device, had to admit the QPL system could no longer be used for such devices, and a process approach was necessary. This revision was quickly overtaken by fast changing events, and illustrates the attempts to preserve the body of knowledge the specification represented. MIL-M-38510 was inactivated on 27 August 1993, but it did not simply go away.[33] The main body of the document was made Appendix A of MIL-I-38535, the specification for integrated circuits. The defense standardization system allowed appendices to contain pretty much anything, because they were considered additional, elaborating material, not compliance material. So, although the information was no longer in the compliance section of any specification, it was not entirely lost. The appendices of MIL-M-38510, dealing with quality assurance plans, statistical sampling, testing, and inspection, rolled over into other appendices of MIL-I-38535. The appendix on custom hybrid microcircuits had already been made part of MIL-H-38534 in June 1989. MIL-M-38510 was then canceled on 15 September 1993.[34] The document that was the key to receiving the "JAN" device marking had been rolled into a performance specification, effectively moving all microcircuit qualification to the QML system.[35]

---

[31] David S. Christensen, Capt. David A. Searle, and Caisse Vickery, "The Impact of the Packard Commission's Recommendations on Reducing Cost Overruns on Defense Acquisition contracts," *Acquisition Review Quarterly*, Summer 1999, p. 252.

[32] Christensen, Searle, and Vickery, "The Impact of the Packard Commission," 252.

[33] Mil-M-38510J Inactivation, Notice 1, Defense Logistics Agency, 27 Aug. 1993.

[34] MIL-M-38510 Cancellation Notice, Air Force Research Laboratory, (RL/ERSS), 15 Sept. 1993.

[35] *The NASA ASIC Guide: Assuring ASICS for Space, Draft 0.6*, (Pasadena, California: Jet Propulsion Laboratory under grant from Office of Safety and Mission Assurance, National Aeronautics and Space Administration, 1993).

Not everyone was pleased with the move. QML addressed a significant problem with the QPL approach to qualifying ICs, ASICs, and hybrids. Still, there were flaws the Acquisition Reformers might have been able to address, but did not.

In 1993, the Semiconductor Industry Association's Government Procurement Committee (SIA/GPC) surveyed the state of the industry, reviewing cost accounting, offshore manufacturing, military performance specifications, and the use of commercial components in military systems.[36]

The group took a somewhat controversial stand, based on their definition of commercial parts, leading to a recommendation to do away with certification documentation associated with defense contracts. More will be said on certification documentation shortly, when the Process Action Teams are discussed.[37]

The report recognized economic and competitiveness issues forced IC manufacturers into a difficult position. While defense and national security contracting called for use of domestically manufactured parts, the electronics industry was becoming global. Manufacturing of ICs and other components was moving to places where the most favorable costs of all the manufacturing factors could be had. That meant Taiwan, Singapore, and elsewhere. For U.S manufacturers, having to maintain domestic facilities to support military and national security needs meant having to pay higher facilities and labor costs, making them less competitive globally. Many had already moved their commercial manufacturing overseas, which meant higher prices for military and national security customers simply because these were becoming the only markets for domestic manufacturing. Recognizing this reality, the DoD had recently allowed some ICs compliant with the MIL-M-38510 JAN requirements to be manufactured offshore.[38]

Not surprisingly for an industry body, the SIA/GPC endorsed the continued and broadened incorporation of commercial parts into military systems. Their motivation is easily understandable. The fewer kinds of manufacturing lines needing to be maintained, and the larger the volume from those remaining, the more competitive manufacturers would be. However, because of the odd way the SIA/GPC had defined commercial products, they were able to endorse "continued use of military-grade ICs produced by suppliers employing 'best commercial practices' in accordance with MIL-I-38535 or a commercial equivalent." Such a statement would have been incomprehensible only a few years before. Commercial products had been clearly different from military and space quality products. The industry's evolution since the early 1960s emphasized the inability to "test in" quality because commercial parts were made differently from the outset. In the IC world, however, the processes had always been different. So different, in fact, the SIA/GPC was able to say with a straight face the only difference between commercial and military parts was additional testing. They went along with the acquisition reformers, ignoring the differences required for space qualified parts.

---

[36] Rick Cassidy, "Maintaining Military Superiority in an Era of Cost Constraints," (Santa Clara, California: National Semiconductor, Oct. 1993).

[37] Ibid.

[38] Ibid.

Instead, the SIA/GPC claimed an important problem with QML was the lack of a standard parts numbering system, reminiscent of World War II, when the original JAN standard created a common vacuum tube numbering system. At the time, allowing best commercial practices meant there was nothing to say one company's part numbering was any better than another's. Lack of a standard system allowed the same industry part number from different manufacturers to have different functions, or to provide the same functions but from different pins. Standard Military Drawings had been mandated for ICs in 1991 but did not standardize ICs. In the SIA/GPC's words, "The purpose of product data sheets today is to establish competitive advantage, not to provide any standardization between manufacturers."[39] This was another industry concern that affected the path of Acquisition Reform in a positive way, as the drawings supporting QML ICs were shortly to be renamed.

Since the Packard Commission, the juggernaut of commercial parts entering military systems had been gaining speed. A fundamental and ultimately very damaging flaw was no one actually agreed on what "commercial" actually meant. Each of the studies, from Packard on, clearly understood what each had meant as "commercial," though somewhat differently and not complementarily.

As the case in point, ICs were not subject to any uniform commercial electrical specifications. In fact, under QML, fabrication processes were driven to increase yields in each production run. So, the process that produced a prior batch of ICs need not be the same for producing the next. Why would anyone expect anything different in an environment where costs and competitiveness were the main forces shaping the IC marketplace? If yields began to decline, a process had to be tweaked to reverse the decline, so changing processes was the norm. "They can change molding compounds, lead finish, soldering specifications, marking permanence requirements, and many other critical characteristics without notifying anyone other than their own QA departments."[40] Anathema to military specifications, this approach to process controls was part and parcel of "commercial best practices." Such variability also means best commercial practices were not common across manufacturers, since all were varying processes independently.

Such changes could (and would) radically alter the suitability of an IC for use in military or space environments. No commercial parts were guaranteed to meet the military temperature extremes of $-55°C$ to $+125°C$, though some might be so qualified through additional testing. However, additional qualification testing raised their costs and also put them into another DSB hit list category. Other commercial parts, such as plastic encapsulated parts, had been unsatisfactory for a variety of reasons, not the least of which was their lack of suitability for such temperature extremes. Broad-stroke admonitions to use more commercial parts in military systems could not sweep away the physics that made such materials unsuitable.[41] No one who fully understood the consequences of such a broad range of temperatures, and the lack of suitability of commercial products to meet these extremes,

[39] Ibid.
[40] Ibid.
[41] Ibid.

could realistically contend commercial products were adequate. Yet, that was precisely what had been said since Packard, and would continue as the mantra for several more years. This was well understood in 1993 on the eve of the acquisition reform movement.

Under the QPL, industry and government created an extensive infrastructure necessary to support military standard testing, such as testing over the temperature extremes. When MIL-M-38510 rolled into MIL-I-38535, overseas manufacturing had been allowed. The overseas facilities, though, did not have the same extensive testing infrastructure, since they geared for commercial markets and global competition. This raised a potential further problem. For potentially suitable products made overseas, compatibility with the military and space environments would be very expensive to demonstrate, and therefore unlikely to occur.

Another example in favor of commercial parts being good enough to meet military needs was the success of the automobile industry's producing vast numbers of specialized ICs. But like most of the acquisition reform claims, that too was only a skin-deep analysis. A contemporary analysis took this example apart:

> *Some proposals have suggested the use of automotive ICs for military purposes. However, there is no "standard" automotive IC. Each auto manufacturer maintains several screening flows, each tied to a different set of applications. Those flows focus on large volumes of a small number of device types, many of them custom designed for unique automotive functions. Their reliability goals aim more at the minimization of customer inconvenience than at the elimination of failures. The low prices typically associated with automotive ICs are based on enormous volumes—typically millions of a given part type per year. Screened in the volumes typical of military applications, they would be priced comparably with military parts. One key point is that almost none of the parts now used in the dozens of military systems currently in production are available from any of the limited lines of automotive products.*[42]

*   *   *

In spite of evidence to the contrary, reforming acquisition moved forward more briskly than ever. Secretary of Defense Les Aspin and his chief acquisition reformers continued the crusade. In January 1993, before Deutsch chartered the DSB summer study on defense acquisition reform, the Acquisition Law Advisory Panel (known as the Section 800 Panel for the affected part of the U.S. Code) reported its findings to Congress. Their report cited the places where laws needed to be changed to implement the ongoing initiatives of acquisition reform. When Colleen Preston assumed her duties in the newly created office of Under Secretary of Defense for Acquisition Reform in June 1993, for the first time a senior defense official was solely responsible for reforming the DoD acquisition process.

By October 1993, the Federal Acquisition Streamlining Act (FASA) was passed, putting into law a vast number of the changes needed by acquisition streamlining, the Cheney Review, and subsequent acquisition reform studies. FASA was to be followed a year later by FASA II to catch up with a large number of subsequent

---

[42] Ibid.

changes. FASA II was needed because in February 1994, a major change occurred. Deputy Secretary of Defense William Perry, so influential on so many prior acquisition studies, panels, task forces, and commissions, replaced Les Aspin as Secretary of Defense.

The changes came fast and furious as Perry made sure reform would take hold. The same month he took the reins, he issued *Acquisition Reform, A Mandate for Change*. Known widely as the Perry Mandate, he made clear no opposition was going to occur. Defense acquisition was going to change along certain specific lines. He knew, from his lengthy experience and the lack of overall progress to that point, there were key elements to making such a major cultural change. Support from the top was easy, since he brought that to his position as Defense Secretary. But cultural change had to be both broad and deep. Working with other reformers already in place, such as Preston, he started filling acquisition positions with people whose mission was reform.

Making his views clear, and to serve notice in the Department, he sent a letter to all senior DoD leaders with *Mandate for Change* to ensure all had an opportunity to read it and understand what was coming. That covered the change being broad. Next came deep change—the *coup de grace*, which no one before him had been able to accomplish.

Chapter 18

Hardly Standing PAT

The *coup de grace* was the report of the Process Action Teams (PATs).[1] In April 1994, just two months after Perry took over, the PATs' *Blueprint for Change* was released. *Blueprint* provided the depth, and, some might say, the heart, of Acquisition Reform.

Blueprint covered many important items. For military standards and specifications, the *Blueprint*'s Process Action Team on Military Specifications And Standards was the first report to cover a vast array of desired changes.

> *Specifications and standards reform is an integral part of the acquisition reform vision, a vision intended to revolutionize the way the government does business. At the root of the problem are 31,000 military specifications and standards. Over the past 20 years or so, it has been an uphill, and not always successful, struggle to keep these up-to-date in a world of continuous and planned obsolescence. As DoD's budgetary and manpower resources are reduced, however, there is little hope that military specifications and standards can be kept either technically current or on track with commercial practices, products, and processes. The greater the divergence between the commercial and military sectors, the less the likelihood that military products and systems can be purchased from or produced in commercial operations.*[2]

Here was a great deal of truth about military specifications and standards. Within the high reliability electronics world, the realization had hit SD several years earlier that military standards' keeping pace as electronics evolved was becoming unworkable. SD had wrestled with the issues in the face of the VHSIC effort. Technology was advancing at a far greater rate than the military standards could change. Clearly, some alternative approach had to be found.

---

[1] Altogether, five Process Action Teams existed: Electronic Commerce/Electronic Data Interchange, Military Specifications and Standards, Oversight and Review, Procurement Process, and Contract Administration Process.

[2] Process Action Team on Military Specifications and Standards, *Blueprint for Change*, Office of the Under Secretary of Defense for Acquisition and Technology, April 1994, p. 1.

Implosion: Lessons from National Security, High Reliability Spacecraft, Electronics, and the Forces Which Changed Them, First Edition. L. Parker Temple III.
© 2013 the Institute of Electrical and Electronics Engineers. Published 2013 by John Wiley & Sons, Inc.

The Specifications and Standards PAT went on to say:

> *DoD cannot afford to pay an increasing "defense-unique" premium for the goods and services it buys. It does not have the wherewithal to subsidize increasingly inefficient defense operations which do not have a self-sustaining market base. As these defense companies downsize, convert, or fail, DoD will lose a significant portion of the industrial base once capable of producing to its specialized requirements.[3]*

Once again, the *Blueprint* struck a reasonable chord with an eminently clear truth. While true as a global statement, everything depended on its implementation in the specific.

> *There are only two ways out of this dilemma. The first is to convert overly prescriptive military specifications and standards into nongovernment standards (NGS), commercial item descriptions (CIDs), and performance-based specifications and standards—the kinds of documents that will allow suppliers to optimize production capacity and DoD to buy from a unified national production base. The second alternative is to face the prospect of an industrial base that is incapable of sustaining our forces in two major regional conflicts simultaneously.[4]*

Again, as a global statement, the *Blueprint* may have been correct, but the tone was clearly "one size fits all" for acquisition. "My way or the highway" can be an effective way to serve notice one is serious, but such an attitude cannot be appropriate for every situation when dealing with the vast range of acquisition efforts within the DoD.

Making military procurement specifications defer to commercial "best practices" was a major conceptual change relying on the belief that commercial practices were better than military standards. The precedent for this change had been set much earlier, when gate array devices (which had become a form of ASIC) were added to the QPL under slash sheet 608. That slash sheet had not been like others, because it left many aspects of design and fabrication open for the manufacturers to choose their own practices. This meant slash sheet 608 was not tied to a specific design and thus could keep pace with industry.[5] This was the genesis of renaming Standard Military Drawings in MIL-I-38535 as Standard *Microcircuit* Drawings.

The PAT wisely recognized limits to what could be addressed:

> *Unfortunately, there are no universal solutions or overnight panaceas that will convert the military specifications and standards program into a "commercial friendly system." It is an extremely complex system spanning acquisition needs ranging from nuclear weapons to chocolate chip cookies. There will probably never come a time when all defense needs can be satisfied by commercial operations. However the goal of the Process Action Team (PAT) on Military Specifications and Standards is to maximize the overlap between DoD needs and commercial capabilities. How DoD defines those needs is a critical determinant of the potential for overlap.[6]*

---

[3] Ibid.

[4] Ibid.

[5] *The NASA ASIC Guide: Assuring ASICS for Space, Draft 0.6* (Pasadena, California: Jet Propulsion Laboratory under grant from Office of Safety and Mission Assurance, National Aeronautics and Space Administration, 1993).

[6] Process Action Team, *Blueprint for Change*, 1.

There seemed to be a ray of hope in the words that allowed for maximizing the overlap, leaving room for retaining some specific unique elements. As though addressing high reliability space qualified electronics, the *Blueprint* said:

> *Thirty years ago, military specifications and standards defined the state of the art. Today, they trail best commercial practices increasing the cost of defense procurement, and create a firewall between the commercial and military sectors.*[7]

The Reformers intended to field the best, most technologically advanced systems possible. During the Cold War, technological superiority was the only way for the United States to counter the Soviet Union's superior numbers. That was etched in stone, never to be questioned. Yet, the PATs could have seen the underlying premise— the need to defeat numerically superior Soviet forces—had been overtaken by events. What greater proof was needed than the United States and some of the former Soviet Union's republics collaborating in Desert Storm? The world was indeed different. So why not also view technological superiority as malleable? The most advanced technologies were expensive, and not necessarily universally necessary if capabilities were considered instead.

That is, if high reliability space qualified electronics were a generation or two behind commercial standards, there might be a way to live with that. High reliability space qualified electronics purchases might have to work with end-of-production, lifetime buys of some technologies as the commercial suppliers moved on to another generation, allowing specific standards to be followed despite their being somewhat behind the latest available gadgets on the street. This option never came forward.

The Specifications and Standards PAT comprised six focus groups: systems acquisition, replenishment, training, management and manufacturing standards, improving specification and standard content, and automation. Each group had representatives from all of the Services, other DoD agencies, and from the Office of the Secretary of Defense's large number of acquisition-related branches. "Taken together," the report continued, the recommendations from all of those groups were the roadmap "to major cultural changes in the military specifications and standards process."[8]

As the PAT report got to the meat of the subject, the writers began to talk down to their audience.

> *Specifications and standards are a complicated problem to understand, much less reform. One of the first barriers is the confusion caused by similar sounding terminology. For example, industry uses the term "standards" to describe both products and processes. In DoD parlance, "specifications" are generally used to describe products, material items, or components, while "standards" generally describe methods, processes, or procedures, i.e. ways of doing something. Performance-based specifications and standards describe what is needed.*[9]

[7] Ibid., 11.

[8] Ibid., 12.

[9] Ibid., 17.

Performance-based specifications exemplified the QML approach for process standards, which the PAT considered a valuable commercial practice. Quality assurance, certification, and qualification focused on processes rather than end-items.[10]

The way to keep up with technology, then, was to get out of the way, and let the commercial world dictate the next steps.

> *Process based specifications and standards describe how to achieve it. In reality, though, all specifications and standards include some mixture of performance and "how-to" instructions. . . . Similarly, the difference between specifications and standards is fundamental to an understanding, both of the problem and the recommendations offered in this report. Specifications and standards are used by every quality supplier in the world and by every major buyer seeking quality products. Specifications and standards are the unseen glue of modern civilization. They ensure that plugs from different appliances fit into the same electrical outlet and that light bulbs fit into standard fixtures. They ensure that the mustard you buy isn't just a yellow-colored substance, and that the vacuum cleaner doesn't give up a week after purchase.*[11]

Obviously, the authors were enjoying themselves, but at what cost to the industry they were going to affect? Satellites are not yellow like mustard, nor were they expected to last only as long as a vacuum cleaner. If defense acquisition were about bandages, tissues, and toothpaste, the discussion would have been appropriate.

> *The difference between DoD and other major buyers, however, is that military specifications and standards do not always stop at specifying what is required. Frequently, they also describe how to make a product, indeed, the one acceptable way to make it. Those detailed process prescriptions often diverge from commercial practice. Where the standard describes processes like how to set up a quality assurance system, how the product must be tested and inspected, or how the work must be measured, DoD loses access not only to a commercial product, but also to the whole commercial facility. Overly prescriptive process standards impede DoD access to commercial operations, which prevents it from acquiring defense unique items from flexible, multipurpose manufacturing operations. The dilemma for policy makers is—as DoD's declining procurement budgets can no longer sustain a separate base of producers who have set up their operations to comply with these military-unique specifications and standards.*[12]

Another generalized truth masked the important alternate question: Could the high reliability space qualified parts industry survive without these?

As though taking aim at the SD JAN Class S stockpiling initiative, the PAT moved to standardization, which, it said,

> *is an entirely different process with a different rationale. One benefit of standardization is that it facilitates centralized purchasing. If every buying agent at every command bought their own supplies, they would pay retail prices and not*

---

[10] Ibid., 175.

[11] Ibid., 17.

[12] Ibid., 17.

volume discount prices. In order to buy centrally, there has to be some commonly accepted description of what needs to be bought. The exception to this rule is where there is no need for volume purchasing.[13]

Yet the PAT report did not reflect a deep understanding or concern about narrow markets such as those that supported satellites. For the PAT, standardization allowed maintainability—a reason that harkened back to the American Revolution with Eli Whitney's standardization of rifle barrels for mass production and interchangeability. Except in this case,

The primary reason to standardize, however, has more to do with the special problems of trying to field many advanced systems which have to perform under the stress of combat. If an M-1 tank is disabled on a battlefield, the Army wants to ensure that there are not five different versions of the spare part needed to make it functional again. That would force the Army to identify which of the five alternatives is needed to repair that particular tank. Army technicians would have to be trained to work with all five versions of the part, and the Army would have to stock each of the five different spare parts to ensure that adequate supplies were available when needed (and that would depend on a crystal-ball calculation of which tanks are likely to be disabled). Clearly, lack of standardization creates a logistical nightmare, one that would only be multiplied if each Service were to stock different versions of the same component for each of their systems.[14]

But what about systems where maintainability was impractical? Although dreamers conceptualize replacing standard parts on satellites 22,000 miles away, the reality was quite different. It seemed the PAT focus teams might have understood that, since a major implementation recommendation was for the Deputy Secretary of Defense to issue a

policy memorandum prohibiting the use of military specifications and standards in all ACAT Programs except where authorized by the Service Acquisition Executive or designees. . . . Exemption may only be granted for performance-based specifications, truly military-unique specifications and standards, no acceptable alternative, or not cost effective.[15]

Early warning, navigation, reconnaissance and other satellites were based on "truly military-unique specifications." Weren't they?

Apparently not to the PAT focus team's satisfaction. After wading through less than 3,000 of the 31,000 specifications and standards, they produced a "Prohibited List of References in Military and Federal Specifications and Standards, Bulletins, or Commercial Item Descriptions."[16] They went into detail, explaining that the

---

[13] Ibid., 17.

[14] Ibid., 17–18.

[15] Ibid., 52.

[16] SecDef William Perry memorandum, "Specifications and Standards: A New Way of Doing Business," 29 June 1994. In Perry's memo, precise numbers are unclear, but the number reviewed may have been as low as 1,600.

following is a list of the types of documents that shall not be cited as requirements in military or federal specifications, standards, bulletins, or commercial item descriptions. . . . Typically, reference to these documents is inappropriate because they more properly belong in the contract, they are policy documents directed at government and not contractor personnel, they inhibit use of commercial products or processes, or they represent traditional management approaches that inhibit more creative and effective risk management alternatives.[17]

Thus, the PAT's broad sweep did away with:

- All directives, instructions, regulations, or other types of policy documents Military Handbooks. These documents are useful for guidance purposes but shall not be cited references in military or federal specifications.

- Data item descriptions (DIDs)—which are the means of describing the form and content of data to be delivered on a government contract

- Management oversight specifications and standards, including but not limited to:
  - MIL-STD-1541, Electromagnetic Compatibility Requirements for Space Systems
  - MIL-STD-1543, Reliability Program Requirements for Space and Launch Vehicles
  - MIL-STD-1546, Parts, Materials, and Processes Control Program for Space and Launch Vehicles.[18]

MIL-STD-1541 was essential for increasingly complex satellites, since it had captured the hard-won lessons about how multiple payloads coexisted on the same satellite without interfering with each other. MIL-STD-1543 was the key document governing the reliability program practices in design, development, fabrication, test, and initial operations of satellites. It mandated participation in GIDEP, and set forth the necessary features for Failure Modes and Effects Analysis. This latter analysis was essential to be able to interpret anomalous behavior, and possibly determine how to save a spacecraft mission when something stopped working properly. MIL-STD-1546 provided the necessary guidance for an effective parts management program. While the PAT recommendations covered many more than the three standards listed here, these three were among the core of the high reliability space qualified electronics upon which the highly reliable national defense and national security space systems were based. The core of high reliability for space applications, then, was about to be swept away.

Their rationale depended on the prime contractors stepping up to new responsibilities for success of their spacecraft. After all, the thinking went, who would know better than the contractors exactly how to do the design and construction of spacecraft? Surely the government had no business telling contractors how to do

---

[17] Process Action Team, *Blueprint for Change*, 57.

[18] Ibid., 57.

what they did so well. Therefore, the government needed to let the contractors loose to use the more beneficial best practices.

Colleen Preston later recalled:

Probably the best example of that strategy in action is Military Specifications and Standards. Dr. Perry could have, very early on when he was the Deputy Secretary of Defense, issued a memorandum saying that, "From this day forward you will not use Military Specifications and Standards." He believed and I believed that it was very critical, instead, that we have a Process Action Team made up of people who were dealing with the issues on a day-to-day basis and let them make recommendations on how to implement or achieve this objective. Use of commercial specifications and standards had been pushed for many, many years. In fact, it had been in law for five years that there was a preference for commercial specifications and standards. As the team worked, we came to realize that the more important thing was the preference for performance standards rather than simply shifting from military specifications and standards to commercial.[19]

Implementing the PAT recommendations, Perry issued "Specifications and Standards—A New Way of Doing Business" on 29 June 1994. This new mandate replaced military standards with "a performance-based solicitation process and the expanded use of nongovernment standards." Perry did not obliterate military standards, however. His memo allowed for some brave souls to attempt to use such standards. "If military or federal specifications or standards are necessary, waivers must first be obtained."[20] Of course, the waiver had to *prove* performance-based approaches would not work. In the absence of prior performance-based acquisitions, such proof was unlikely to be forthcoming. Instead, a whole generation of weapon systems had to be developed and fielded—taking five or more years from inception to fielding before the proof would be available. When the proof began to arrive, in the form of failed satellites, it would be too late to go back or do anything but live with the consequences.

Perry claimed great progress by the time of his newest mandate:

Since June 1994, DoD has adopted an additional 1,200 nongovernment standards, raising the total number of nongovernment standards adopted by DoD to nearly 7,000. This represents a growth from about 17 percent to nearly 20 percent of the total for all specifications and standards adopted by DoD. Additionally, every military specification and standard is being reviewed to ensure that it supports acquisition reform principles. Industry and private sector organizations are helping DoD decide whether to cancel a military specification or standard, convert it to a performance-type document, replace it with a nongovernment standard, convert it to a guidance-type document, or retain it.[21]

This was the moment when the former designation system for military standards and specifications changed. What had formerly been known as MIL-I-, MIL-H-,

[19] "Colleen Preston on Acquisition Reform," PM: Special Issue, Jan.–Feb. 1997, p. 25.

[20] Perry, "Specifications and Standards." The Defense Standards Improvement Council was created to carry out those policies.

[21] Ibid.

or MIL-M- became MIL-PRF-xxxx, indicating they were performance-based specifications.

By the time of his memo, the Defense Standards Improvement Council had identified what they considered the top 107 cost-driver standards. Of these 107, some of which were key to the high reliability space qualified electronics industry, nearly

> *half have been canceled or declared inactive for new design, 20 percent will be converted to use for guidance only handbooks, and 10 percent are being retained until an adequate nongovernment standard becomes available. The rest will be converted to performance-type documents or retained.*[22]

Many of the very specific electronics standards were converted to "guidance only handbooks," making any contractual implementation or enforcement based on the goodwill of the parties involved.

One illustration can serve for the overall effect. MIL-STD-1540, "Test Requirements for Space Vehicles," established a uniform set of definitions and ground testing requirements for space vehicles. It covered the full range of testing, excluding the testing of piece parts, concentrating on higher levels of assembly known generally as environmental testing (the V diagram's ascending right side). Its tests were crucial to successful spacecraft, dealing as they did with vacuum, thermal, and vibro-acoustic testing, simulating as closely as possible the launch and orbital operations of the spacecraft. Revision B, released in October 1982, had 93 pages of tests, which were "a composite of those tests currently used in achieving successful space missions." Revision C followed in September 1994, adding more requirements and incorporating other standards such as electromagnetic compatibility testing. It had grown to 135 pages. Revision C's scope explained "The test requirements focus on design validation and the elimination of latent defects to ensure mission success. The application of these test requirements to a particular program is intended to result in a high confidence for achieving successful space missions." Revision C's release came right in the midst of Acquisition Reform. When it was next revised, in the wake of the Perry initiatives, it had shrunk to twenty-eight pages. The information it contained, accumulated from decades of experience, was not folded into another standard, as MIL-M-38510 and MIL-I-38535 were. This information was simply gone.[23]

Perry said, in a later interview, his DoD Specifications and Standards Reform memorandum of June 1994 was a major step forward in Acquisition Reform. In terms of the high reliability space qualified electronics industry, it was the key reform, and its consequences only began to become clear later. Aside from the obvious impacts to new contracts having lost the basis to ensure their quality and

---

[22] Ibid.

[23] MIL-STD-1540, Test Requirements for Space Vehicles, Revision B, 10 Oct. 1982; Revision B Change Notice 1, 31 July 1989; Notice 2, 8 Feb. 1991; Revision B Change Notice 3, 12 Feb. 1991; Revision C, 15 Sept. 1994; Revision D, 15 Jan. 1999. Of course, the information as not completely gone, as people retained copies of the more rigorous version C, and it would eventually find its way back into use as problems and the need arose.

reliability, the loss of standards rippled through organization and cost reduction efforts over the next few years.

Reform became a government-wide crusade as the second wave of FASA was enacted. Paul Kaminski, another of the key reformers, assumed the office of the Under Secretary of Defense for Acquisition and Technology in October 1994. Working with him, Colleen Preston led 11 different teams along with other government agencies as the overarching federal procurement laws, Federal Acquisition Regulations, Defense Federal Acquisition Regulations, and a host of other implementing policies and directives were changed at a furious rate. Taking a clean sheet approach to rewriting the regulations, eleven integrated product teams (nine led by DoD) ferreted out necessary changes and implemented them as quickly as possible.

FASA II, enacted on 13 October 1994, cleaned up items left over from the 1993 law and added some new changes driven by the Perry initiatives. FASA II repealed competitive prototyping and competitive alternative sources requirements. This reduced required documentation and reporting, removed statutory detail from several other required reports, and implemented other, similar actions. Although meant to reduce the glut of paperwork associated with defense contracting, it had an insidious effect. In addition to the FASA's deletion of deliverable data, the FASA II eliminations forced acquisition programs to move from "oversight" to "insight." The Oversight and Review PAT defined these two terms in a special way. "Oversight" was the *continual process* of monitoring program execution between phases, which stood for the PAT's understanding of the "old" culture of defense systems acquisition. By contrast, they were making even that level of involvement difficult, favoring instead "insight," which was a minimal level of "oversight," allowing only limited flow of information concerning program risk, complexity, funding, and problems.[24] Neither corresponded to the reality of national security satellite systems acquisition, and both would have caused problems.

Several unintended consequences of paperwork reduction were to ripple across the acquisition community for decades to come, and more will be said about this shortly. For the moment, to be sure, some of the standard data deliverables could have been deleted without impact, as they produced reports that mainly sat on shelves. Data deliverables and government data rights were keystones of government acquisitions. More to the point here, data rights and deliverables also allowed information sharing on defective parts and related problems, ensuring that all affected organizations were aware of issues.

From this point on, restriction of data flow would be used for competitive advantage. Manufacturers became less concerned about customers finding out about problems, and more concerned that competitors could use such problems to shift market share. Support for information sharing forums such as GIDEP waned. That

---

[24] "Oversight" was the *continual process* of monitoring program execution between phases, and "review" was the *discrete* process of gathering and evaluating information to make a decision about a program. The PAT intended redefined oversight to be a tailorable entity, based on several criteria. The reduced form of "overview" became "insight," but not as either term is normally understood.

being the case, manufacturers could keep buyers of defective parts unaware of the extent of the problem. The external correcting function afforded by parts buyers working in concert was seriously impeded. This might have been workable had they not also cut the resources necessary to check and test incoming parts and ensure manufacturers were properly audited.

The problems did not end there for the FASA II measures. Until FASA II, suppliers of parts to the government had to certify they were either the original equipment manufacturers (OEMs) or authorized distributors. In the latter case, an authorized distributor handled some of the sales logistics for the OEM, particularly when the OEM did not want to handle that end of the business. Shortly after the passage of FASA II, however, the number of brokers of all kinds (authorized and not) mushroomed. For profit margins from 10 to 30%, the market was a real economic stimulus for legitimate small businesses. Paperwork reduction allowed anyone with a contractor code to become a broker. As reductions in auditors and government insight had been streamlined out of acquisition processes earlier, the result was a temptation for unscrupulous parties. This paved the way for what has since become an epidemic of counterfeit and fraudulent parts.[25] In addition to the damage done to defense acquisition sources, counterfeiters undermined the confidence and trust in an industry where these had been the norm. In a business where parts problems had formerly resulted from honestly made mistakes in a business where the source could be trusted, counterfeits would eventually emerge to undermine the confidence in sources of supply from all but the original manufacturer or their licensed/franchised distributors.[26] By 2009, some market analysts estimated "the legitimate electronics companies miss out on about $100 billion of global revenue every year because of counterfeiting." As of 2010, 46% of surveyed manufacturers producing discrete parts ran across counterfeits of their products. Microcircuit companies found an even higher rate of counterfeiting, with 55% having such issues.[27]

Data reduction's broad brush saved some unnecessary expenses, but went too far by removing essential data in an environment that also removed process controls. When problems were uncovered a decade later, reversing the situation was out of the question. The annual cost of counterfeits alone more than outweighed the savings.

---

[25] *Business Week*, "Dangerous Fakes—How Counterfeit, Defective Computer Components from China Are Getting into U.S. Warplanes and Ships," 13 October 2008. This was the DoD Activity Code (DODAC) which had been required to participate in the JAN Class S stockpiling effort. The use of recycled electronic parts as counterfeits was enhanced by the lack of paperwork and the European initiative known as Waste Electrical and Electronic Equipment.

[26] Even the terms "licensed" and "franchised" were eventually undermined by unscrupulous individuals claiming greater authority from the original manufacturer than was the case. The key was the test of whether a part purchased would carry with it the original manufacturer's warranty. But even that has begun to erode as manufacturers move their capabilities overseas.

[27] *Defense Industrial Base Assessment: Counterfeit Electronics* (Washington, D.C.: U.S. Department of Commerce Bureau of Industry and Security, Office of Technology Evaluation, Jan. 2010), 8; Diganta Das, "Keeping It Real: A Researcher Educates Industry on Perils of Counterfeit Electronics," *Metrics*, A. James Clark School of Engineering, University of Maryland, Fall 2009, p. 3. Das was a member of the Center for Advanced Life Cycle Engineering at the university.

The Systems Acquisition Process PAT reported its recommendations in December 1994, resulting in major changes to defense systems acquisitions management. Contracts became performance-based. No longer would government acquisition offices actively participate and have deep understanding of what was happening in the development of their weapon system. Instead, the contractors would handle things, letting the government know if anything was going wrong. With the paperwork reduction, conversion to performance-based contracts, and use of commercial standards, many space program acquisition specialists and engineers got the message their expertise was no longer needed. Experience trickled away slowly at first but soon became a torrent.

The widespread changes required another revision of the DoD 5000 acquisition regulations. Acquisition Reform overwhelmed the now relatively minor Acquisition Streamlining changes. Updating DoD 5000 began in earnest in March 1995. The changed Federal Acquisition Regulations followed September 1995. The effort had been intense, broad-ranging, and, in bureaucratic terms, implemented in the blink of an eye.

*   *   *

Converting the military standards to the new "performance-based" approach was neither simple nor straightforward. Military standards, whatever their perceived ills may have been, were a treasure trove of lessons learned from past problems and directives based on first-principles of physics, chemistry, and thermodynamics. Discarding this wisdom rankled with some participants, resulting in a slow and painful process of conversion. In a few cases, the wisdom was conserved in handbooks or appendices to the new performance standards.

Thus, DESC was unable to issue an initial draft revision to MIL-I-38535 to convert it to MIL-PRF-38535 until 4 January 1995. The changes were far from cosmetic. Standards language that amounted to edicts on how to do things was no longer permitted. The government could no longer direct manufacturers how to produce or test the parts destined for use in military systems. Instead, manufacturers were to use their own "best commercial practices." The expected results of such changes included putting the best available technology into military systems sooner. Also, costs would be correspondingly lower, approaching those of the commercial marketplace.[28]

It is difficult to overstate the sweeping nature of the revisions and their implications. For instance, to address concerns about the global marketplace and international competitiveness, QML microcircuits were no longer required to be domestically built. For such overseas production, no country of manufacture needed to be cited. Parts could come from multiple countries, and be assembled in another. In fact, design, fabrication, assembly, and test had no requirement for being under the control of a single process owner.

---

[28] Joe Barrows, "QML Specification To Become Performance-Based," *EEE Links* 1, no. 2 (April 1995).

The consequence, in many cases, was that purchasers lost traceability of device heritage (handling, environments, manufacturer, place of manufacture, etc.). Gone were the days when a historical jacket followed a device, documenting its manufacture and handling. The long tradition of historical jackets dated to June 1969 when they were made mandatory by SAMSO at the release of *Assembly-Through-Flight Test and Evaluation Guide*. Failures and process deviations were an inevitable part of manufacturing, even under the best of circumstances. Under this new regime, there was little possibility of tracking down the source of the failure and correcting it.[29]

Class S suppliers still had to meet some requirements, included in an appendix to MIL-PRF-38535. However, the requirements for qualification were deleted under the process-oriented scheme adopted in the performance standards. Qualification data was no longer necessarily available, although some manufacturers would make some data available to potential customers as a marketing tool. A considerably looser continuing auditing process was instituted as well.[30]

As DESC released the new draft of MIL-PRF-38535 to the joint EIA-JEDEC meeting in January 1995, the members of that group were assured the implementation and intent of the QML system remained unchanged. One observer reported:

> MIL-I-38535 has always allowed vendors the option to optimize device processing and testing through the vendor's technical review board. However, the proposed changes to the document will now essentially allow vendors the flexibility to create their own processing, environmental screening and evaluation programs. DESC will only include recommended guidelines for designing, processing and testing microcircuits. There will be very few (if any) strict requirements.[31]

*   *   *

In the flurry of reinventing defense acquisition, other decisions flowed in part from canceling military standards. The nation began to lose far more than intended. The high reliability space parts industry had combined efforts from the government in terms of Rome Laboratory and other government facilities, the Aerospace Corporation (for space systems), the prime contractors, and major subcontractors. The military standards caused a proliferation of high technology testing equipment to be present in all of these places. Whereas Rome Laboratory might emphasize failure analysis or failure modes and effects research, contractors were analyzing parts they had developed to understand their manufacturing processes better, or as screening and receiving inspections for parts received from others. With the elimination of the standards and specifications requiring these, expensive maintenance contracts for high technology inspection equipment were not renewed as contractors became more

[29] Ibid.

[30] Ibid.; author interview with Thomas Turflinger, former parts engineer with Navy CRANE, July–Aug. 2010. Turflinger explains that another element of the change was a tension between the avionics grade, Class B, and the space grade, Class S, camps. The main focus followed the size of the market, which meant Class B, and only the necessary lip-service was paid to proponents of Class S.

[31] Barrows, "QML Specification to Become Performance-Based."

cost conscious and price competitive. Obsolete equipment was not replaced. Some inspections could be contracted out to independent third parties. The industry lost extensive expertise as engineers and specialists who had grown up with the space industry saw they were no longer necessary. Many with the most experience simply retired.

Without directive military standards, there was no need to keep expensive government organizations whose mission was writing and maintaining them. In 1995, the Base Realignment and Closure Commission (BRAC) drew a bead on Rome Laboratory. In the cold calculus of the BRAC's decisions, the closure of Rome Laboratory was justified because the "Air Force has more laboratory capacity than necessary to support current and projected Air Force research requirements. The Laboratory Joint Cross-Service Group analysis recommended the Air Force consider the closure of Rome Laboratory."[32] The Commission had taken a high level look at breaking up the laboratory and dispersing its functions and capital equipment, all of which assumed people would simply follow their functions to Hanscom AFB, Massachusetts, and Fort Monmouth, New Jersey.

Not surprisingly, some in the Air Force initially disagreed. James Boatright, Deputy Assistant Secretary of the Air Force for Installations, as recently as 1993, had announced publicly "the Air Force has no plans to close or relocate the Rome Laboratory within the next five years." Five years was simply the planning horizon for the defense budget process, so, in other words, Rome Labs was covered in the budget.

One of several Air Force "superlaboratories," Rome Laboratory was a critical mass of facilities, equipment, and most important of all, expertise, concentrating on the future of the Air Force, DoD, and industry. In 1995, as the BRAC considered its decision, Rome Labs was a Tier 1 (top) Laboratory within the Air Force.[33]

The BRAC seemed unimpressed. They asserted breaking up Rome and collocation at other facilities would "reduce excess laboratory capacity and increase inter-Service cooperation and common C3 research."[34] Not everyone saw it the way the BRAC did, and some of those who disagreed had far more expertise in the subject than anyone on the Commission.

For instance, on 5 May 1995, the BRAC Commissioner received a letter signed by five former Chief Scientists of the Air Force: Drs. George R. Abrahamson, F. Robert Naka, Michael I. Yarymovych, Joseph V. Charyk, and H. Guyford Stever. They included a former Director (and founder) of the National Reconnaissance Office, a former Deputy Director of the National Reconnaissance Office, and the scientist who predicted the rise of personal computers and the revolution they would bring in military and civilian life. They had "grave concerns regarding the Department of Defense recommendations to relocate most of Rome Laboratory to Hanscom

---

[32] 1995 DoD Recommendations and Justifications, Rome Laboratory, New York. This was part of the overall decision package of the Base Realignment and Closure Commission in 1995.

[33] Letter, James Boatright, Deputy Assistant Secretary of the Air Force for Installations, to the Base Realignment and Closure Commission, 7 May 1993.

[34] 1995 DoD Recommendations and Justifications.

Air Force Base and Fort Monmouth."[35] They understood the impact of closing Rome Laboratories in both capability and human terms. The Laboratories were home to many world-class scientists who would not move to wherever the Laboratory's functions were dispersed. They explained "Rome Laboratory is a unique and irreplaceable resource; movement will severely damage that resource." In the area of Command, Control, Communications, Computers, and Intelligence (C4I), the "Lab undertakes some unique and outstanding activities that ought to be preserved." The BRAC's mistake was in simply moving functions, personnel slots, and some equipment and related items, using obscure models and equations to demonstrate cost savings (as though moving from a high cost area to a low cost area was not sufficient on its own to demonstrate savings). The Chief Scientists explained Rome's value was not floorspace, equipment, and personnel slots, but people and the system they supported. At the time, Rome Labs' C4I effort relied on key "experienced scientists . . . recent recipients of doctoral degrees, to doctoral candidates."[36] In other words, Rome Labs had a built-in sustaining system. It was filling current needs uniquely while providing the training grounds to maintain the health and viability of this work through close ties with Cornell and Syracuse Universities.

Beyond the irreplaceable human resources, what, specifically, were Rome's unique activities? Rome Laboratory had evolved its functions extensively in keeping with the necessity to stay up with (and ahead of) advancing electronics technologies.

At the time of the closure and dispersal of Rome Laboratories, their efforts comprised extensions of the efforts discussed earlier. The loss of the Electronics Reliability Division had the longest term implications for the high reliability and space qualified electronics areas.

Prior to their deactivation, the Electronics Reliability Division contained the Reliability and Diagnostics Branch which was the

> *DoD's premier test and analysis facility for analog devices. This group pioneered the evaluation of analog devices—especially Monolithic Microwave Integrated Circuits (MMICs)—used in advanced AF and DoD systems. . . . In addition, the group includes one of the world's experts for the testability and fault tolerance concerns of microprocessors and other complex devices. This work has led to the design and development of the RH-32, a radiation hardened, fault tolerant, 32 bit computer for space applications. . . . Finally, this group is leading the investigation of the reliability of photonic devices.*[37]

The Design and Diagnostics Branch was the digital device equivalent to the analog activities of the Reliability and Diagnostics Branch. "This group manages DoDs most sophisticated tester for digital devices—the . . . Teradyne tester. It can

---

[35] Letter, Dr. George R. Abrahamson et al. to Alan J. Dixon, Chairman, Base Realignment and Closure Commission, 5 May 1995.

[36] Ibid.

[37] Rome Laboratory Command and Control Systems Fax, Subj: Back-up Data, Programmatic Impacts of SECDEF Recommendations, 21 April 1995, p. 2.

test the most complicated and highest speed integrated circuits and multi-chip modules built today."[38] The significance of this work, as an example, was its foundation for the standard processes of MIL-STD-883. Like so much of the work Rome Laboratories did, the knowledge and wisdom gained were captured and turned into sentences, paragraphs, and footnotes within various military standards. That was why the Teradyne tester was not just "the only facility of its kind in DoD," but was also key to advancing governmental and commercial microelectronics by capturing this information. The group led

> *DoD efforts in the rapid prototyping of signal processing architectures—crucial to the design of advanced systems for air and space platforms. And the group provides automatic test technology that reduces costs for logistics support by an order of magnitude. The systems avoid lock in to contractor proprietary test equipment and allow test vectors to be generated directly from high level equipment descriptions.*[39]

Fundamental research was the function of the Reliability Physics Branch:

> *Basic research that investigates the influence of materials and interfaces on the reliability of silicon-based and compound semiconductor devices. Fundamental work in electromigration in thin films—an increasing reliability problem as device geometries become smaller and smaller. Group develops improvement in semiconductor processing to desensitize products to this failure mechanism. Area also works on the failure mechanisms of simple test structures which can be used "on-chip" for cost-effective in-line screening. In addition to electromigration, evaluates hot-electron degradation and time dependent dielectronic breakdown. Currently researches the R&M impacts of the use of plastic encapsulated microcircuits in defense systems which offer large potential cost savings, but have little reliability data in defense uses. Efforts support all AF systems especially air and space platforms. In addition to AF customers, supports ARPA, NASA, NRL and the electronics industry.*[40]

Some reliability problems had nothing to do with physics of devices, but stemmed from design issues. That was the province of the Design Analysis Branch, which

> *[d]evelops simulation tools for the Air Force and DoD to evaluate the mechanical, thermal, and electronic performance of devices and components before they are built and to investigate failures after the devices are fielded. Recently, these tools were instrumental in an investigation of problems in Travelling wave Tubes (TWTs) at*

---

[38] Ibid. It has not been possible to further identify the Teradyne tester model. Teradyne was a large Boston-based manufacturer of test equipment, and what distinguished this one in the minds of the Rome Laboratory people defending themselves against the BRAC is not clear. Teradyne testing machines were in other government laboratories. Perhaps what was meant was the personnel who could run the machine properly for unique purposes were themselves unique.

[39] Ibid. The irony of the situation was that Rome was behind some of the commercial industry's advances, which the DSB and others were trying to force the military to use. The problem was that the advances Rome Labs helped with were for high reliability systems, and these migrated to support commercial lines, while the DSB and Packard were trying to reverse that flow.

[40] Ibid., 3.

[Warner Robins Air Logistics Center]. The simulation tools pinpointed the problem in the thermal design of the tubes and was able to definitively indicate which tubes should be scrapped and which could be saved—returning a substantial investment of TWTs to the inventory. Group has developed a multi-chip module (MCM) thermal analyzer that allows design evaluation of these complex devices in software before committing to hardware production. The analyzer simulates the full electrical and thermal performance of the devices including the interactions between thermal and electrical properties. This pioneering work will greatly reduce the costs and schedule for advanced systems which use MCMs. This work supports all product centers and logistic centers but has special significance to the space community. Air Force efforts have been greatly leveraged by ARPA funding in this area.[41]

The mention of the funding by ARPA underscored the multiagency interest in and support for the capabilities of Rome Laboratories, for the benefit of all.

This also highlights another shortfall in the BRAC Commission's recommendations and the appeals against their actions. The work being done for space programs occupied a narrow space within the breadth of the C4I work done at Rome Laboratory. The breadth comprised several key DoD, inter-Service, and interagency responsibilities that the Air Force did not defend effectively. While covering the forest breadth, the pending loss to space program trees was lost in the mix.

The former Chief Scientists went further, pointing out that

[the] implications for the Air Force are profound: No other function ranks as highly as C4I in the eyes of senior military and civilian leaders, as evidenced by the repeated statements to that effect made by the Commanders in Chief of the unified commands, the Chairman of the Joint Chiefs of Staff, the Secretary and Under Secretary of Defense, and the Director of Defense Research and Engineering, as well as in resolutions made by both houses of Congress.[42]

They warned the BRAC Commission that the "damage will take years to rebuild." The BRAC had "suggested that any difficulties encountered will be justified by a reduction in administrative costs and by the benefit of new synergies that will develop among the services and with the universities surrounding the Hanscom and Fort Monmouth sites." Such an eventuality was not going to happen overnight. It might not ever happen, given the loss of the human resources overlooked by the BRAC. The former Chief Scientists were not alone in their serious concern. They told the BRAC, "Evidence of the truth of the above statements includes the Navy's declining to participate in the proposed action, and both the Army and Navy declining to participate in other . . . proposed relocations of C4I capability."[43] The other Services understood the impacts and the inadvisable breakup of the Rome Laboratory. Sadly, the BRAC's spreadsheets did not take into account Rome's real strengths.

---

[41] Ibid.

[42] Letter, Dr. George R. Abrahamson et al. to Alan J. Dixon, Chairman, Base Realignment and Closure Commission, 5 May 1995.

[43] Ibid.

Those spreadsheets claimed a return on investment in four years. It went as so:

The total estimated one-time cost to implement this recommendation is $52.8 million. The net of all costs and savings during the implementation period is a cost of $15.1 million. Annual recurring savings after implementation are $11.5 million with a return on investment expected in four years. The net present value of the costs and savings over 20 years is a savings of $98.4 million.[44]

Dick Helmer, the BRAC Commission's Cross Service Team Senior Analyst for Rome Laboratory, said several key questions needed to be answered about the closure of Rome Laboratory: "(1) Should the Services have accepted the Laboratory Cross Service Group's proposal to consolidate C4I Functions . . . ? (2) Does it make sense to lose the lab and realign its functions at two different locations?"[45] These were addressed by the Chief Scientists, whom the BRAC Commission found uncompelling. Perhaps the Chief Scientists' appeal was lost in the flood of adverse reactions, from New York Governor George Pataki's claim of bipartisan support to retain the Laboratory and the impact on the community and the state, Senators Daniel Patrick Moynihan and Alfonse D'Amato's assertions the costs and benefits were not accurate and that "everything is to be gained by leaving Rome Lab in place," to Congressman Sherwood Boehlert's criticism there were no cost savings in closing the lab. Politically, all of these criticisms were the expected standard reactions of politicians over the loss of a government facility. The BRAC Commission ignored input from people who should know more specific detail than either the commission staffers or the politicians. In addition to ignoring the former Chief Scientists, the commission ignored others such as General Robert Mathis, former Air Force Vice Chief of Staff (and former Rome commander), who said "splitting Rome Lab will prove impossible."[46] Since the BRAC Commission looked only at numbers of dollars and people, it believed such claims were overstatements of the situation and emotional reactions. Time would only tragically demonstrate the BRAC Commission's deep misunderstanding.

Laboratory equipment could be dispersed, facility square footage replicated, and missions combined with other organizations. But when world-class experts decided not to move upon Rome Laboratory's dispersal and chose either to continue teaching or just retire, the nation lost in many important ways.

*   *   *

In the spirit of conversion to commercial standards and practices, acquisition reform caused a new twist for the move from QPL to QML. Under QML, of course, any parts on a qualified line would then be allowed to claim whatever quality level (such as Class S) for which the processes were certified. The immediate impact was minimal, since the necessary equipment for testing, and the processes themselves,

---

[44] 1995 DoD Recommendations and Justifications.

[45] Base Visit Report, Rome Laboratory, Griffiss Air Force Base, Rome, New York, 5 April 1995.

[46] Ibid.

were based on the former, tightly controlled system, which complied with military standards. In fact, it seemed like a good idea given the realities of the changes in the electronics industry. In effect, Rome Laboratories had read the handwriting on the wall, and proposed a solution to take high reliability military grade parts (Class Q), and, with additional testing requirements, allow them to be considered space level quality (Class V). Experience in the early days of high reliability electronics had underscored the fallacy of doing precisely this, but with adequate controls in place, the new approach to space quality parts was claimed to be nearly as good, assuming enhanced Class Q high reliability processes.

However, acquisition reform changed the method of enforcement to one of occasional insight when audited by Defense Supply Center Columbus (DSCC), successor to DESC. Over time, all processes stray, and occasional audits, while important, proved insufficient. For instance, advancing technology caused some subtle effects in testing. Older test equipment was no longer suited to the new, smaller feature sizes of the rapidly advancing technologies. Without the constraints imposed by independent technical oversight, some manufacturers claimed their experience indicated a certain test or number of test cycles were excessively costly, without resorting to first principles analysis. Thus, they claimed their processes could be "optimized" by discontinuing or reducing such tests. Optimization in this sense, though, was not in regard to assuring the quality of the products, but in terms of profit margins and competitiveness. Performance standards gave little technical guidance on which to base approvals and disapprovals beyond what the suppliers provided. Such "test optimizations" required a government laboratory to verify the deletion or modification was allowable, but the principal laboratory had fallen under the axe of the BRAC.[47] Gone was the ability to independently determine the impacts of such optimizations on reliability and failure rates.

* * *

Among the first space systems acquired under acquisition reform was the Space-Based Infra-Red Satellite (SBIRS). SBIRS itself had a high orbiting component as well as a low earth orbit one, but the high system provides the best illustration of what was happening. SBIRS-High was to improve upon and replace the venerable Defense Support Program, which itself had succeeded the Midas program of early warning satellites. These space systems were responsible for the earliest detection of missiles launched against the United States. This mission was fundamental to getting the national security and defense space programs started in the 1950s.

[47] Evaluation of the Defense Supply Center Columbus Qualified Products List and Qualified Manufacturers List Program, Department of Defense Office of the Inspector General, Report No., D-2002-090, 14 May 2002. An indicator of the problems of maintaining quality, due in part to the increasing workload of inspections and certifications, was that "512 (42.8 percent) of the required 1,196 manufacturing line audits scheduled during 1999–2000 were not accomplished. Some of the manufacturing lines have gone 8 years without certification." Author interview with John Ingram-Cotton.

SBIRS-High began in 1996 with a plan for five geosynchronous satellites combining both wide area surveillance sensors and the ability to stare at particular points of interest. Two additional satellites were to perform similar missions from highly elliptical orbits covering the north pole. The first geosynchronous satellite was to be delivered in 2003 with a launch expected in 2006. In 2000, the contracting officer described SBIRS-High as "a consolidated, cost-effective, flexible space-based system that in time will meet the United States' infrared global surveillance needs through the next several decades.[48]"

But SBIRS-High's acquisition was based on Total System Performance Responsibility (TSPR). Specifically, TSPR was actually a contract clause, stating that

> *the contractor agrees to assume TSPR in accordance with the terms and performance requirements of this contract, and to furnish all necessary effort, skills, and expertise within the estimated cost and award fee pool of this contract.*[49]

Limiting the contractor's liability in such a way meant that, were any cost overruns to occur, the contract would be little different from a cost plus contract from the government's perspective. The number of military standards and specifications for the program was reduced from 150 to only two; everything else was based on "commercial best practices" and international standards.[50] The program began amid a wide range of cost expectations. Four years after initiation, some degree of optimism still remained about the program's acquisition approach. That was about to change.

*   *   *

Trends, of course, are not immediately obvious. When trends involve critical national assets, they become obvious only after some serious damage has been done to national security and national defense.

In January 1997, a Delta suffered the failure of a solid rocket motor while launching a GPS satellite. A national security satellite was lost in August 1998 when a Titan IV lost control due to electrical shorts. Later that same month, a Delta failed as a result of a control system design deficiency. Eight months later, a DSP early warning satellite was placed in the wrong orbit due to an upper stage failure. In April of 1999, another Titan IV failed launching a Milstar communications satellite. A month later, a Delta failed. These were only the failures; nine other events involved notable anomalies.

---

[48] David N. Spires, *Beyond Horizons: A History of the Air Force in Space, 1947–2007* (Peterson Air Force Base, Colorado: Air Force Space Command, 2008), 328; Major Henry P. Pandes, *A Quest for Efficiencies: Total System Performance Responsibility* (Maxwell Air Force Base: Alabama: Air Command and Staff College, 2001), 22. Pandes based this description on an interview with Terrence D. O'Byrne, the Lockheed Martin Space Systems Company contract administrator on the SBIRS High contract.

[49] Pandes, *A Quest for Efficiencies*, 23–24.

[50] LtCol Wesley A. Ballenger, Jr., *Acquisition Reform: Where We've Been; Current Legislation and Initiatives; and Where We're Going* (Fort McNair, Washington, D.C.: Industrial College of the Armed Forces, 1995), 14. Ballenger later managed the Global Positioning System program.

The dollar value of these launch vehicle and satellite losses ranged into billions.[51] The monetary loss, though, did not compare with the impacts to national security and national defense. Satellites are not built quickly, so the nation suffered years of diminished or lost capabilities as replacements were built.

In keeping with the traditional approach to failures, the situation was thoroughly analyzed. The Department of Defense directed a Broad Area Review (BAR) of launch vehicles. Similar to Bradburn's 1976 Ad Hoc Study Group on Space Launch Vehicles, the BAR's examination reached back to 1985 to include all catastrophic failures in U.S. launches of potential relevance.

The BAR found the most recent failures suffered from a lack of engineering design robustness. Additional information indicated more thorough independent analyses prior to launch might have helped avoid problems. Such findings were in keeping with prior failure reviews, and served to underscore the importance of the basic engineering and systems engineering processes that had built the highly reliable space programs. Importantly, they also saw "[a]t least six failures of systems developed on 'commercial practices' are attributable to the non-application of lessons learned (as documented in Commander's Policies, which have been cancelled, even for DoD programs, since 1995)."[52] Translated into this history's terms, the BAR concluded at least six of the failures tied directly back to the consequences of acquisition reform's move from military standards to "best commercial practices."

Sadly, the Launch Vehicle BAR had uncovered only the tip of the iceberg. Satellites take longer to build than their launch vehicles. The consequences for orbital systems took two more years to surface. By the early years of the new century, no more launch vehicle problems occurred due to the changes immediately imposed following the BAR. Instead, the failures began to occur in satellites in orbit. Nationally critical satellites and commercial satellites began failing. Again, the cost quickly rose to billions of dollars in cost impacts, and further damage to national defense and national security.

The immediate thinking was the orbital failures reflected problems undiscovered during integration and testing, the last stages of a satellite's building (the ascending right side of the classic V diagram). Perhaps the testing was no longer adequate. Perhaps tests were not perceptive enough to detect problems, or the wrong tests were being done, or not enough of the right tests were being done. In an effort mirroring the Launch Vehicle BAR, an extensive analysis of satellite failures examined testing programs. Almost immediately, the study participants found problems that went well beyond testing.

---

[51] William F. Tosney, *Proceedings of the 4th International Symposium on Environmental Testing for Space Programmes*, Liege, Belgium, 12–14 June 2001; Michael C. Sirak, "Our Share of the Pie," Airforce-magazine.com, 29 Feb. 2008; Alex Liang, *Task Force Launch History Panel Final Report* (El Segundo, California: Aerospace Corporation, 1 Oct. 1999).

[52] Liang, *Task Force Launch History Panel Final Report*; William F. Tosney, *Faster, Better, Cheaper: An Idea Without a Plan* (El Segundo, California: Aerospace Corporation, 5 Oct. 2000).

Testing problems did indeed exist. Testing early in the development of satellites had been cut back to save costs. The gamble was that any problems existing in units would be caught in the later phases of integration and testing. The problems were not solely attributable to this deferral of testing. Failure rates of parts and subsystems began rising in the late 1990s. Starting with piece parts, testing had become inadequate in terms of the quality, quantity, timing, and types of tests conducted.

The underlying confidence in the quality of manufacturing processes contributed to the latter problems. Such confidence was unjustified because cutting early testing was based on a continuation of high quality processes that had not been maintained. The entire industry was caught counting on the inertia of the highly reliable pre-acquisition reform practices. As the study progressed, dozens more problems were identified across the spectrum of space system acquisition. While immediate impressions were tied to the loss of standards due to acquisition reform, the list of causes was top heavy with a litany of other changes also directly resulting from acquisition reform.

Understandably, actions began immediately to improve testing and reinvigorate the appropriate military standards. As experience began to mount and further analysis continued, acquisition reform became synonymous with all the problems uncovered. That was only largely true, but not the complete story.

Many other forces were found to be at work. Some of these had been in the background for years prior to acquisition reform. Others had been accentuated or exacerbated by acquisition reform. Many were direct consequences of acquisition reform, though others were unintended consequences. Of the unintended consequences, some were foreseeable. Further in-depth reviews of the rise in all failures uncovered even more systemic problems unrelated to acquisition reform. These problems did not exist in isolation, but were as interrelated as each thread in a fabric.

Expertise had been lost as people retired, affecting both the government and industry. A generation of program managers was taught the principles of acquisition reform and the supposed evils of military standards. Such things were not easily unlearned after ten years of preaching acquisition reform and demonizing the former acquisition processes. Pushing back against the gospel of reform required considerable effort and education over years. Some new disciples of acquisition reform refused to accept the findings of the failure reviews claiming there was no data to make such findings. But that was a self-fulfilling denial. Acquisition reform's lack of deliverable data led to an industry where data was not released to the government because the government had no data rights, and had lost control of its own acquisitions.

Even the flow of information about quality problems had slowed to a trickle. GIDEP had been the information exchange forum for sharing issues about electronic parts and materials, and had been beneficial for participating programs. The competitiveness of the commercial marketplace had been swelling in the 1980s, and once freed from government and FFRDC oversight, adverse information about parts quality could be hidden by invoking proprietary rights. While lawyers worked and reworked the wording of a part recall or problem notice, the homogenized end results were delayed beyond relevance. Not all suppliers took refuge behind proprietary fences to hide problems and protect market share, but the loss of such information undermined the benefit for many high reliability programs. After all, if a problem was not revealed until after launch, not much could be done to mitigate its impact on a satellite. The resulting situation allowed questionable and defective parts to enter supplies, not to be discovered until very late in a program development, when the cost of finding and replacing defective parts was very high. Litigiousness overtook the team spirit of cooperation among the organizations participating in the process of making high reliability satellites.

Acquisition reform replaced standards with trust in the contractor's processes. Acquisition reform's new gospel of "best commercial practices" completely turned on its head the whole process of quality and reliability controls. The government no longer oversaw the practices used, which began a rather fast decline in some areas. Competitive market pressures forced reduction or elimination of expensive tests, test equipment maintenance, and high reliability parts. As though Economics 101 had been eliminated from the curriculum, the government introduced a new market dynamic that had no mechanism to maintain high reliability beyond trusting those whose metrics were based on return on investment and shareholder equity.[53] Detailed understanding on the part of the government was gone. The government could not even do as President Reagan had suggested with the former Soviet Union—trust but verify—due to the absence of deliverable data. Relationships with parts suppliers had undergone a change after the loss of the standards and the technological underpinnings provided by Rome Laboratory. Lacking adequate engineering support as test optimizations proliferated, DSCC approved most such test eliminations with little in-depth review. Because everything starts with parts, space systems were being affected by this reduced technological insight and engineering at the most fundamental level.

On top of all these problems, and many more specifics, came the realization that the Packard Commission, Cheney Review, and acquisition reform had not saved any money, but only deferred and magnified expenditures.

> *Initiatives based on the recommendations of the Packard Commission did not reduce the average cost overrun percent experienced on 269 completed defense acquisition contracts evaluated over an 8-year period (1988 through 1995). In fact, cost performance experienced on development contracts and on contracts managed by the Air Force worsened significantly.*[54]

Whether the approach was called TSPR[55] or Faster-Better-Cheaper (in NASA), the cure was far worse than the perceived ills. So much had gone wrong by the end

---

[53] Author interview with John Ingram-Cotton.

[54] David S. Christensen, Capt. David A. Searle, and Caisse Vickery, "The Impact of the Packard Commission's Recommendations on Reducing Cost Overruns on Defense Acquisition Contracts," *Acquisition Review Quarterly* (Summer 1999): 251.

[55] Federal Acquisition Regulation, Part 14.201-1, 1 Jan. 2000, p. 242.

of the 1990s, in fact, by the time the dimensions of the problems became clear, several studies recommended sweeping reforms to return as closely as possible to the pre-acquisition reform days. Sweeping reforms, however, cost money. Resource prioritization limited the scope of the corrective reforms to a few important and essential efforts. Besides, some of the changes had been so fundamental with critical infrastructure dismantled, and struck so severely at the core of the prior approach to acquiring high reliability spacecraft, that there was no going back.

By 2000, the SBIRS-High program manager was claiming "TSPR has been successful in allowing the contractor to determine approaches for interfacing . . . however, it has not shown to be successful in meeting acquisition program baseline parameters for delivery of the first increment."[56] The delays cost the contractor its incremental award fee at that point, and as the program manager further observed, "[t]his program has shown that the road to success is not an easy one despite inclusion of TSPR in the contract."[57]

Shortly thereafter, though, in 2001, SBIRS-High exceeded its Nunn-McCurdy cost growth limit for the first time. The program was recertified to Congress as remaining essential for national security. Such a recertification, though painful to those involved, was obvious: the nation remained threatened by others with credible long range ballistic missile programs, and no other program was available to replace the aging DSP satellites. After going through the internal bureaucratic steps, in May 2002, Undersecretary of Defense for Acquisition, Technology, and Logistics Pete Aldridge issued the recertification, which required a program restructuring to bring down the cost, in part by reducing the number of satellites.[58]

At that time, a marked contrast existed in the cost of the program. From a bid of $1.6 billion in 1996, the 2002 expectation for cost at completion had quadrupled (see Fig. 18.1). But that was not the end of SBIRS-High's problems.

The importance of the program, and the degree of its problems, led to an important analysis of space systems acquisition. This time, the study was actually a joint DSB and Air Force Scientific Advisory Board (AFSAB) review, led by Dr. A. Thomas Young. Young led an extensive team of highly qualified engineers and acquisition experts in what was, at that time, the largest and most talented group of such individuals to examine such a topic. They examined all the existing space programs that had a TSPR clause in their contract or were being managed under the TSPR "philosophy." Their conclusions were not kind, but reflected the seriousness of what had happened to critically important space acquisitions.

One overall observation was that "'[c]ost' has replaced 'mission success' as the primary objective in managing a space system acquisition. . . . The task force unanimously believes that the best cost performance is achieved when a project is managed

---

[56] Pandes, *A Quest for Efficiencies*, 24–25. Pandes was quoting interview responses from LtCol Michael J. Wallace, the SBIRS High program manager, and Terrence D. O'Byrne, the Lockheed Martin Space Systems Company contract administrator on the SBIRS High contract.

[57] Ibid.

[58] Spires, *Beyond Horizons*, 328.

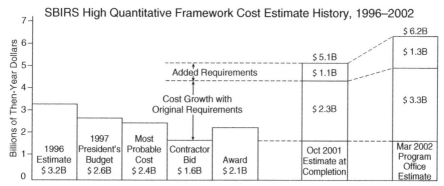

Figure 18.1. An example of satellite cost growth. Source: Booz-Allen-Hamilton Study: Space Systems Development Growth Analysis Report, as adapted by A. Thomas Young, *Report of the Defense Science Board/Air Force Scientific Advisory Board Joint Task Force on Acquisition of National Security Space Programs* (Washington, D.C.: Office of the Under Secretary of Defense for Acquisition, Technology, and Logistics, May 2003), 22.

for 'mission success.'"[59] That was not to say they had a formula that would solve cost growth or avoid problems—these occur because advanced programs contain some element of technical or other risk.

However, TSPR was shown to be deficient, in mild terms, for the acquisition of space systems. The Young report called for a return to the basics of space systems acquisition, with "strong government involvement and control throughout the development and procurement process"—which was to say, a return to pre-acquisition reform practices. While their language about returning to the basics was a bit muted, they were less restrained in their conclusions about SBIRS-High.

> *In short, SBIRS High illustrates that while government and industry understand how to manage challenging space programs, they abandoned fundamentals and replaced them with unproven approaches that promised significant savings. In so doing, they accepted unjustified risk. When the risk was ultimately recognized as excessive and the unproven approaches were seen to lack credibility, it became clear that the resulting program was unexecutable.*[60]

Even that was not the end of the ramifications of acquisition reform. It took years to get into trouble, and would take years to get out. By early 2007, SBIRS-High had come up against Nunn-McCurdy provisions four times, the original five geosynchronous satellites had been cut to three, and the launch date slipped into late

---

[59] A. Thomas Young, *Report of the Defense Science Board/Air Force Scientific Advisory Board Joint Task Force on Acquisition of National Security Space Programs* (Washington, D.C.: Office of the Under Secretary of Defense for Acquisition, Technology, and Logistics, May 2003), 13.

[60] Ibid., 30; Spires, *Beyond Horizons*, 333–334.

2008. In April 2007, the GAO remained unconvinced and "expressed reservations that despite recent rebaselining the 'program still faces considerable risks.'"[61]

The application of TSPR in national defense space programs was demonstrably a serious problem. The problems were not isolated to national defense space's TSPR, or NASA's "Faster, Better, Cheaper." The joint DSB-AFSAB Young report examined classified programs such as the Future Imagery Architecture (FIA). For reasons of classification, their report could only say, "The task force found the FIA program under contract at the time of our review to be significantly underfunded and technically flawed. The task force believes that the FIA program—thus structured—is not executable."[62]

Similar findings of flaws applied to the Evolved ELV program, which was the sole means of access to space for both national security space and commercial space programs. These were serious red flag warnings reminiscent of the all too recent launch vehicle failures of the 1990s.[63]

David Spires, a generous assessor of what was happening, gave additional accounts of the status of all national defense space programs in his comprehensive 2007 history, *Beyond Horizons: A History of the Air Force in Space, 1947–2007*. The problems for the National Polar Orbiting Environmental Satellite System (NPOESS) and the Advanced Extremely High Frequency (AEHF) communications satellites were extreme.

NPOESS was to have launched in 2006. Each satellite was to have fourteen essential instruments for

> climate, oceanographic, atmospheric, environmental, and solar observations. . . . By the spring of 2007, NPOESS costs had increased by 34 percent, precipitating a Nunn-McCurdy review in 2006, subsequent recertification, and program restructuring. That led to removal of six of the satellite's 14 "essential" instruments, which, together with the instrument removed in 2005, left the program with half of the original configuration. . . . The first NPOESS satellite launch date by this point had been extended from 2006 to early 2013.[64]

Similar costs and delays affected the AEHF, the successor to the Milstar system. The latter system had been one of the launch failures in 1999 leading to the initial concerns that something was amiss in space systems.[65]

By the early years of the 21st century, no one producing highly reliable spacecraft had remotely the same relationships with each other as they had less than ten years before. The years from 1994 until past the turn of the century were the spacecraft development equivalent to the moments after the lookout yelled to Captain Smith there was an iceberg ahead. It takes time to realize the dimensions of a disaster. Captain Smith and the passengers of the *Titanic* simply found out faster.

---

[61] Spires, *Beyond Horizons*, 328–329.

[62] Young, *Report of the Defense Science Board*, 31.

[63] Ibid.

[64] Spires, *Beyond Horizons*, 329–330.

[65] Ibid., 331.

* * *

Unlike the movie of the name, the high reliability spacecraft "perfect storm" did not come together in a few days or hours. A Darwinian evolution produced highly reliable spacecraft in an extremely optimized system comprising elements from government, FFRDCs, prime contractors, parts suppliers, acquisition policies, standards, technology advances, technology understanding, and laws, all related through complex interactions. Just as Darwinian evolution took time to create modern flora and fauna, decades optimized the complex interdependence of high reliability spacecraft acquisitions in place in the early 1990s. The high reliability spacecraft perfect storm resulted in the near simultaneous convergence of many different forces of change. The most devastating of these were the 1993–1995 policies, directives, and laws collectively known as acquisition reform.

The Darwinian forces affecting production of highly reliable spacecraft had been altering for years. Without acquisition reform, the failure rate would have inevitably increased due to other forces at work, and the system would have addressed the problems, correcting some and adapting to others. That is what happened before, such as when General Phillips set teams to study and understand the underlying causes of reliability problems and failures in 1969–1970. The processes would most have corrected and adapted to some new optimum that would have been distinctly different from the one forced by acquisition reform. Seeking and adapting to a new optimum is the nature of a complex enterprise describable by a production function, more about which will be said in a moment.

The key elements dismantled by acquisition reform truly changed the culture and constituent elements of the prior optimized production system. The acquisition reformers had to change the culture, and in that they succeeded. Meanwhile, with the best of intentions, they dealt a crippling blow to a system that produced high reliability spacecraft that were the envy of the technological world.

They attacked what they thought were the key elements of costs in weapon systems acquisition. Without meaning to, they aggregated acquisition policies and directives to such a general level that high reliability spacecraft were lumped together with tissues and toothpaste. They tampered with a system whose details they only barely understood. They set in motion changes that resulted in the loss of Rome Laboratory's expertise, for example, while overlooking the views of people who had greater understanding than they. The evidence was at hand, but ignored in the zeal of reform.

Part V

Resetting the Circuit Breakers

Many constituent elements came together to make well-intended change become a disaster. Understanding the forces of change, both controlled and uncontrolled, helps define the means to recover at least something of the former capabilities.

Although the period of reform had been focused and relatively short, its measures had been effectively and efficiently emplaced and made to take root before the consequences became obvious. The complexity exceeded contemporary understanding, and unexpected consequences were widespread. Dealing with the aftermath depended considerably on understanding what had happened. With the failure of an electronic part in orbit, a Failure Review Board, armed with an understanding of the physics of operation of the part and the physics of its failure modes, might be able to determine what happened and how to prevent it from ever happening again. Could the same thing be done to repair the system that produced high reliability spacecraft?

What is the current state of affairs? More important is the ultimate question: Can things be set right again?

Implosion: Lessons from National Security, High Reliability Spacecraft, Electronics, and the Forces Which Changed Them, First Edition. L. Parker Temple III.
© 2013 the Institute of Electrical and Electronics Engineers. Published 2013 by John Wiley & Sons, Inc.

Chapter 19

Brewing the Perfect Storm

A perfect storm does not arise from a simple set of circumstances. We have ranged broadly across topics bearing on the final outcome. Before drawing final conclusions about what happened (and continues to happen), or about corrective actions, we need to discern the factors that converged between 1993 and 1995. These should by now be familiar, but it is important to see is their complex interdependence. Also, although they may be familiar, this section ties together and summarizes how each actually worked.

This history has covered eighteen trends or forces of varying effect to the overall system producing high reliability spacecraft. Many, but not all, were direct consequences of acquisition reform, others were unintended consequences, and others were independent of acquisition reform. There is no order, significance of impact, or precedence implied here:

1. Long lead time Of JAN Class S parts affecting cost and schedule;
2. Minimum-buy requirements for JAN Class S;
3. Greater reliability demonstrated begets greater reliability demanded;
4. Cost and schedule constraints cause false shortcuts;
5. Rising consumer electronics market volume and competitiveness;
6. Rise of commercial space;
7. Cost of the high reliability parts infrastructure;
8. Loss of sponsorship for basic research;
9. Complexities of the high reliability parts marketplace;
10. Loss of engineers in acquisition;
11. Loss of standards;
12. Loss of industry data sharing;
13. Loss of insight to contractor activities;
14. Loss of in-factory government plant representatives;

Implosion: Lessons from National Security, High Reliability Spacecraft, Electronics, and the Forces Which Changed Them, First Edition. L. Parker Temple III.

15. Lack of pragmatic standards approach across device types;

16. The end of the Cold War and the "peace dividend";

17. Increasing device complexity; and

18. Push for the most advanced technology.

These forces waxed and waned over time, interacting in complex ways. Some acted subtly, others as fearsome bludgeons. Each added its unique contribution to the ultimate chaos made even more devastating by their unintended temporal convergence. Their coincident interaction magnified the overall impact, as the whole was far greater than the sum of the parts. Each of these is examined below to clarify how they contributed.

LONG LEAD TIME

During the Second World War, the United States became tied to the notion that only the latest technology should be put into U.S. systems. The technology lead was what set U.S. systems apart on the battlefield, especially *before* the battle, which might be avoided by the American methods of keeping track of its enemies. Sorting out the most advanced technologies was not painless—by their nature, highly advanced technologies mature on unpredictable schedules.

Lengthy delays increase program costs and put programs at risk. Advanced technology contends with program schedule length and attracts unwanted attention. The history of space systems contains multiple examples of this, but few are more notorious than McNamara's cancellation of the Dyna-Soar spaceplane in 1963. He did so based on Director of Defense Research and Engineering Dr. Harold Brown's claim far less costly alternatives existed: robotic reconnaissance satellites. Dyna-Soar had taken too long, and its technologies remained elusive. Just prior to the cancellation, Brown asked the Dyna-Soar program manager, William Lamar, "What can you do for me that SAMOS [an early reconnaissance satellite] can't?"[1] Reconnaissance satellites had already won the high technology competition by being a bird-in-hand, having become critically important to the National Command Authorities. The incident highlights how pushing for the most advanced technologies can sometimes leave one without any.

The move to Class S parts reflected a different conviction, that higher technology would produce more reliable parts. But long lead times of JAN Class S parts affected cost and schedule. Except for stockpiled parts, most parts required orders well ahead of time, and the lead time could get excessive. For more complex parts, such as analog-to-digital converters, lead times became as long as four years. Long lead time items required purchase of such parts early in the acquisition of a spacecraft, thus locking in technology that was out of date by the time of the launch of the spacecraft.

---

[1] Richard P. Hallion, *The Hypersonic Revolution* (Wright-Patterson A.F.B., Ohio: Special Staff Office, 1987), vol. 2, p. xvii.

The desire to put only the newest technology into space systems (in particular) was not an easy thing to do for other reasons as well. The extensive testing and qualification necessary to understand a new technology's failure modes delayed availability. That same delay of testing also considered whether the technology would work suitably in the space environment. These were essential differences in the highest reliability applications that were of lesser concern in most terrestrial uses. Thus, with thousands of hours of device time necessary to characterize failure rates (which are fundamental to reliability and lifetime calculations), the lead times for simple parts, but of advanced technology, could also be lengthy.

Consequently, related to the push for the newest technologies, the advancement of greater complexity, and greater difficulty in qualification of individual parts, long lead times provoked significant pressures for ways to reduce schedules and cost.

MINIMUM-BUY REQUIREMENTS

Initially, solid state device lot sizes and low yields kept the supply of parts comparatively low. In the early ICBM period, demand for large numbers of parts for qualification testing and actual use exceeded the industry capacity. As yields increased and lot sizes grew, high reliability parts buyers could specify all of their parts were to come from a homogeneous (same lot date code) batch. Homogeneity allowed statistically valid acceptance testing of a small sample from the lot. That would not last forever, particularly for larger and more complex devices. Lot sizes and yields increased as demands for the highest reliability parts slowly decreased inversely with the longer lifetimes of satellites. As the highest reliability parts became ever smaller portions of a manufacturer's overall business, manufacturers had less motivation to sell partial lots of parts. That was especially true for SMD parts. Minimum-buy requirements for JAN Class S reduced the attractiveness of partial-lot buys, since manufacturers did not want to produce a whole lot but sell only a portion of it, thus being left with unsold inventory. Minimum-buy quantities became the lot size of a production run taking into account any reductions due to process yields. Minimum-buy quantities moved the excess parts problem from the manufacturer to the purchaser. Since the purchasers were the spacecraft manufacturers, this increased the costs of spacecraft.

Such problems were program management issues because they affected schedule and cost. They also reflected a trend among the high reliability parts manufacturers to be supportive of customer needs, but less tolerant of impositions made by the high reliability customers. Contrarily, as the national importance of the piece parts critical to space systems became clearer, the importance of the highest reliability parts within the marketplace declined. One characterization of the chain of events was that the manufacturers moved from welcoming the business to continuing the business out of a sense of patriotism, national pride, duty, and eventually as an obligation. This view, however, does not really fit the continuation of production of high reliability Class S parts, which remain a small but profitable portion of the overall market. Lot sizes and partial lot buys began as a nonissue, and evolved to be an irritation that would rise to an annoyance.

Class S stockpiling was an interesting response to contending pressures. The need to stockpile Class S parts resulted from creating Class S in the first place. Class S was the necessary and obvious path to follow to improve the quality and reliability of satellites. It offered the better value for the taxpayers' investment in satellites by avoiding the high costs of late development test and orbital failures. But Class S's downside decreased the demand for high reliability parts due to longer satellite lifetimes, and the subsequent demand for even greater reliability and longer lifetimes.

Consider further, however, the need to stockpile was an obvious consequence of the small market created when Class S was defined. That small market was not viable because higher reliability trends resulted in lower volume demands. High reliability parts were a small portion of the overall U.S. demand for electronic parts, and Class S was a very small sliver of the high reliability market. Within little more than a decade, the downward pressure on the demand for Class S reflected the immense success of Class S, and also made its continuance problematic. Had acquisition reform not happened, one might reasonably conclude more than simple stockpiling would have been needed to sustain the Class S market. Moving all parts to QML, rather than just the complicated and technologically advanced ones, would probably not have been the solution.

Acquisition streamlining and acquisition reform took hold as JAN S stockpiling was getting up a head of steam. As it turned out, stockpiling did not solve the partial lot purchase issues, because the military standards that forced buying high reliability parts disappeared. It may be this was a false start, but it may also be JAN S stockpiling was simply a means to stave off the inevitable. The small JAN Class S marketplace was the root of the problem, driven by longer-lived satellites, the declining need for these satellites, and, hence, parts. The stockpile would have reached a point of diminishing returns; the stocks might never have been depleted while the parts themselves would have become obsolete technology (in most cases).

The success of Class S, its intricacies, and its consequent downward moving market illustrate the level of complexity and subtleties that acquisition reformers really ought to have understood before acting.

GREATER RELIABILITY DEMONSTRATED BEGETS GREATER RELIABILITY DEMANDED

Demonstrating greater reliability begat demands for even greater reliability. If a satellite designed for an 18-month mission duration ends up with an operational life in excess of seven years, the expectation for the next generation of spacecraft became more like the demonstrated seven-year lifetime. Longer lifetimes reduced the demand for spacecraft, driving down the number of spacecraft purchased, affecting workforce sizing, skill retention, training, and so on and driving up the cost of space launch (due to the same decreases in volume and timing).

Ever greater reliability was a dead end, as longer lifetimes decreased the need for more satellites, and hence demanded fewer parts. Although initially slightly offset by greater satellite complexity, the dominant effect came from the longer lifetimes. Fewer satellites and the longer times between satellite builds created the same sort of feast-or-famine purchasing of high reliability parts seen in the earliest ICBM programs. Such peaks and valleys hurt manufacturers and their willingness to deal with such variability. Class S stockpiling might have been able to smooth out the peaks and valleys a bit, but parts obsolescence would inevitably have been an issue for many of the stockpiled parts.

Such diminishing returns, at some point, without acquisition reform, would have forced a crisis in the marketplace. In fact, that point was probably reached in the late 1980s or early 1990s. Large block buys of satellites, especially the most complex ones, simply were unsustainable, the sole exception possibly being GPS.[2] The length of satellite lifetimes reached and exceeded a decade—meaning the need for a replacement discouraged the ability of primes and subcontractors to keep trained and skilled workforces in place between builds. That, in turn, resulted in each satellite becoming a new start, in essence. While some might have worked on a prior version, significant numbers of new labor meant workforce inexperience and workmanship problems. With the loss of strongly prescriptive standards, variation among prime contractors increased, and led to wide differences in subcontractors' products. Essentially the same kinds of units would be required to be built according to some different set of commercial standards, depending on the prime contractor. "Best commercial practices" proved illusory, along with adherence to commercial standards. These differences in acceptable practices exacerbated problems in training the workforce at the lower levels of assembly. Absence of the military standards contributed to this, but was not at the root of the problem. Greater reliability demonstrated and demanded became its own worst enemy.

COST AND SCHEDULE CONSTRAINTS CAUSE FALSE SHORTCUTS

Cost and schedule constraints are the norm in any large acquisition program. These constraints must be taken into account. At times, but certainly not universally, shortcuts obviating the benefit of the high reliability space qualified parts (either JAN Class S or SCDs) must be taken. When JAN Class S parts were made mandatory, they were not universally mandatory. Some programs, such as short duration space experiments and tests, were allowed to use lower quality parts. Then, since the lower quality parts were used on a prior spacecraft, the bounds of acceptability became blurred. Additionally, during the transition to JAN Class S parts, space program budgets developed before imposition of the mandatory JAN Class S requirement

---

[2] Even GPS, with its long view of committed production, ended up hitting a major hurdle when changing from Block II to Block III. Continuing "more of the same" came face to face with the need for "more and better." Long production runs of other kinds of spacecraft have never been sustainable.

were underfunded due to the increased cost of the new parts. The difference in total program cost was small for a new start (less than 5% of the overall budget), but sometimes much higher if the change forced a redesign. To fit within the budget, upscreened military quality parts were the earliest solution. Further exacerbating this particular forcing function were the long production lead times of some JAN Class S parts, forcing programs to use the more available military (and, later, commercial) parts.

Programs using JAN Class S parts amply demonstrated the value of these highest reliability parts. Nevertheless, differences of opinion persist about whether lower quality parts are really an unsatisfactory compromise. To a limited extent, other factors of larger design margins, de-rating, production, and even testing can partially compensate for lower quality parts. The evidence comes from the earliest days of the U.S. space programs. At first, testing began at the box level. If the box performed as required, the design, parts, and workmanship were considered acceptable. While this may sound like an endorsement of "testing in" quality, in fact, it is the opposite. A decrease in parts quality often results in more testing failures later in satellite development, which hurt cost and schedule much more the program would have experienced had it begun with Class S. The budgets and schedules limiting affordability of the highest reliability parts are the same ones that also limit the extent of testing, which is even more expensive than buying all JAN Class S parts from the outset.

Economizing has been a part of the normal acquisition process since long before the beginning of the space age. Economizing and budget cuts annually or biennially have been routine. However, strong economizing pressure beginning with acquisition streamlining was greatly pushed by the "peace dividend" at the end of the Cold War (despite being interrupted by Desert Storm). Economizing culminated in the great slash of acquisition reform, optimistically creating inadequate budgets and schedules.

RISING CONSUMER ELECTRONICS MARKET VOLUME AND COMPETITIVENESS

Rising consumer electronics market volume and competitiveness shunted the high reliability market aside to a volumetrically insignificant niche. The decrease in overall market share was somewhat offset as a limited specialized market for high reliability parts where high profits remained. The primary and niche markets never operated independently of each other. As consumer electronics exploded, a variety of market pressures increasingly pushed high reliability parts closer to commercial. In the early days of solid state electronics, the high reliability parts positively affected and improved consumer electronics. Increasingly, though, high reliability parts became more affected by the way consumer electronics parts were being made.

Tight process controls on high reliability parts production lines made them less attractive to the volume-hungry commercial market, where cost was a significant (but not the only) factor. The emphasis on high yield was often at odds with the reliability needed for long life. Transistor radios were a novelty at first, quickly becoming necessities of modern life, so the cost of transistors dropped as sales rose rapidly. That was the harbinger of things to come. The surprising explosive rise of consumer electronics in the 1970s with the introduction of hand-held calculators was followed quickly by the ubiquitous personal computer. The seemingly unquenchable thirst for new consumer electronic gadgets caused manufacturers to develop priorities other than military and space parts sales. With a focus on meeting the demands of the commercial marketplace, low-volume high reliability lines occupied plant floor space that could be replaced by high volume commercial lines. This pressured industry and government to find a *modus vivendi*, in terms of process relief, materials, and more. Had volume been the driver from the start, the industry and the market would never have developed high reliability electronics. The early and extensive investment by the military, mainly by missile programs, was an important kick start that could not sustain the industry because of the military's infrequent needs for large numbers of high reliability parts. Commercial applications sustained the industry. Military investment was often beneficial, such as the VHSIC program's investment in advanced technologies, but such efforts still fell by the wayside quickly as the industry's advancement (following Moore's Law) accelerated away. High reliability electronics, especially those associated with space programs, moved from the dog's collar to the farthest tip of the dog's tail.

High yield processes reduced the cost of electronic parts so extensively that a new class of behavior developed. Who today would look for a cell phone repair store when he or she drops his or her cell phone and it no longer works? It is not just economically unjustified to repair a broken cell phone; the advanced technology of newer cell phones makes it attractive to simply throw away the dead phone.

Another factor affected high reliability spacecraft, somewhat through the back door. The rise of what became information technology accompanied the rise of the personal computers. Software developers and computer technicians became highly sought after as the tech boom took off. Salaries and profits in the information technology field attracted many who might otherwise have been interested in the engineering disciplines essential to high reliability satellite development and operation. The number of domestic entries into engineering disciplines had been declining, and took a severe hit with the rise of consumer electronics and information technology.

Another consequence of the consumer electronics explosion came from entrepreneurial producers. Startup companies came into existence as lean as they could be, at least initially, in order to produce parts competitively with the larger manufacturers. Through careful selection of niches, applications, and technologies, many became very successful as the technology bubble of the 1990s continued to expand. For some of these startups, quality was not as important as cost control and volume of production. None of these new companies began by making Class S parts. Class S certification and qualification required a lengthy initial period in military avionics grade high reliability parts. However, some of these companies aimed at Class S as a long-term strategy. By the time of acquisition reform, such companies could be pointed to as examples of successful businesses that did not need, or have any heritage in, military standards. Thus, some of the entrepreneurial successes of the burgeoning consumer electronics industry indirectly supported attempts by others to eliminate military standards and specifications. Only after they were used this way by the acquisition reformers did the "tech bubble" burst, leading to the failure of some and the sale of others that could not survive. (See further discussion under Item 11, Loss of Standards.)

RISE OF COMMERCIAL SPACE

The movement of commercial space from communication satellites into commercial launch services introduced competitive pressures in unexpected ways. While the communication satellite market had one business model, launch services developed a different one. Both became very much bottom-line businesses. U.S. satellites had to compete with foreign satellite service providers and eventually with fiber-optic cable businesses. Launch services got off to a rocky start. The domestic market for commercial launch services decreased in the wake of the Space Shuttle's advent and the rise of overseas competition from the European Ariane launch vehicles. Commercial space launch rapidly became a global market in which the United States competed with the economically uncompetitive government-subsidized Space Shuttle. All of that changed when the Shuttle Challenger was lost. The small effort to develop a prudent backup to the Shuttle in case of exactly such a loss quickly emerged as a robust set of domestic launch services with overall capacity far in excess of actual needs. Economy and economies of scale meant survival.

Commercial satellites and launch providers faced different high reliability challenges. Both demanded high reliability, the former achieving it based on maturing high reliability manufacturing processes that allowed workforce cross-flow between military and commercial communications satellites. A few new approaches to traditional space missions were attempted, such as replacing geosynchronous satellites with proliferated low earth orbit satellites, but these continued only in niches of their own.

Launch services, on the other hand, experienced a surge of venture capital start-ups lacking the experience of their older ELV rivals. For the latter, lack of prior experience was a badge of honor since they could advertise they were unfettered by the "outmoded" ways of the past. Such a message found a receptive audience about the same time acquisition streamlining began with those convinced old systems acquisition methods were archaic and in need of extensive reform.

The resulting "us versus them" atmosphere portrayed the high reliability spacecraft and launch vehicle builders as dinosaurs too set in their ways and unwilling to get with new, upbeat, consumer-oriented commercial space practices. The pendulum for both swung too far away from previously successful practices, as both began to experience high failure rates. At first, both covered losses with commercial insurance, which rapidly dried up as claims mounted.

High reliability approaches to satellites and launch vehicles had never been "least cost" approaches. They were simply tried and proven processes known to work successfully. Such a line of reasoning, however, did not affect the acquisition reformers. Before commercial services started failing, their "streamlined" commercial best practices were held out as the model the "outmoded" acquisition practitioners should adopt. What had been a difficult case to make in the first place (to use the tested and proven practices) became an easily dismissed argument under acquisition reform.

COST OF INFRASTRUCTURE

Infrastructure cost is a frequent target for cost cutters. Infrastructure can rarely be tied to a specific product or outcome, and thus, the argument goes, less is better. In the case of the high reliability spacecraft enterprise, however, that was not the case. New technology qualification and extensive testing facilities were the part of the national infrastructure that became targets.

Extensive testing facilities, equipment, and workforces grew in government laboratories (such as Rome Labs), within prime contractors, parts manufacturers, and independent testing houses. These testing capabilities supported new technology development, ongoing production, and problem resolution.

Advanced technologies introduce new ways components fail. New failure modes arise from smaller feature sizes, different materials and material combinations, and packaging techniques associated in new ways. Rome Labs played a key role in preparing specifications for new technologies and monitoring the processes once in place to maintain understanding of failure modes and effects. The same capabilities that helped qualify new technologies for use in space could also be used to chase down an anomaly's root cause. Root cause determination was a key to feeding lessons learned into the next revision of standards and specifications. As feature sizes decreased and complexity grew, however, staying up with technology became increasingly expensive for all involved.

When, for other reasons, cost cutting became the order of the day, extensive testing capabilities in the government became targets of interest. The rationale was industry had to have these capabilities anyway, so why should the government pay to duplicate them? The spurious rationale about unnecessary redundancy was implicit in the late DSB studies and especially in the BRAC hearings deciding the fate of Rome Labs. Faulty reasoning weakened support for the government's unique infrastructure.

As military standards and specifications began to move toward process rather than piece part qualification, industry testing capabilities began to fall away. Cutting cost became more important than cutting edge.

Further undermining the continuation of the extensive testing infrastructure was the testing cost itself. Satellites had to be tested as assembled systems late in their development prior to launching. Thus, the reasoning ran, tests early in development, at the box and subsystem levels, could be reduced or eliminated. Problems would be caught in the later integrated systems tests, or so thinking ran. But not all tests are created equal.

Saving on testing when it was comparatively inexpensive (early) was short-sighted, because it deferred detection until repairs and rework were very expensive (late), or let problems slip through to be found in orbit (far too late). The rationale was even more flawed in light of other trends and considerations in testing. The classic V diagram provides more than just an orderly activity flow. With the underlying testing standards, work moved from qualification in the early stages, ferreting out bad parts and similar basic flaws, on to higher levels of integration of complex vehicles, where functional testing discovered flaws in workmanship, design, and assembly. Some testing determines latent defects within a part (very early) whereas thermal/vacuum or vibration testing can find workmanship problems in assemblies (late). Furthermore, as expensive later testing was also streamlined, reduced, or eliminated, the likelihood of compensating for early test deletions decreased. Eliminating early unit tests did not simply defer discovery of problems until later systems tests. It reduced the possibility of later discovery because later tests discriminated different problems.

The rationale also overlooked the growing complexity of spacecraft and the difficulty in actually testing and exercising all of their systems. Even when early testing was included in proposed budgets, when cuts inevitably came, early testing was easily eliminated with no immediate pain. Eventually, the contractor infrastructure for this portion of the testing infrastructure also fell under the axe.

Finally, government acquisition policies pulled back from deep levels of understanding of every facet of their spacecraft's development. In place of understanding the government's former mandatory practices, contractor judgment substituted financial reasoning, trimming costs where possible. Contractor testing, as an element of overall cost (thus affecting profit), got slowly whittled away. Market forces are powerful.

LOSS OF SPONSORSHIP FOR BASIC RESEARCH

The cost of infrastructure affected more than the support for the Rome Laboratories. Decreasing expenditures on basic research within the Air Force had a subtle effect. Investment by the Air Force in science and technology fell precipitously (in terms of constant dollars). From a high above 2.5% of the Air Force budget before 1989, expenditures fell to 2% in 1990, and leveled off at about 1.5% in 1992. This was the era in which the acquisition reforms and Goldwater-Nichols began strongly emphasizing the combatant commands' advocacy responsibilities.

That advocacy, however, focused on near-term capabilities that overlooked promising technologies not obviously connected to immediate problems. One advantage of the government's support and involvement with industry had been its resulting capacity to advance promising technologies without having to worry about bottom line profits. The resulting synergy was not usually something that had a visible or identifiable product in terms of specific weapon systems. Consequently, long-term development of promising technologies was always a hard sell with warfighters concerned with the here and now. This was the end of a trend begun in the early 1960s as the cost of the Vietnam War began to take a toll on basic research. According to former AFSC Commander General Marsh, "Air Force investment in the technology base in constant dollars . . . declined since the early 1960s. Except for a short period of modest growth—4% a year—from 1982 through 1986, it is still declining. . . . I say that our store of technology on the shelf is . . . sparse."[3]

Government laboratory funding reductions affected their ability to perform their traditional tasks. The combatant commands advocated (for the most part) specific weapon systems, but basic research was unassociated with any specific weapon systems acquisition. Consequently, when AFSC (and its successor Air Force Materiel Command) reduced their role in advocacy, the laboratories became one step further removed from the critical support of the ultimate arbiter of the annual budgets. Supporting the characterization of new technologies, audits, qualification of product lines, development and maintenance of standards, and other fundamental tasks was simply not an operational combatant command concern. Laboratories' support for the high technology electronics industry withered and died just as acquisition reform was forcing cost cutting of allegedly extraneous functions.

COMPLEXITIES OF THE MARKETPLACE

Changing bureaucracies is hard. Effective improvement of large bureaucracies is even harder. Consider, now, large bureaucracies in government and industry tied to each other by the complexities of market forces. Few members of the global marketplace can actually exercise changes that are sufficiently extensive and also constructive. Most changes are simply well-intentioned meddling in the complexities of the marketplace.

Zealous reformers created unintended consequences that were worse than the ills they wanted to solve. Top-level acquisition reforms implemented without understanding the implications at the intermediate or lowest levels of acquisition processes cannot succeed except by chance. Public policy and political science studies underscore that the original intent of a policy or law can be completely frustrated by the way in which it is implemented.[4]

The acquisition reformers knew their task was going to be difficult because it involved changing culture. In a two-year span, with dedicated reformers in key positions, they implemented changes that accomplished the culture change. They succeeded, if the metric was overturning a prior way of doing business. They did not take into account a number of factors, as we have discussed, not the least of which was their inability to foresee how the larger, global marketplace of electronics would adapt.

[3] *Shortchanging the Future—Air Force Research and Development Demands Investment*, Air Force Association Special Report of the Science and Technology Committee (Arlington, Virginia: Air Force Association, January 2000), 17, 20–22.

[4] Jeffrey L. Pressman and Aaron Wildavsky, *Implementation* (Berkeley, California: University of California Press, 1984). This is an excellent reference for understanding how well-intended policies go awry through the difficulties of implementing them.

To put it another way, adding to the complexity of changing the acquisition bureaucracy was the fact that the government bureaucracy was only one of the many bureaucracies involved in acquiring high reliability satellites. Changes in the government bureaucracy would ripple through the others. Unexpected and unintended consequences should have been no surprise, but their extent was.

By the mid-1980s, the nation's high reliability satellite enterprise—the mixture of government, FFRDC, and prime and subcontractors—was becoming part of a global market. Changing the marketplace for high reliability parts had become, by that time, a highly risky business, and more likely to miss its target than to hit it.

A critical mistaken impression contributed to the problems affecting high reliability satellite acquisitions. The acquisition reformers counted on contractors maintaining the same level of performance using equally rigorous (albeit commercial) processes when the motivation for compliance was removed. Marketplace forces drove economizing to least cost despite increased risk. Production of high reliability satellites, among the most complex machines made in any part of the U.S. economy, was the result of a complex mixture of many different entities. Trying to fundamentally alter one portion of this complex mixture was playing with fire, since the outcome could not be predicted.

Making policies about acquisition is one thing if left at a sufficiently high level. The problems arise when implementation takes place. The implementers of any broad policy are never the same ones who came up with the policy because implementation is a lower level function than policy making. Some policy makers may be among the implementers, but organizational considerations make this unusual. Changing a large, complex bureaucracy that is situated in a web of complex interdependence requires implementation measures quickly exceeding the grasp or control of the original policy promulgators. New motivations and different understanding of intent creep in. Within a moment, implementation gets out of control. When that occurs, all that has been accomplished is equivalent to meddling with a mechanism that is generally poorly understood. In fact, any attempt to correct something that is itself as poorly comprehended as the complexity of acquiring high reliability systems is rarely anything other than well-intentioned meddling.

Acquisition reform was certainly well intentioned, no matter how flawed its implementation. However, the Perry and Burnett letter to Fowler demonstrates how fundamentally they misunderstood the highly optimized acquisition of space quality parts.[5] Without so intending, they undermined essential elements of the high reliability and space qualified parts acquisition system, which went well beyond the government organizations to interdependencies in the parts manufacturers, their commercial and other military manufacturing, and even international considerations.

They succeeded in their drive to change the acquisition culture because they occupied all the key management positions within the Department of Defense and coordinated well with Congressional reformers. The reformers could set in motion irreversible, far reaching, and widespread changes.

---

[5] James R. Burnett and William J. Perry, Letter to Charles A. Fowler; "Acquisition Reform, Implementing Defense Management Review Initiatives," General Accounting Office, NSIAD-91-269, August 1991, 11.

So, is the message that all such meddling should be avoided? On one level, perhaps so, but some improvements were undoubtedly needed. For instance, the military standards system of specifying toothpaste and Band-Aids needed to be revisited. Also, hybrids and microcircuits were moving forward too fast for the traditional military standards process. Their move to performance standards was warranted and already under way well before acquisition reform began. The same was not true for standards related to discrete devices such as capacitors and diodes. These latter devices could have continued with the military standards system that had served them so well.

Perhaps the acquisition reformers should have learned from acquiring and operating the highest reliability space systems, rather than giving them superficial consideration (if at all). One such lesson is when a big change is planned, especially within a system that is already operating in space, a back-out plan is needed. That is, at each incremental step, before proceeding to the next, everything must be assuredly working well enough to justify moving to the next step. If not, then some way of returning to the *status quo ante* should be provided—that is, to back out and rethink. Anyone who has ever participated in the development of such a plan can testify to the intensity of the effort of many minds forging a coherent and well thought out change.

This implies two things. First, the thing being changed must be well understood (not necessarily fully understood, since complexity is the enemy of full comprehension). Second, such understanding informs implementation. Implementation must be well thought through to whatever level of detail is necessary to allow stopping and returning to a safe prior state if anything goes awry. Changing machines or processes allows this sort of prudent implementation. Public policy implementation, which is far less reversible, must therefore be thought through even more carefully. It almost never is.

When Packard was so difficult to implement and of limited effectiveness, two possible major conclusions could have been reached. The first could have been the changes had not gone far enough, so what was really needed was not a small change but a fundamental change of culture. Alternatively, the lesson could have been the desired changes were not working for some important reasons, and more study and understanding was needed. Government bureaucracy revisionists almost never reach the second kind of conclusion unless their motivation is to stave off change by delaying and studying the problem some more.

Opportunities existed to avoid marketplace meddling and dangerous unintended consequences. As Packard was found to be only partly implemented, reasons were found, and they were good ones, indicating more understanding was needed. Instead, different conclusions were drawn, and the Cheney reforms attempted to bring home what Packard had intended. With a change of administrations, but with no less zeal for reform, the Perry initiatives aimed at a fundamental culture change in the huge bureaucracy. Each of these efforts proceeded from studies and analyses. But the studies, such as the DSB, fell short of good service to their masters by trying to bolster a preconceived outcome. Contrary opinions were poorly received. In effect, all the changes were "green light" changes—no back-out was ever conceived or considered necessary. Each would proceed to completion. Only then the outcome would be observed. The expectation was the results would surely be the ones their studies had indicated.

None of this means attempts to improve bureaucracies should not be attempted, or that such attempts are immediately doomed. However, the literature on the subject is immensely rich with all the different considerations needing to be taken into account before attempting anything of the sort. Superficially informed changes do not stand a chance of improving the outcome.

LOSS OF ENGINEERS IN ACQUISITION

Acquisition of highly complex, high reliability satellites is an intense engineering challenge. It combines engineering disciplines of virtually every kind at different times. High technical expertise was foundational to the early success and continued improvements in high reliability satellites from the 1950s.

Sputnik was not the actual driver of the infusion of engineers into systems acquisition. That infusion was built on the highly experienced engineers who came out of World War II and transitioned into the emerging field of long range rocketry. *Sputnik* created a new demand for more engineers who would sustain the space enterprise after the World War II group retired.

Thus the years 1969–1973 contained a crucial turning point. Public excitement began to wane once the moon landings occurred repeatedly. The gap between MOL's cancellation and the end of Apollo before the start of the Space Transportation System was a bad time for retaining a skilled engineering workforce. NASA's engineering force drawdown benefited a few other programs, but the overall decrease, especially at entry and senior levels, was severe. No new horizons appeared to capture imaginations and attract new entries. The experienced senior engineers from the 1930s and 1940s were retiring. As consumer electronics took off, more technically inclined people were attracted in that direction due to the promise of interesting and lucrative careers.

The Reagan Administration partially restored a highly technical workforce with challenging programs such as SDI. The government's engineering acquisition workforce was once more a robust and interesting place to work, competing well for new entries and retaining the highly skilled older engineers.

Then acquisition reform caused two important changes. The standards were mostly eliminated, and with their loss came the loss of deliverable data. This data was the grist for the government's engineering mill. Absence of both standards and the data on which engineers operate left little for government engineers to do in high reliability satellite programs. Many moved on to other things, relinquishing the field to acquisition specialists without engineering degrees or experience. The government's loss of engineers further undermined the technical underpinnings of even the loosest of the new military performance standards.

One of the reasons why the TSPR approach worked in the pathfinder aircraft programs in the 1970s and 1980s was the Air Force had a highly skilled technical

(A)   (B)

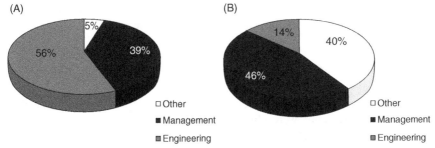

Figure 19.1. Changing skills in acquisition programs. The figure reflects Air Force System Program Office officers whose highest degree is in the area indicated in (A) 1974 and (B) 2001. Source: LtCol Steven C. Suddarth, "Solving the Great Air Force Systems Irony," Air Power Journal 16, no. 1 (Spring 2002): p. 11, Fig. 1.

workforce. These people, trained in the disciplines necessary to acquire advanced technology systems, were essential to the early apparent success of streamlined acquisition. Their knowledge, experience, discipline, and skill skewed the results of early streamlining efforts, and made streamlining appear to save considerable money and time. When these people left defense acquisition, nothing behind them compensated.

And that is what happened.

The 1990s workforce changed, as shown in Figure 19.1. Many of those leaving space system acquisitions found the rising information technology sector challenging, fun, and profitable. The talents and abilities they developed while acquiring space systems were turned to other areas as the streamlined techniques made defense systems acquisition less challenging and interesting. For similar reasons, engineering was less attractive for new entries.

The acquisition reformers recognized changing the culture meant changing people and their skills. As one insightful analyst put it:

> *Finally, and perhaps most destructively, the new management ceased to understand the importance of human capital and replaced it with process. This management took a minimalist approach: don't hire the best; hire only whom you need, and use the process to ensure whomever you hire can do the job.[6]*

The exodus of engineering expertise also made any reversal less likely, since technical issues became more difficult to fully understand for the acquisition specialists. That meant when an adverse trend developed, such as increased satellite and launch vehicle failures, it would only be recognized years after the engineering expertise had left. Fewer people who had to take or approve the appropriate corrective actions would understand in detail what needed to be done. Exacerbating this

[6] LtCol Steven C. Suddarth, "Solving the Great Air Force Systems Irony," *Air Power Journal* 16, no. 1 (Spring 2002): 10. LtCol Suddarth's article is an excellent précis of the changing themes in acquisition management, carrying several thought provoking insights, despite its narrow focus.

problem, the elimination of the lessons-learned feedback systems meant future decisions would not be guided by the wisdom gained from the past.

LOSS OF STANDARDS

To be precise, not all standards were eliminated. Certainly, those most critical to high reliability spacecraft were eliminated because they were considered overly prescriptive, with too much "how to" and not allowing for alternative (assumed to be less expensive) methods. Some of the remaining ones became performance standards, just as ICs and hybrids had, because the tight controls placed on military standards precluded the standards' keeping pace. What was good for the goose was not necessarily good for the gander in the case of all the remaining standards. Not all microelectronics technologies moved at the pace of ICs, and did not need to become performance standards. The shift of a few standards to a performance focus should not have been the precedent for all.

Foremost, perhaps, was the standards' basis (for the most part) in hard won lessons learned and first principles in chemistry, physics, and so on. The rationale and basis of these lessons was lost with the emphasis moved to "best industry practices."

The whole concept of "best industry practices" or "best commercial practices" was muddled. "Best industry practices" assumed such practices were not originally based on directive military standards, and the workforce skills based on experience with the military standards practices. In some cases, entrepreneurial startups seemed to prove success could be had without a military standards basis. The Acquisition Reformers glommed onto these new practices before time allowed them to be disproved. Success was often overstated, as seen in the wake of the bursting technology bubble in the late 1990s well after the Acquisition Reform revolution took place. The DSB studies searched for those few examples where some manufacturer had a robust commercial enterprise to show commercial acquisitions could teach the military how to acquire better systems or do so more inexpensively. But the DSB got it backwards. They cited successful commercial acquisitions that had actually benefited from the manufacturers' military acquisitions. The real "best" commercial practices were the ones derived from disciplined application of the military standards. When these rigorous processes became a routine part of a company's manufacturing culture, they paid for themselves in terms of higher yields, lower reworks and returns, and fewer claims.

Sadly, that was not universally true, and glaring exceptions existed. When military standards were imposed to provide uniformity across manufacturers of toothpaste and tissues, commercial processes clearly were less expensive than the military standard products. The military standard system overreached the range of items actually benefiting from rigorous standardization. So, when acquisition reform advocates cited examples in these categories, they made the further mistake of assuming the methods that produced the lowest cost toothpaste could also work for complex, high technology items such as aircraft and spacecraft. The studies produced no justification for such a claim. It was a matter of faith.

Specifically, for the most complex high reliability parts, the QPL way of qualification had surpassed its useful life by the mid-1980s. The QML response was a reasonable one, but should never have been the universal way of dealing with high reliability electronics. When the word of the day was reform, however, the QML approach that seemed to address complex devices so well naturally became a consideration and then a mandate for other types of devices.

Breaking apart the military standardization process undermined support and continuity for virtually every other part of high reliability satellite acquisitions. While many elements combined to decimate the high reliability satellite acquisition process, standards were the glue holding the enterprise together. Once they were gone, virtually every other negative acquisition reform consequence could be tied back to their loss.

Recall, for instance, that one of the very earliest and most basic reasons for standards, not surprisingly, was to standardize electronic parts. JAN-1, the first electronic military standard, was put in place to provide commonality, interchangeability of parts, common stockpiles, joint inspection, and overall efficiency. With reliance on chimerical commercial standards came the proliferation of dissimilar corporate standards that would be imposed on electronics parts manufacturers and subcontractors. Workmanship inevitably suffered as individual workers in a parts manufacturer had to do things one way for one customer, and, for the very same job, differently for another customer. The result was lack of skill transfer, lower quality, and increased costs due to repairs and replacements. In response to the proliferation of varied commercial standards for essentially the same product, some manufacturers were forced to make their parts for the lowest common denominator. Either the parts had to be designed into satellites as is (essentially as commercial parts with some limited additional testing), or parts had to be acquired by more expensive Source Control Drawings. Either way, without the extensive government–industry testing infrastructure and high levels of engineering expertise among the government and industry, reliability took a further decline as designs were themselves to become another uncompensated source of errors.

Doing away with standards and cutting budgets because of the "savings" left no alternative but to try to "inspect in" initial quality, and then hope later testing would catch any escapes. But that idea should have died by the mid-1960s. Quality, expressed as high reliability, could only be built in from the very start. Such things were demonstrably true and had been proven since the early 1960s, as space system acquirers worked to progressively reduce the sources of low reliability and on-orbit failures, evolving an optimized series of standards and processes. Acquisition reforms, from the Packard report onward, eroded and finally eliminated the foundational standards containing what had been learned, undermining the centers of expertise supporting the standards and irreversibly damaged national assets. All to replace processes known to work with unproven and ephemeral "best commercial practices" to achieve imagined cost savings.

We have already explained how "best commercial practices" were illusory, but we need also to recognize the metrics by which such "bests" were determined. For instance, the commercial satellite business has been very successful for a long time. Surely, then, the commercial satellite manufacturing sector can teach national defense and national security space a thing or two about how to get better systems faster and less expensively? The metrics for determining reliability in each sector are different. In a national system, outage or partial failure of a payload is a significant impact and treated very seriously. Any such failure represents the loss of critical information, as this history has emphasized throughout. On a commercial communications satellite with several dozen transponders, loss of a single transponder does not have the same impact. So long as customers of the relayed data do not see an outage or loss of their connectivity, a commercial communications satellite is successful and, from the standpoint of the end user, reliable.

So, is the problem simply inappropriate metrics applied to national defense and national security satellites? Absolutely not. Success metrics must focus on what is important to users. Commercial reliability is measured appropriately in terms of continuous service without interruptions or down time. Such metrics for a national defense or national security spacecraft have far different and more serious consequences. These latter systems are far more complex, by at least an order of magnitude in terms of number of parts (for instance), and payload redundancy is not as easily provided as in the case of transponders on commercial communications satellites. An outage that misses a hostile nuclear missile launch and its intended target, for instance, is a different order of consequence than shifting from one dead transponder to one of multiple backups.

In space systems acquisition, possibly more than other technological endeavors, old wisdom won by hard lessons rarely became outdated. Capturing wisdom in the form of lessons learned incorporated into military standards was the method used to perpetuate this knowledge. Of course, space systems can be built without such knowledge—there have been hundreds of small satellites built without following the old ways. But none of them approach the complexity, capabilities, performance, and lifetimes of satellites built with Class S parts before acquisition reform.

Interestingly, the part numbering schemes are the lasting legacy of the original intent of military standards development.

LOSS OF INDUSTRY DATA SHARING

Space systems acquisition never had a perfect information sharing program for parts, materials, and processes problems. GIDEP came closest with varying levels of support from its participants. GIDEP was at times mandatory on some contracts, and voluntary on others. One of the key roles of the FFRDCs was to facilitate the cross-flow of problem information among space programs to head off cost and schedule problems, and ultimately avert orbital anomalies. Data, and access to it, was the basis of both proper engineering and independent evaluation as a check to ensure the best possible outcome. Acquisition reform, by eliminating many data deliverables, permitted contractors to retain information and protect themselves against sharing data on proprietary grounds. Widespread use of proprietary markings staunched the flow of quality information, and raised the specter of litigation for violations. Previously, when the government paid for key technical information data rights about its satellites, from the highest to lowest level of manufacture, problems were uncovered and solved early. The move to commercial practices made any problem a potential advantage for competitors to exploit in the marketplace, so such data had to be protected. Before acquisition reform, with involvement of the FFRDCs, problems were found early and treated effectively, without recrimination, because of the teaming among government, industry, and contractors. Such teaming was yet another casualty of acquisition reform.

LOSS OF INSIGHT TO CONTRACTOR ACTIVITIES

Closely related to restricted sharing of information was the loss of the government's overall access to satellite technical data. Government purchased data rights gave insight into contractor activities, so the government could see the same things the contractor did in the development of a spacecraft. The result has been that some problems are hidden to the point at which they cause major impacts to cost and schedule, often affecting multiple programs, when they could have been reasonably dealt with earlier. When space system failures rose in the late 1990s and continued into the new century, the failures were not found on the factory floor, but in space or on the way to space. Access to technical data with independent analysis by an FFRDC would have identified most of the problems much earlier, when they could still be corrected. Coupled with the absence of military standards, loss of data produced some very expensive failures and was directly attributable to acquisition reform.

Lack of insight into the activities of second- and lower-tier contractors raises issues about the quality measures at those levels of manufacture. With loss of the government's demand for delivery of data about the satellite's development came the prime contractors' similar lack of pass-down to their subcontractors for the same information. The result was an environment open to substandard parts, inappropriate parts, counterfeiting, inadequate materials, and a vast array of incompatible manufacturing processes, all hidden from the view of those responsible for integrating the final products into a satellite.

Loss of the free flow of problem information and the virtual end of delivered technical data affect nearly all of the other forces cited here. Recognition of problems tends to remain at the lowest possible levels, with little information flowing up the chain from manufacturers to vendors to subcontractors to prime contractors and, eventually, the government. Some serious problems remain hidden for so long they become severe impacts to a single program just because of the effects of delay. Another phenomenon has arisen, known as "balkanization." A problem may affect multiple purchasers of a part or device, but the original manufacturer can use proprietary exclusions and other means to keep the purchasers from realizing they are not the only ones affected. Once again, mounting an effective and less costly corrective effort is hampered by balkanization, which effectively ends only when the degree of the problem becomes so severe as to be impossible to contain. Whether this happens in automobiles, computers or satellites, the effects are devastating and incredibly expensive.

LOSS OF IN-FACTORY GOVERNMENT PLANT REPRESENTATIVES

Less obvious than the lack of data deliverables was the major reduction of the in-plant government representatives. The Minimum In-Plant Surveillance (MIPS) program saved some short term money, but committed space system (and other) acquisitions to pay for problems discovered much later in acquisition. Presence of in-factory government plant representatives provided continued assurance of stability of manufacturing processes and early warnings of problems. Such personnel were complementary to the delivery of data and early problem reporting to GIDEP, and served as yet another screen to weed out all possible sources of failure. In a sense, they were redundant, in that they were looking for problems, ensuring processes, and helping to eliminate bad parts and devices. But, just as system thermal vacuum testing finds certain kinds of errors, but not the ones that unit qualification testing discovers, the apparent redundancy was meaningful. What appeared to be superficially redundant and therefore unnecessary was actually a source of uncovering different kinds of problems than other measures uncovered. Moving from insight to oversight reduced or eliminated this presence.

LACK OF PRAGMATIC STANDARDS APPROACH ACROSS DEVICE TYPES

By the 1980s, the situation had become clear: the kinds of standards needed to properly qualify a capacitor or diode for use in space were very different from those for unique, application specific or very highly complex device types. The rate of technological change was fastest among the most complex device types. The methods of qualifying each individual integrated circuit, application specific integrated circuit, or hybrid device were incompatible with the dynamic microcircuit market.

Several analyses pointed to the use of SCDs as evidence of the inability of the military standards to keep up. Depending on the analysis, SCDs were either an expensive and unnecessary evil or the most efficient method to ensure the highest quality part. Much of the adverse criticism of SCDs inappropriately grouped together abuses of the SCDs with the legitimate uses, and in any case misapplied conclusions about SCDs.

For instance, the prevalence of SCDs could indicate the parts being driven by the military standards did not meet the specifics of the contractors' design. Alternatively, one might conclude the contractors could drive a unique piece to be manufactured to increase its price to the government and hence improve profitability. Both criticisms have been levied against SCDs. But, as in each case in this history, peeling back another layer reveals the problem is not so simple.

While both criticisms undoubtedly have a basis in reality, the extensive use of SCDs does not seem to have been a widespread abuse of the system. A standard part is exactly that—a part consistent with all its siblings. By its very nature, a standard part can also be considered the lowest common denominator among all potential customers' uses of that part. While a designer can attempt to use standard parts to the maximum extent possible, inevitably in space systems complexities arise that force some additional feature or qualification to meet the specifics of the operational end use. The SCD is the only way to bring the standard part in line with its intended use in a specific application.

Upon seeing a large number of SCDs in use, some concluded the cost of parts was out of control and the SCD system was being abused. Early in the era of the JAN Class S parts, an analysis revealed SCDs were still more prevalent than the mandatory Class S parts, to the chagrin of the SAMSO and SD commanders. As considerable effort was devoted to development of the JAN Class S qualification criteria, their hope was JAN Class S devices would be the standard parts used on spacecraft. But a thin layer below the SCD statistics was the revelation that Class S parts either did not go far enough in specifying necessary tests or needed to be adapted to a specific use. Many of the SCDs in that analysis were actually modified Class S parts.

Similarly, the ASIC is not an abuse of a standard integrated circuit, but an efficient and effective way of meeting a specific need. Standard Microcircuit Drawings provide generic kinds of integrated circuits for such complex but widespread ends as analog-to-digital converters. An ASIC might use an SMD as a departure point to adapt it to a specific use, or might start with a wholly clean design. One cannot tell which merely by use of the ASIC term any more than one can tell what an SCD's basis is.

Once acquisition reformers latched on to the number of SCDs versus the standard parts, two conclusions could be reached. As said above, some considered the large numbers of SCDs to be an abuse to drive up the price charged to the government. Others took the numbers to mean the Class S standards had driven the industry too far by overly prescriptive specifications that were ineffective, forcing designers to have to use SCDs.

The large number of SCDs, even after the JAN Class S mandate, supported the contention that acquisition processes needed reform to end this apparent costly abuse.

THE END OF THE COLD WAR AND THE "PEACE DIVIDEND"

The anticipated "peace dividend" from the end of the Cold War allowed cost cutters to scale back the size of the defense industry on a broad front. Government actions became motivated more by the need to draw down than to preserve essential niches.

Fuzzy reasoning hopefully conflated commercial electronic parts with high reliability and space qualified parts in the DSB studies, Packard Commission, Cheney Review, PATs, and FASAs. Authoritative counter-arguments, such as the Air Force chief scientists' letter to the BRAC, were dismissed as special pleading. When studies came forward with conflicting data or without drilling to the necessary level of detail (such as the Rome Laboratories BRAC decision), bad budget cutting decisions were made and smacked of premeditated conclusions needing to be justified.

Budget cuts, especially when intended to be dramatic reductions, have to be somewhat arbitrary. Full insight into budget cut effects over an enterprise as widely varied as defense systems acquisition is difficult at best. Space systems were expensive. As critical as they were, and as much as they had contributed to the end of the Cold War, they had to take their share of cuts. Budget cutting was widespread and on a short deadline, leading to decisions that handled priorities as well as could be done under the circumstances. The budget cuts were premature, because the end of the Cold War was followed quickly by other challenges in a world that still has hostile actors.

INCREASING DEVICE COMPLEXITY

Greater capabilities entailed greater complexity in ASICs and hybrid microcircuits. Such spacecraft devices represented very small lot sizes (a handful at any one time) incompatible with the military standard space qualification processes. Complexity complicated qualification, which translated as long lead times. The pace of commercial technology meant long lead times, which resulted in older technology being built into spacecraft. The modern synthesis of QML and qualifying manufacturing lines was necessary to avoid the large and increasing cost of qualifying a few unique parts, but remained insufficient to ensure the latest technologies went into spacecraft.

The complexity of ASICs and hybrids thus contributed to the valid criticism that out-of-date technologies were going into America's front line systems. This particular heresy was meant to be corrected by Packard and Cheney's acquisition streamlining. All too often, the scale of systems complexity was not used as a meaningful metric when analyzing the overall topics of reforming acquisition. When complexity was mentioned, it was usually cursorily, without appreciation for its real implications. For instance, simply noting that cell phones are made affordable and disposable can be readily contrasted with satellites that, at the size of a Greyhound bus, are not disposable. That's easily grasped. But complexity also drives the difficulties of manufacturing and testing as the machine is built up. Complexity also affects the skills and training of the workforce because of dependence on a certain level of expertise and workmanship.

Dealing with complexity is itself complex, since it implies intertwined specialties, schedules, and so on. This history is an illustration. The consequences of changes in public policy and their implementation illustrate the exigencies and intricacies of complexity. This chapter also illustrates a facet of that complexity, because most of the forces in motion and discussed here are related in some way to one or more others. By altering one, the consequences are rarely local and often have unanticipated ripple effects.

PUSH FOR THE MOST ADVANCED TECHNOLOGY

For a very long time, the U.S. military has been at the forefront of advanced technology, a key advantage used in its favor when facing the overwhelming numbers poised against it in the Cold War. Technological superiority depends on the idea the United States can innovate and insert advanced technology faster and more effectively than its adversaries, basically outrunning their ability to keep up. This has been cited as one of the contributors to the final collapse of the Soviet Union, whose economy could no longer bear the expense of competing with the Americans.[7] In that view, the Reagan presidency's fascination with extreme high technology was the straw that broke the Soviet camel's back. Anything less than the latest technology was anathema.

However, that was not a truly global good for American systems.[8] Satellites were long considered the epitome of high technology. While that may have been true for some sensors and other advanced devices, it was not really necessary for each individual part. Full characterization of a spacecraft's parts was key to ensuring the satellite would work reliably for a long time in the space environment. Full characterization timelines grew longer as technology advanced, eventually requiring more time than the release cycle for the following advanced technology allowed. By the mid-1980s, the more natural and practical attitude toward satellite technology would have been to stay as advanced as possible, but drawing the line at some point to ensure the next generation of satellites would be built on well-understood and mature technologies ready for use in space. Space systems would be a generation, or perhaps more, behind the technology going into cell phones, while still providing advanced capabilities.

In fact, this was happening, as acquisition programs began to set dates for technology freezes—no technology developed after a certain date (usually well in advance of the satellite's start of assembly) would be used. The capability available,

---

[7] Martin Anderson and Annelise Anderson, *Reagan's Secret War: The Untold Story of His Fight to Save the World from Nuclear Disaster* (New York: Crown, 2009), 37–42. This insightful book tells how Reagan worked to stop nuclear intimidation and ultimately bring down the Soviet Union.

[8] David E. Lockwood, Cost Overruns in Major Weapon Systems: Current Dimensions of a Longstanding Problem, Report No. 83-194F (Washington, D.C.: Congressional Research Service, Library of Congress, 1983), CRS-25. Lockwood reports that a staff paper, "Defense Facts of Life," apparently no longer available, developed for internal use in the DoD by Franklin C. Spinney, 5 Dec. 1980, anticipated another aspect of the high technology drive. "His thesis is that the increasing complexity of weapons has led to sharply rising costs which, in turn, have had an immediate effect on the number of weapons that can be afforded. The decline in numbers produced is accompanied by a second effect, namely a decline in performance reliability." That change apparently affected products such as aircraft more than spacecraft, because spacecraft acquisition had always been a low-volume activity.

then, reflected technologies that were somewhat out of date when the satellite went into orbit, and there was truly nothing wrong with that.

The mistake, if it can even be called that, was assuming advanced space capabilities *required* the most advanced technologies, and the advanced technologies for sensors (for instance) were also needed in the more mundane housekeeping functions. Early in the history of the national security and defense space programs, obtaining the most advanced technologies for all spacecraft systems meant greater capability in orbit. Through the 1960s, it was clear that further electronics miniaturization increased reliability through more redundancy per payload pound. While that remains somewhat true, it no longer has the same impact as it did when some satellites needed vacuum tube technology to perform their missions.

Acquisition streamliners and reformers wanted American systems kept at the forefront of new technology. Without appreciating the distinctions drawn here, that led to shortcuts in qualifying new technologies, which raised failure rates as a consequence.

Chapter 20

Summing the Parts

I hope by this point we can see that, clearly, a large number of trends and forces played a part in this story. Equally clearly, they are not a separable list, as you may have concluded. The list points to the complex interdependence of the constituent trends and forces affecting high reliability satellites, the electronics industry, and their relationships to national security and national defense space programs. Together, working at times in reinforcing concert, and at times in destructive disharmony, these constitute what may be called a technology trajectory. Specifically, they comprise and guide the technology trajectory of high reliability satellites. It is time we collect our thoughts and draw some conclusions to guide future actions.

One major force might appear to be missing. Wasn't this whole story about rapid technological change? Undeniably, rapid technological change runs rampant and accelerates throughout. The forces cited generally fall into direct consequences or mitigations of rapid technology change. Like many topics discussed here, rapid technology change is neither good nor bad. It's what people *do* about such change that matters.

Rapid technology change is relative and situational. The change rate considered rapid in 1947 was easily dealt with by 1960, was tepid by 1970, and irritatingly slow by the 1980s. An important consideration, then, is not whether change was rapid, but whether existing policies, processes, and other mechanisms could handle the *rate* of change. Several times we have used a Darwinian analogy to illustrate key points. Consider rapid technology change as merely Darwinian natural selection within the confines of this history. Everything else follows from it.

If rapid technology change was being selected for, what then was the source of the random mutations on which it operated?

The trends and forces summarized above reflect either fact-of-life, physics-based realities, or were the consequence of circumstances in which the best intentions led to unintended consequences. While some political and personal agendas undoubtedly crept in, this perfect storm owed its convergence to happenstance, not malice. This is not to say the result was inevitable.

At first blush, these trends and forces might seem to provide the agenda to recover high reliability spacecraft. Studies in the early 2000s (responding to the

Implosion: Lessons from National Security, High Reliability Spacecraft, Electronics, and the Forces Which Changed Them, First Edition. L. Parker Temple III.

failures of some critical national security and national defense satellites) identified some solutions as just such candidates. While these studies were correct, they were incomplete. For instance, one early and obvious correction was reinstatement of the space-related military standards. Replacing the missing standards was important, but all this could do was create new conditions within which the marketplace would reoptimize but not necessarily achieve the desired goal.

The problem is that the forces described above, and others, are part of what economists call a "production function." A production function describes and relates *all* the factors of production, linking inputs to outputs. The production function was what became optimized. When a change perturbs a production function's optimality, its interacting and competing processes seek and settle onto a new optimum. Darwinian rapid technology change ensured the optimum was never static. That new optimum may be unpredictable; as related in this history it certainly was.

So, if only one factor is repaired, no matter how important, the result does not necessarily have a major effect on the overall production of high reliability spacecraft. In the case of high reliability spacecraft, the production function is incredibly complex, dynamic, poorly understood as a complete entity, and certainly undescribed in any practical sense. The acquisition reformers never understood how all the factors of production related to the outputs, thus creating serious unintended consequences.

No agency or set of agencies, contractors, suppliers, laws, or policies can solve the problems in the short term. The reason is the production function is not owned (much less understood) by anyone. The Darwinian source of random mutation was the constant interplay of the contending parts of the overall production system. Changes to the high reliability space parts industry, for example, are important, but are insufficient to solve the overall problems, and in themselves have unexpected and unforeseen consequences elsewhere, such as in foreign competitiveness, balance of trade, export controls, and more.

Without acquisition streamlining and acquisition reform, market forces would have caused a new optimum—rapid technological change would have ensured that. New optimums had always happened in the past because of the basic nature of markets. Whatever that new optimum might have been, it would not resemble the current situation because it would have been incremental, not fundamental. Ongoing movement of commercial production to overseas facilities and the continuing need for only small lots (and, as lots grew, only a portion of a production lot) for space programs would force a new industry–market optimum. Government acquisition policies would have to change, just as they did when the cost of partial lots caused creation of the Class S stockpiling program. The difficulty of continuing the QPL approach for ASICs and hybrid microcircuits would have resulted in the QML approach, or something very like it, aside from any attempt to streamline or reform acquisition. These were the realities of acquiring high reliability satellites. It might be amusing to consider what the current state of affairs might have been had the efforts to streamline and reform acquisition never occurred. Amusing, perhaps, but pointless, since they did occur.

The central thesis in this book was that high reliability—as an overall technological bin comprising many individual technologies—was on a technology trajectory. The idea of a technology trajectory appears in the literature of technological change. It is usually taken to mean the inevitable course along which a new technology continues to evolve. These are countless examples, leading to the conclusion that technology trajectories do, in fact, exist. What is less certain, though, is why. The idea of a technology trajectory has many facets, two of which concern us here.

To begin with, the trajectory described in this book covered the applications of the first solid state electronics. This trajectory continued into the early 1960s. High reliability satellites were part of an overall system, comprising elements from the electronics industry, government, and, to a lesser degree, public consumers. The events described in this book follow that system through various attempts to reform systems acquisition. The system accommodated the changes from a variety of organizational imperatives and acquisition reforms, seeking a new optimum after each perturbation. Thus far, the motivation for seeking a new optimum has been alluded to in passing as Adam Smith's "invisible hand." But Smith's "hand" and this explanation are little more than hand waving. There was no single intelligence behind the changes, but there was an overall plan. And that plan is the principal cause of the technology trajectory.

In his book on the development of inertial guidance and navigation, Donald MacKenzie makes some important observations about how increased accuracy in long-range ballistic missiles was not driven by any identified need. He uses the idea of a technology trajectory to explain:

> *That turns the potential anarchy of technological change into a persistent trajectory is the way social interests are created in its continuing to take particular form. It is not simply that people want it to take that form—though they sometimes do—but that they build into their calculations that it will take that form. They invest money, careers, and credibility in being part of "progress," and in doing so help create progress of the predicted form. Finding their predictions correct, they continue to make them with confidence.*[1]

Is that what happened to high-reliability satellites?

In the case studied here, we have laid out a wide variety, though probably not all, of the factors and forces at work shaping and changing high reliability satellites. But if we step back to the beginning and look at the biggest picture—that of the solid state revolution—we see the reason for a particular technology trajectory. Moore's Law. There is no industry body enforcing adherence to Moore's Law. Nevertheless, there have been several key changes in solid state electronics and radical alternations to integrated circuit designs in order to keep on Moore's trajectory. MacKenzie's conclusion about the circular predictions, extrapolations, changes, and

---

[1] Donald MacKenzie, *Inventing Accuracy: A Historical Sociology of Nuclear Missile Guidance* (Cambridge, Massachusetts: MIT Press, 1993), 168.

reinforcing achievements of predictions is certainly correct for Moore's Law. Many times has this "law" been used to set the next goal in feature size.

In this book, we have discussed the drive to increase reliability and shown how more reliability was initially extremely important. Its effect on force-sizing, in-service rates, and strategic targeting of missiles was an essential element behind government investments in electronics reliability leading to "Minuteman quality" parts. These in turn, as we showed, led to the ultimate Class S parts, which also continued to increase in reliability.

How much reliability was enough? It seems like the same question MacKenzie asked about how much accuracy was needed, but it is not. The question of the extremes of reliability and what constituted them was actually open-ended. Increased accuracy reached an end essentially with a "good enough." Reliability reached an end because of something inherent in the way Moore's Law has been tracked.

To this point in our discussion, at various places, we followed the early techniques to reduce feature sizes of solid state electronics devices. That sustained Moore's Law for a while. Eventually, though, other techniques had to be invented, such as a move to multiple core integrated circuits. To some, Moore's Law was taken to stand for something other than the number of transistors per unit area and had more to do with the number of bit flops. Sustained by the kind of underlying mechanism MacKenzie described, the electronics industry had to make other compromises to align with Moore's Law. Economics certainly had a major and probably the most significant impact. But technology changes such as were necessary to achieve greater speeds had other effects. For instance, as feature sizes change, the response to various kinds of radiation changes. More to the point, though, was the effect on margins.

Once on a sound footing, solid state electronics made a major advance in lifetimes of electronic parts. Models predicted some part lifetimes of several thousand years. In practice, once some parts were powered on, they would seemingly run forever in conventional, commercial temperatures and environments. High reliability improvements extended these parts' abilities to operate in the extremes of the military temperature ranges and radiation environments, which fed back to the electronics industry and provided further improvements. However, as feature sizes shrank and other technological changes were made to materials, lifetime margins began to shrink. For parts with lifetimes of several thousand years, shorter lifetimes of a few percent were imperceptibly different.

Fractions of a cent make a difference when making things like cell phones. Their low cost is one of the reasons no cell phone repair stores exist. As margins were sucked out of commercial electronics, the lifetimes of these parts came down drastically. That was not the trajectory on which high reliability parts could survive.

Minuteman quality and Class S parts beat down the causes of early mortality failures on the left hand side of the classic bathtub curve. The solid state electronics industry's earliest competitive advantage over vacuum tubes was their lower failure rates. The solid state electronics industry benefited from the government's investments in driving down the early failure mechanisms. Once turned on, the length of time before some end-of-life failure meant some electronics would last a very long time. Now, however, some industry analysts see the end-of-life failure mechanisms (the right hand side of the bathtub curve) have become dominant as the duration of the middle, nearly static failure rate mechanisms has nearly disappeared. This inward movement of the life-limiting failure mechanisms is not due to one change, especially in light of the wide range of electronics technologies now available. It was, however, a consequence of the "invisible hand" of an industry driven to follow Moore's Law for want of a better or different technology trajectory.

It would be unfair to blame Moore's Law as the sole underlying cause. There was a self-fulfilling mechanism in it similar to MacKenzie's explanation of the social aspects of technology trajectories. The need for longer lifetime satellites was in response to the demand for dependable, available national security and national defense satellites in time of crisis and conflict. Once started down that path, as satellite lifetimes grew dramatically longer, they drove down the need for more satellites. This happened right at the time the Space Shuttle was being proposed to have a 60 flight per year launch rate based largely on the idea that satellites were inherently unreliable. So, high reliability satellites undermined Shuttle economics, but also undermined themselves. As the industry supporting high reliability satellites found, longer lifetimes meant fewer satellites were needed. Fewer satellites needed meant lower demand for piece parts and other critical technologies. Lower demand meant these were less attractive to more manufacturers, just as the consumer electronics market began its explosive growth. As industry analysts projected the course of the consumer electronics technologies and the industry overall, high reliability electronics became increasingly irrelevant to the corporate bottom line. As they became increasingly irrelevant, their technology trajectory became very different from that of their consumer electronics cousins. High reliability piece parts made today reflect more of the processes supporting consumer electronics a few years ago than vice versa. That trend will continue.

Will moving to some newer technologies, such as nanotechnology, spintronics, or photonics, remedy this? Maybe not. As these or other alternatives emerge, new early mortality and end-of-life failure mechanisms will affect them. These will lead to their own technology trajectories, possibly independent of Moore's Law. The one thing that is changed forever, though, is the influence, and perhaps even the consideration, of the extremely high reliability satellites. There may be a narrow window for government investment, just as in the earliest days of the sold state electronics industry, to influence reliability of such new technologies. But the will to do so is probably lacking in both government and industry. Cost and rate of technology change are moving government procurement to adapting and adopting commercial technologies rather than influencing their development. That's where Adam Smith's "invisible hand" comes back into play, because current trends reflect the economic realities that will eventually make high reliability electronics a memory of an older generation.

But surely defaulting to some ineffable "invisible hand" overlooks another key part of the story presented here. In sociological terms, we can define an institution as any "stable, trans-individual pattern of social behavior."[2]

[2] MacKenzie, *Inventing Accuracy*, 386; Paul L. Berger and Thomas Luckmann, *The Social Construction of Reality* (Harmondsworth, UK: Penguin, 1971), 65–109.

In this history, the institution has clearly and readily defined members. Specifically, we can cite the long-term stable institution comprising at least universities (innovation and basic physics), Rome Laboratories (standards, industry guidance, physics of failure), industry (innovation, packaging, implementation), SAMSO in all its forms (systems design, integration, testing, and operations), FFRDC (institutional memory, continued pressure for improvement, lessons learned feedback), and operational military commands (pressure for reliability and dependability). One might be tempted to add Congress to the list, but the roles played by Congress were not stable over the period of interest. In fact, it would be a mistake to include Congress as a monolithic entity. At the same time, parts of Congress were pushing for greater efficiency in acquisition; controlling budgets; improving reliability to keep costs down; keeping local district defense contracts alive and expanding; and legislation regarding standards, FFRDCs, specific space systems, federal acquisition regulations, and so on. While there was some consistency among certain members of Congress, one could never use the term "consistent" with any other part of the congressional role regarding high reliability spacecraft.

There is another way to use the idea of technology trajectory to understand this history. The technology trajectory had to do with increased reliability, which translated into longer lifetimes. Ancillary to that was greater capability.

The initial capabilities of satellite systems were impressive and in many ways a quantum leap over their terrestrial counterparts. The very features of solid state electronics that embodied their greater reliability compared to vacuum tubes also enabled greater capability in orbit. The space business in all its variety is still very much a utils per orbital pound enterprise. As feature sizes shrank, of course, greater redundancy was possible, enhancing reliability which expressed itself in longer satellite lifetimes.

At the same time, though, more sophisticated versions of early capabilities were enabled. Sophistication could be expressed in wider bandwidths, greater resolution, faster processing speeds, and so on. For a while, increased sophistication contended with redundancy for inclusion in the next version of each satellite. That is, with a fixed weight and power budget, early on the alternatives were to expand a sensor or transceiver's capabilities, or provide a redundant telemetry package, for instance. That contention for weight and power was transitory as feature sizes shrank, essentially enabling both paths to be followed. Further enabling both redundancy and capability enhancements was the growth of launch vehicles.

One example was the Titan launch vehicle, sized to be able to put a large structure such as a Dyna-Soar spaceplane in orbit. That same launch vehicle provided an immense opportunity to expand (or, should I say, an opportunity for immense expansion?) for other satellites. The launch vehicles themselves underwent constant tweaking to squeeze a few extra pounds of performance. This all came down to enhanced capabilities.

Enhanced capabilities, however impressive, were not a guarantee of long-term support from the principal national defense beneficiaries if reliability concerns were not met. Starting in 1969 and intensely for the next four years, continuing with some bumps in the road until the formation of Air Force Space Command in 1982, the majority of national defense space users converted to a belief in the reliability and dependability of space systems. That bumpy road alluded to in this book is covered in detail elsewhere, but that conversion would not have been possible were it not for the already ongoing technology trajectory of space systems toward higher reliability and consequent longevity. What happened in 1969–1972 was decisive and, some would say, game-changing decisions were made to keep that trajectory moving forward.

The fact of the ongoing technology trajectory of spacecraft reliability should not be viewed as technological determinism. Determinism of that form is a fallacy. There were many ways in which spacecraft and acquisition practices could have evolved. (See the Epilogue for an overview and discussion of these alternative paths.) Many of these possible evolutionary paths have been resurrected recently to correct the dysfunctions arising from the 1993–1994 acquisition reform and related changes.

To some extent, longevity was not the only needed expression of high reliability. A case in point is the Corona film-return photoreconnaissance satellite. This technology was chosen because it offered the earliest means of obtaining sufficiently high quality imagery to satisfy crucial national security needs. The U-2 was a diminishing asset after Harvey Stockman's first overflight of the Soviet Union. Only 25 flights later, the U-2 overflights of the Soviet Union were history never to be repeated. The immediate need for the U-2 flights was ascertaining the extent of Soviet bomber and missile acquisition and deployment. Carmine Vito's flight, the second to overfly the Soviet Union, was understood to definitively answer the issue of either bomber or missile gaps—neither existed. The confidence in the conclusions drawn from just these two flights, particularly from Vito's, was not based solely on the imagery returned. One might have argued, based on a few hundred square miles photographed and the relatively few bombers and missiles found, that either there were few bombers and missiles, or the photographs were taken of the wrong places. Other information, including intelligence, available at the time resulted in targeting of these flights for the greatest payoff.

Photography was incredibly important in the assessment of the relative status of forces, gaps, deployment locations, and much more. But photographs are no longer current once the camera shutter has shut. Bombers that were not at an airbase an hour ago might be there now. Ongoing monitoring was no less important than the first confirmatory pictures. But the U-2 program could not be the source of continuing monitoring. That had to come from Corona, initially, and later from the rest of the Keyhole series of satellites. The first Corona mission lasted only seventeen orbits before the capsule was retrieved over the Pacific. That is hardly a longevity problem. Corona did not need solar arrays for electric power, because the duration of the mission was short.

The reliability needed for Corona was, in a sense, the same kind of reliability sought for the ICBM fleet. Both had to work the first time with extremely high confidence. Each picture taken was a national treasure, not to be squandered or frivolously expended. Each mission had to take as many cloud-free pictures as possible to maximize the return on investment of national resources. So, for the single bucket Corona missions, reliability did not mean longevity. Corona represented a compromise between competing interests to replace the U-2 as soon as possible with a system of comparable performance and to achieve greater access with a limited means of obtaining images (the film-return capsule). Pressure mounted to add more capsules. A second film-return capsule on each satellite, then, meant double the imagery (both in terms of number of images and of the areas covered) with less than double the cost. It even allowed the possibility, or at least the option, to do an early return of a capsule if images of a time sensitive nature were thought to have been captured, because the satellite still had more capsules aboard. This same line of thought eventually grew the number of film-return capsules to more than double. But film-return, as explained in this history, was not responsive to crises.

A crisis response system, such as the near-real-time electro-optical system, had to have the same reliability to work when demanded that was present in Corona and its successors. It finally transitioned the national security imagery reconnaissance systems to the other sense of reliability expressing itself as satellite longevity.

For decades, correcting changes and variations from desired levels of high reliability has been a disciplined application of systems engineering processes, as discussed in this history at length. Those processes, when focused on the goal of mission success, are known as mission assurance. Mission assurance processes are the means by which, over time, some of the lost ground may be regained. Mission assurance improvement was the first response this time as well.

Epilogue
Can One Ever Truly Go Home Again?

The broad sweep of trends played greater or lesser roles affecting the factors of production, setting a new state of play. Unlike previous problem corrections, this time was different. No prior reform policy implementation had so successfully undermined an existing culture. Its implementers succeeded in changing the culture so substantially, it could no longer be sustained or recovered. They left no back-out option. The long lead time for new systems acquisitions assured that discovery of the problems would take many years. By then, a new cohort of acquisition specialists, having no ties or understanding of the old, would learn the new culture and nothing else. Acquisition reformers committed government acquisition to a new course amidst a swarm of other factors they did not, and could not, recognize and take into account. None of the new generation learned the untenable grounds on which the new culture was based. When problems inevitably arose, they attempted to solve these in the context of the new culture.

That begs the question about where things stand. Before considering any way forward, we must summarize the current interim state.

A key study in 2002 held acquisition reform responsible for the failures of launch vehicles and satellites. Acquisition reform was the largest and most obvious difference in the previous ten years. So acquisition reform, as the root cause, had to be overturned for the high reliability applications. This initial response concluded five major areas of corrections were needed.

First, the critical standards and specifications had to be rejuvenated. The most important standards had been canceled about a decade earlier while electronics technology continued its rapid advance. So, rejuvenating standards meant a significant effort aimed at updating the content before the standards could be made useful again. But which specifications and standards were needed? In 2005, a short-term

Implosion: Lessons from National Security, High Reliability Spacecraft, Electronics, and the Forces Which Changed Them, First Edition. L. Parker Temple III.

effort, named the Mission Assurance Improvement Task Force, began a review of the former standards with the intent to develop a set of approximately—seventy to eighty documents. The thinking was once these were brought up to date and made contractual compliance documents, they would force a return to the formerly successful disciplined practices. The standards and specifications had never been intended to be applied *in toto* to any development effort, since their individual scopes were often quite broad. The standards especially had always been tailored to fit a specific system acquisition. Tailoring, as might be imagined, meant specific modifications of the document due to lack of applicability or other reasons. However, in the decade since the standards and specifications had been compliance documents, few people were familiar enough with them to properly tailor the documents. Many whose only experience was post-acquisition reform found it hard to accept that standards produced anything useful, so resistance was fierce. Worst of all, the decade or more of lost data from industry, the former Rome Laboratories, and anomaly lessons learned meant updating the standards had to be done in a data-poor environment.

Reinstatement of MIL-STD-1546 and -1547 was a large step to reinvigorating the general Class S level of assurance. It is one thing (a rather easy thing, as it turns out) for a standard to say what should be done. It is quite another to implement it and keep manufacturers interested in providing such parts. As they currently stand, the new military standards strike an uneven balance between prescriptive and descriptive content, direction- versus performance-based approaches. This leads to compromises for manufacturers, designers, and ultimately the users of the final products.

The practical reasons for such compromises are a consequence of acquisition reform. As discussed earlier, when the standards were eliminated, many manufacturers did not retain their testing equipment and facilities. As pieces of test equipment were sold off, broke and were not repaired, or simply became obsolete with advanced part technology, test optimizations eliminated the need for such tests. By the time of reinstatement of MIL-STD-1547, some manufacturers no longer had the ability to do the required testing. Most are unwilling to make heavy investments in such equipment and testing facilities, preferring to contract out such services. The market forces, once again, do not favor returning to high reliability Class S parts as they existed in the late 1980s.

For instance, current military avionics or Class S parts are not the equivalent of their ancestors of the 1980s. The oversight of the high reliability parts industry by DSCC has, based on the scantiest of pretexts and little data, allowed "test optimizations" to undermine parts reliability. Test optimizations were intended to reduce the cost of making a specific part when a manufacturer could prove some aspect of required tests (number of tests, extent of tests, etc.) was excessive for their part and could be eliminated or reduced. This approach is abused and undermines specific Class S parts being accepted with confidence. Whether any Class S part meets a particular design requirement necessitates a full understanding of the test optimizations applied to that specific part. In short, the buyer must be aware of what is *not* included in the specific manufacturer's qualification to compensate with additional testing.

It is even possible to have space qualified parts unsuitable for operation in some space radiation environments. This apparent contradiction in terms arises because radiation hardness testing is not required for space qualification under QML. Radiation testing is an additional set of tests with respect to an intended orbital regime's environment to determine radiation hardness. Radiation hardness levels are important because ITAR restrictions preclude overseas sales of the highest radiation hardness parts.[1] To avoid ITAR restrictions, some manufacturers perform radiation tests only stringent enough to stay under the ITAR limits, allowing them access to the international parts marketplace. Complexity and interdependence of this sort abounds, and cannot be brushed aside easily without deep understanding, indicating market forces will continue to dominate any proposed solutions.

Maintaining or increasing the level of assurance in space qualified parts is a more herculean task in the 21st century than it was when Class S was introduced. Space quality parts are even more of a niche market now than they were in 1978 when Aerospace Corporation Vice President William Leverton predicted there would always be someone willing to profit by making such parts. What he did not see was there are conditions placed on the willingness of manufacturers to provide such parts resulting in the whole range of test optimizations.

The second thrust was systems engineering. Systems engineering covers such a wide range of disciplines there should be no surprise that the largest number of implementation actions was tied to it. Acquisition reform's emphasis on trusting the contractor to do the right things without independent verification meant systems engineering had atrophied. A cursory examination of the failures in the early 2000s revealed stronger systems engineering by the government could have ameliorated some of the problems of acquisition reform. Government systems engineering organizations began receiving more attention.

Third came testing. Testing was crucial to successful spacecraft since the inception of the space age. When acquisition reform replaced mission motivations with the cost bottom line, important tests early in spacecraft development were eliminated. Moving testing emphasis away from parts placed more reliance on subsystem and integrated system testing. Testing at higher levels of integration, however, only discerns some kinds of problems. Furthermore, many discriminating tests were not performed or were performed inadequately. Parts-related problems were slipping past such tests only to be discovered in orbit. The space acquisition community had to relearn the lesson that "testing in" quality did not work. Test methods and procedures were supposed to draw on first principles of engineering, chemistry, thermodynamics, physics, mechanics, statistics, and, as experience mounted, on failures

---

[1] Specifically, under the Arms Export Control Act, 22 U.S.C. 2778, authority to control the export of arms was vested in the President. Executive Order 11958, Administration of Arms Export Controls, delegated that to the Secretary of State, which the International Traffic in Arms Regulation implements. In the ITAR, Section XV covers Spacecraft Systems and Associated Equipment, describing the allowable radiation hardness limits. These limits do not translate well to actual operational space environments.

and lessons learned. Thus, rejuvenating testing, with a push to move testing earlier in program development, rose to a prominent place in the recovery from acquisition reform.

In a similar vein, especially with the intervening rapid advances in computer technology, software development and software systems engineering needed attention. Both computers and software had increased in complexity, resulting in the need for improvement in testing and other areas to eliminate unnecessary complexity.

Finally, and certainly not least, the entire range of parts, materials, and processes needed attention. Everything may *start* with parts, but everything *relies* on information. The government lost access to critical data. Parts and materials selection lists, as-designed parts lists, and as-built parts lists were no longer available to the government or their FFRDCs. One of the characteristics of the failures of the late 1990s and early 2000s was some problems affected multiple programs. The government and FFRDCs had no means to identify which programs might be affected by any given part problem. In fact, because of the de-emphasis on data exchange and a move to greater litigiousness, GIDEP no longer provided the early warning of parts problems. The possibility existed that parts problems were widespread, unidentified outside the contractors and their suppliers, and the government would be completely unaware of these and their effect on cost and schedule until a crisis ensued in orbit.

Working a triage approach, all five areas received immediate attention in varying degrees. Work began on standards, identifying which were most urgently needed. The parts, materials, and processes standards were among these. Implementation of the range of recommended actions began quickly but slowed just as quickly. In a classic replay of the move to Class S parts in the early 1970s, changing existing programs happened slowly. No programs had budgeted for the additional "new" requirements to return to the disciplined processes of the past. Partial implementation of standards and processes was the best that could be done immediately. Resistance to change kicked in on both sides of the government–private sector divide, so many recommendations did not stand much of a chance of being implemented. Shortly after the rush to recover began, reality set in.

It was one thing to identify which standards were needed, but quite another to ensure they were adequately updated. Staffs were inadequate to handle new workloads. Budgets did not exist for additional people, services, and other requirements. The "savings" of acquisition reform meant the money did not exist to fix its damage. Recovery was not going to happen in the short term.

Improved mission assurance went beyond the five areas identified initially. Just as acquisition reform had fundamentally changed a culture, another culture needed to change. Recovery was a long-term effort.

By 2005, after making organizational, documentary, staffing, contractual, and procedural changes (at least, those that could be afforded), a new realization arose. Not everything which had changed could be laid at the doorstep of acquisition reform. It seems that other forces had been at work, not as obvious, pernicious, or pervasive as acquisition reform, but there nonetheless. And the other forces were not trivial. Rolling back acquisition reform was not the complete answer. In fact, some analysts began to think there might not be a complete answer.

Other observations are important to public policy implementation. Several times we have mentioned the culture change that acquisition reformers sought and achieved. They knew the government acquisition methods and processes were a culture. Changing cultures entails manifold problems, as this history describes. Even the extensive coverage of this history only scratches the surface. Much of the reason is the complexity involved in changing a massive bureaucracy. The size of the defense acquisition community alone, not to mention the rest of the government, means the change was rife with second- and lower-order implications. We mentioned eighteen, but surely our list is incomplete as further time and analysis will reveal, and they serve only to make the point.

Massive changes feed on panels of experts. Those experts have justly earned the respect of their peers and others. Nevertheless, experience becomes dated, especially in a culture facing rapid change. Defense acquisition is a microcosm of the American culture; it draws from the same pool of talents and abilities as any other technical activity. Generational divides may be only poorly understood by the *éminence grises* who may not see as deeply into any particular subject as once they did. The basis for their eminence has, in some cases, been overtaken by their own success. The culture they experienced as mid-managers is no longer the actual situation, and yet that is their perspective. Experts have as many problems dealing with complexity as anyone else. Acquisition reform, as comprehensive as it was, has to be seen as a policy and its implementation amidst a redolent forest of other forces at work. Implementation of policy, then, by offloading responsibility to panels of luminaries must lead to unintended consequences unless the problem being examined is very narrow and shallow. Reforming acquisition is neither. Recognizing the difficulty of policy implementation and the complexity of acquisition reform as affected by all the other forces at work, then, informs our recovery. Recovering is not really like putting the toothpaste back in the tube. It is harder than that. The complexities involved mean recovery will have its own unintended consequences.

Undermining recovery is the length of time it will take. Some in government and Congress are already tired of hearing about the problems created by acquisition reform, and how much it costs to fix them, why problems persist, and so on. The American lack of patience with protracted efforts is already eroding support for a recovery that has only barely begun.

This is only the interim. If we learn nothing else from this history, let us recognize change is the norm. Undoubtedly more problems lying ahead need to be recognized, analyzed, and addressed. The interim is characterized by efforts whose goal is to return as closely as possible to the *status quo ante*, but the *status quo ante* is unachievable. It may also be the wrong goal. For one thing, the production function remains unknown and changing, so it is not possible to decide how closely the *status quo ante* can be approached. For another, things were moving to some new unknown optimum before acquisition reform, and what the result might have been is equally unknowable. How closely the *status quo ante* can be approached is an open question, and may even be unimportant.

The reason is simple. No prior problem recovery returned to its *status quo ante*. Instead, problems had solutions and lessons leading to improvements over the *status quo ante.* After all, something about the *status quo ante* had led to the problems in the first place. Mission assurance was about using current circumstances to best advantage. The ultimate goal must be making mission assurance part of the space acquisition culture—the normal way of doing business. Returning to what *was* is not as important as shaping what *will be.* But the price tag is very high. Recovery is not a zero sum game. The cost to correct for what was lost will greatly exceed any claimed or actual acquisition reform savings. And it will take time.

The perfect storm lasted only a couple of years, but its effects were profound. The thing that was different this time was how completely the basis for high reliability spacecraft had been undermined.

Paradoxically, in an environment of rapid technological change, the damage took over a decade to comprehend because complex, high reliability spacecraft are not built quickly. More than a decade after the first smoke appeared indicating some problems, extensive studies revealed much of what went wrong.

Correcting the problems cannot be short term. Senior officials must not believe a quick fix can demonstrated before their next promotion or posting. Although little will be accomplished at each step, each incremental positive step matters. Repairing the damage requires sustained, focused, and coordinated effort on a broad front without any expectation that changes will be effective in a year, two, three, or more.

The last section questioned the *status quo ante* as a goal. The issue is whether high reliability spacecraft can ever really go home again. Can high reliability spacecraft be built again? Proven methods will work. Probably not as effectively as they did, but work they will. It's not a matter of faith. The truth of history points the way to the future. Just as different paths diverged from the 1969 Space Task Group, circumstances forty years later appear to indicate another fork in the road. What form, then, should the path ahead take? Four basic alternatives exist.

PATH 1: STAY THE COURSE

High reliability space programs are most familiar with staying the course because they are by nature conservative. That is, they conserve wisdom and are averse to risks. Conservatism also means a resistance to change. However, the current state of affairs is still too affected by acquisition reform's influences. National security and defense space programs have not yet returned to the highest degree of reliability. Now is not the time to stop reversing the adversities of the mid-1990s.

Staying the course is about a path of incremental change moving toward highly reliable spacecraft. In addition to improving the quality, reliability, and supply of Class S parts, this includes a wide range of related activities under the rubric of mission assurance. Another way to think of staying the course is the incremental recovery of lost or diminished capabilities. Somewhat like a soft metal, once deflected, it may slowly return to near its initial state, but never recover completely. The current state is still closer to full deflection than its pre-1993 initial state.

Significant obstacles need to be nibbled away at and overcome. Resources do not allow wholesale recovery, so incremental, constant pushing for improvement is what staying the course must be all about. Government agencies and the spacecraft industry need to work together to shore up confidence in Class S. What will that entail?

Changes need to cover a broad front from the lowest level, because everything starts with parts. A change as simple as moving to Class S parts took an extended time because of the difficulties of changing ongoing programs. Marketplace pressures never favored Class S parts as a long-term mainstream enterprise. Everything may start with parts, but what kind of parts?

As discussed in the previous chapter, Class S is not what it used to be. All of this suggests market and practical limits to the degree of assurance attached to high reliability Class S parts. In order to maintain the actual high reliability demands of this approach to spacecraft design, a few options exist within "staying the course." All Class S parts need to be understood based on their individual test optimizations. Any shortfalls in testing have to be conducted either by the purchaser or by third party specialized test houses that can determine whether any given lot of parts will meet the appropriate standards. Other such options exist, but all increase the cost of obtaining the high reliability Class S parts that were formerly the accepted norm.

Part of staying the course requires improving the supply of Class S parts. Not all device types are available. To compensate for necessary parts' unavailability, the United States might take the approach it did in the early 1960s, and pay for some "captured" manufacturing lines. While such an expense may seem to be egregiously large, the alternative is that spacecraft designed for long orbital lifetimes will suffer early and devastating failures. Third parties might be engaged to do this function and perform it profitably by taking over obsolete manufacturing lines for the government.

A compromise variation on this path is allowing lower qualification limits based on the actual intended usage. The space environments in low earth orbit vary considerably from those at geosynchronous orbit, so parts destined for one or the other would have to be qualified only to their specific end use. In other words, qualification for one orbital regime may not apply to a very different regime, so a part qualified to fly in one may not be qualified adequately for the other. This allows a cost-tailored approach to determining whether the device will (or will not) work in the intended environment.

Along these lines, considerable work is being done on use of heritage designs and new technology insertions. The former recognizes nonrecurring engineering costs, a substantial part of satellite design, can be preserved even with newer part types under certain conditions. The latter addresses both truly new technologies (such as new materials) and use of previously flown technologies desired to be used in a new application or new environment.

These ideas represent part of the range of activities and options available by staying the course of recovery from the perfect storm. The *status quo ante* will not be recovered, but may not even be the best end state. The activities mentioned here also carry with them implications for infrastructure and sustainment, so there is no easy path to follow. That points to the need for a cyclical improvement path.

While continuing to push for improvements, analysis of where problems are occurring needs to be done periodically. That analysis should highlight where the most significant remaining problems are, allowing prioritization for the next steps on the course. As resources allow, new initiatives would be started and old ones made more robust. After a couple of years, further analysis would again be needed.

Two things need to be seen in this cycle. First, it is very long term, with years between cycles to allow the marketplace to adjust and other forces to somewhat settle in. This was the historical behavior of a marketplace that became increasingly dynamic, and there is no reason to assume the future will provide stability. Second, limited resources preclude quick fixes. Resources allocated to staying the course of recovery should be commensurate with the level of pain felt by agencies that experience problems, unexpected cost increases due to late development failures, and failures in orbit. Staying the course of recovery would not be the path to be parsimonious.

However, sensible recovery does not mean throwing money at the problems, either. For one thing, only limited human resources are available. Human resources focused on mission assurance can be spread too thinly by trying to do too much without ensuring previously started initiatives are securely in place. Staying the course also implies the human resources clearly understand the path ahead. Staying the course is as much a policy implementation problem as acquisition reform, except in staying the course, the culture being changed is the one created in the aftermath of acquisition reform. The change desired is a recovery that closely approaches the situation before things got off course. A shotgun approach worked to bring down the former culture, but recovery demands well-thought-out goals and objectives that can be implemented correctly. Staying the course, then, carries with it a monitoring function to ensure the initiatives and the overall space acquisition implementers not stray from the intent.

That requires a change in focus. For instance, current discussions often do not really weigh cost versus benefit. Such discussions usually only include anecdotal information without greater supporting data due to the way industry evolved and the lack of capturing lessons learned in useful form. For the purchasers of electronic parts, this means one of two things. Either the problem is in quality issues with parts received, which the purchaser then needs to address (which means more cost), or it affects the manufacturers, whose costs go up because they have greater lot loss (yield problems) and require increased testing. But, in the latter case, greater cost tends to lead to fewer companies willing to manufacture Class S parts. This is mysterious behavior, because it means Class S is not sufficiently profitable. That is where the focus needs to be. Profit also has to include the ongoing maintenance of manufacturing capabilities on the long centers of modern spacecraft build cycles. If a satellite program buys its parts over a year or two early in development, and the satellite will take five to ten years to build, that program may not need more high reliability parts for a very long time. And when the parts are needed, the manufacturers are expected to have hot production lines to meet the need. Curiously, greater unreliability may increase the availability of better parts, because higher failure rates raise the demands for parts.

Second, though, is the question of whether manufacturers will still be willing to continue to make parts under SCDs, which are inherently more costly and therefore more profitable. They seem willing. However, like the first point, this still comes back to the government's willingness to pay.

This line of thought coalesces along these lines: If government programs plan adequately, and prepare budget profiles to include more expensive parts up front, then it matters little from the government perspective whether the parts are from QML, QPL, or SCDs. It is simply the cost of doing business. If SCDs are the way to solve continued supply, their individual nature precludes use of DSCC to qualify them, and puts the burden of independent quality assurance back on the space acquisition organizations. This also becomes a cost of doing business. The questions remaining open are under which approach does the government gain access to the best available parts, and what does it need to do in order to deal with the consequences within its spacecraft contractors? The answers lie in sustained, strong systems engineering practices.

PATH 2: A REPAIR SHOP IN EVERY ORBIT

An alternative path that held a brief flurry of interest in the late 1970s and early 1980s is the maintainability approach. If extremely reliable parts do not support ten-year lifetimes in satellites, then make the satellites so that failed boxes can be replaced. While some orbital repairs have been conducted—notably on SolarMax and the Hubble Space Telescope – such an approach has generally not been economically or physically viable.

This approach immediately falls apart due to the wide range of orbits used for various reasons, and the relatively few spacecraft in any specific orbit (except for GPS). The physics-driven cost of getting to such individual satellites is usually the death knell of spacecraft maintenance and repair. Space launch is not inexpensive. The cost of repairing satellites and of several space launches quickly favors the option of not repairing, but simply doing one space launch of a brand new satellite.

Further problems come from the means by which spacecraft designs are made "maintainable." Were this path to be pursued, spacecraft would have to be designed with more connectors and access to critical components in order to be maintained. Increasing the number of connectors, for instance, reduces the reliability of the spacecraft, as connectors are often the source of failures. Maintainable satellites would require more launches than satellites designed with proven reliability processes. Furthermore, with the retirement of the Space Shuttles, the ability to perform anything other than robotic servicing even at low orbital altitudes does not bode well for this concept.

While limited examples may continue to exist, this potential pathway simply does not support the wide variety of satellites and orbits in use.

PATH 3: GARBAGE IN—HIGH RELIABILITY OUT

Another path involves giving up and giving in. That is, back off completely from space qualified parts and highly reliable satellites, using only the least expensive commercial grade parts. If subsystems of a spacecraft were made sufficiently redundant, which is to say multiply and completely redundant, without any single point failures possible, then a form of long term high reliability systems might be possible.

The problem with this approach may be obvious. The underlying architecture of deeply redundant subsystems is radically different from anything designed to date. The number of backups, then, is driven by the length of time the mission is expected to operate. Instead of taking the approach of long ago, backing up the most critical subsystems, this approach assumes every system will fail and has internal mechanisms for switching to a backup. This approach increases the level of system complexity by orders of magnitude, and is only possible because of the high degree of miniaturization of devices. It calls for an entirely new approach to systems design, employing a system controller that senses when a subsystem is affected by radiation effects or bad parts. The system controller would itself have to be multiply redundant, and self-aware enough to know a backup controller needs to take up its load. Voting schemes have done this kind of delegation in very radiation tolerant designs, but the level of complexity of any such architecture would be well beyond anything done previously. Such complexity would drive design, assembly, and testing costs to an extreme degree.

Furthermore, the trend to control costs in space programs is to use much smaller scale electronics to save weight, making satellites smaller and allowing the use of smaller launch vehicles, thus saving considerable cost. However, the cutting edge of technology feature sizes is what commercial parts possess, and they are not designed to withstand the temperature variations, vibrations, shocks, and radiation in space. In this alternative path, more of the cutting edge, smallest possible feature size electronics would have to be used by greater numbers of redundant backup systems, which would drive satellite size and weight up.

While intellectually interesting, it remains a curiosity more than a real alternative.

PATH 4: HENRY FORD IN SPACE

The fourth alternative path has lengthy historical precedent in both the United States and the former Soviet Union. Until the United States began to pursue its longer spacecraft lifetimes, both nations launched very frequently. Each satellite launched was not expected to last very long, so another had to be in the pipeline for another launch in the near term, often in as little as two weeks. The Soviets mass-manufactured satellites, minimizing the differences from one satellite to the next. What mattered most in the Soviet model was capability, not pushing technology to the nth degree. The resulting high launch rate reduced the cost per launch (but not necessarily the total cost of launches in a year). The assembly line approach also allowed differentiation of manufacturing processes, which favored retention of skill sets and standardized training. Is this the way of the future?

This model favors selection of low cost parts, since the expectation is that the satellite will not last. Concerns about radiation, temperature, shock, and vibration effects inherent in space operations would remain to be designed for and accommodated or such satellites would be vulnerable to becoming the space equivalent of dropped cell phones.

This approach is inherently tolerant of orbital failures due to the spacecraft manufacture pipeline. Commercial or military grade parts would be the normal kinds of part types. Further in favor of this path would be the assembly line type of manufacturing, looking more like a clean-room version of Henry Ford's Model T manufacturing. Variations in workmanship and issues dealing with poor workmanship would tend to dampen out in a long production line environment, correcting another problem with current manufacturing methods.

This radical departure from current approaches and the launch infrastructure would require a heavy investment to be able to support such launch rates. The heavy investment would drive down individual launch costs. Total development, manufacturing, launch, and operations costs might not be significantly different from current costs.

While this path has some attractive features, the realism of the approach is limited to a few orbital regimes. For instance, if satellites were intended for six-month lifetimes, this limited lifetime might not support the kind of robust communications satellite business to which the world is now accustomed. The consequence might be that the United States would cede communications satellite development to other nations, but that also seems unrealistic. Some satellites simply need to live longer. This would tend to drive extremely redundant backup systems for satellites designed with lower reliability parts, and new approaches to fault tolerance in satellite design, but these are enabled by the extremely small feature sizes of current and planned solid state devices.

While one can begin to conceive that this alternative path might be viable—more viable than spacecraft maintenance and repair—fundamental issues remain unsolved. First is whether the current way of doing business could be turned around to this approach in anything resembling affordability. Second, and by no means less important, this approach emphasizes clones of a given satellite design, and adapts poorly to changing mission requirements and needs. Clones for communications satellites might be realistic, but not the most advanced capabilities needed to support national security and defense, where capabilities must advance to keep pace with a hostile world.

* * *

Even if Captain Smith had done everything possible to turn when the iceberg was sighted, could he really ever have changed the *Titanic*'s course sufficiently? The U.S. way of using space was heavily optimized for highly reliable, long-lifetime spacecraft by the early 1990s. Since that time, the emphasis has been on recognizing what was lost. As difficult as it may be to imagine how, the formula for returning completely to that way of producing highly reliable spacecraft has been lost. Many forces broke apart the formula, some irreversibly. Surprisingly, though, while the desire to return to the former system may exist in various agencies and organizations, the will and resources to make it happen have been lost in an age when problems need to be fixed by the end of the fiscal year. A system devastated by a short and intense combination of cultural change and other forces, which then continued foundering from multiple consequences over more than ten years, does not get fixed simply, easily, or in a short time.

This history teaches that poorly understood, radical changes to complex systems are unwise and do not fare well. Although such a statement is only partially supported by studies, it seems pretty clear that trying to steer space systems acquisition away from long-established high reliability demands would be difficult if not impossible to implement. Even then the implementation might be as disastrous as was acquisition reform—radical solutions are just as bad, and for many of the same reasons. Steady, incremental recovery takes resolve, but staying the course makes the most sense.

Some may argue for a hybrid solution, cherry-picking the strengths of the different paths above. However, without providing more of the necessary evidence here, a hybrid would only introduce a wider variety of problems, as it would be impossible to cherry-pick the good without inheriting some of the bad. This history shows how changes made without extensive understanding create serious unintended and unanticipated consequences. Fundamentally, the damage was done by lack of understanding of the critical factors of production and the overall production function. Without such understanding, a hybrid solution would be no better than a shot in the dark.

To this point, we have shown how space systems evolved and squeezed nearly every ounce of reliability out of every aspect of an electronic piece part's lifetime. Everything starts with parts, but parts also have a start. Reliability in the *status quo ante* started with understanding technology and physics of failure, then continued in the inception of each part's design; following standards through preparation for manufacturing; the manufacturing process itself; auditing of the processes; checking and testing the processes; design of the spacecraft and further mission-driven requirements added to the part; the handling of the part from the manufacturer through the whole chain of custody to its assembly in a spacecraft; testing at the box, subsystem, and integrated system levels in vibration, thermal, and vacuum chambers; at the launch site ready for launch and finally in orbit; and through constant feedback about anomalies, failures, and root cause determinations to be folded back into the guiding standards to produce the next part for the next spacecraft. Highly reliable spacecraft demanded the involvement of every participant in the life of each part, performing to a standard of quality assurance that tolerated no compromises.

Clearly, those days are history. As fascinating as the history may be, it cannot be fully recaptured. As difficult to maintain as the *status quo ante* was, staying the course requires great resolve, resources, and a plan. The nation's work is cut out for it.

One certainty in staying the course is the current situation will only continue to decline and fractionate if treated by half measures. It took several years for acquisition reform to wreck the havoc it did, but it will take many more years to recover as much as possible of what was lost. Staying the course implies a national leadership team of government and FFRDCs committed to steady improvement, providing the resources and management attention essential to long lived spacecraft. FFRDCs were created to help the government deal with personnel turnover and as the corporate memory to keep the enterprise on track. Now, just as at the beginning of this history, that role is essential.

This history emphasizes the evolution of handling rising complexity. Making complex systems requires the sort of disciplined, orderly approach exemplified early by the crossover diagram and later by the V model. This is essential, but not sufficient, because high reliability also demands the greatest attention to quality detail starting from the very lowliest of all things in space—each individual piece part. Everything starts from parts. The essence of what was lost in the mid-1990s is the handling complexity from the lowest detail to the highest level of assembly.

While it is true you can never go home again, we can at least visit the old neighborhood. There is no question the highly optimized system of producing extremely reliable spacecraft based on space qualified parts would have come changed to some extent in the mid-1990s. There is equally no question options and opportunities existed that could have, with great effort, kept the U.S. space programs on a relatively even keel. BRAC and acquisition reform, among all the forces tearing down the old edifice, were the most egregious blows. Their direct and collateral effects are the hardest to reverse or mitigate.

If staying the course is generally the right approach, then the requirements the United States could and should address have just been discussed. To date, several studies have examined this subject and made a number of important recommendations, some of which have been partially adopted.

The current danger is that mission assurance is now old news. After all, we heard all about that in the mid-2000s. At this writing, barely five years into recovery, impatience with solving problems created by acquisition reform has been overtaken by intolerance with continuing to point to acquisition reform as the cause of current problems. Others dismiss the claim that acquisition reform really affected as many things as are blamed on it, choosing instead to believe current problems are manifestations of some new problem source.

This is about national security and national defense. The seriousness of the situation warrants a far better response than it has received. Rome was neither built nor destroyed in a day, nor can it be partially restored overnight. The solutions to reversing the fortunes of high reliability, space qualified parts and long-lived spacecraft will take a consistent plan, time, resources, cooperation, and, most of all, steady, dogged *persistence*.

Index

Note: Page numbers in *italics* indicate
figures; those in **bold**, tables.

A-12 Lockheed reconnaissance aircraft, 44
Abrahamson, George R., 260
Acquisition, 152; *see also* Defense, Dept.
     of; Defense Science Board
  concurrency approach, 19, 45
  faster-better-cheaper, NASA's, 269–270,
     272
  and loss of engineers, 290–292, **291**
  space systems, 146, 294
  streamlined, 160
  weapon systems, 206
Acquisition culture, 189, 288–290, 291,
     313
Acquisition Executive, 207
Acquisition Executive Programs, 209
Acquisition Law Advisory Panel, 246
Acquisition Logistics division, SD's, 205
Acquisition management, and technological
     change, 227
Acquisition process; *see also* Defense
     Science Board; Stockpiling
  civilian control over, 192–193
  contractors in, 286
  economizing in, 282
  operating stockpile in, 211
  Sec. Perry on, 234, 235
  Sec. Perry's view of, 188
Acquisition programs
  changing skills in, 291, **291**
  false shortcuts in, 281–282
  and technology freezes, 299
Acquisition reform, 194, 225, 234, 236,
     309, 312, 320
  aftermath of, 275
  basis for, 235

and "best commercial practices," 292
and *Blueprint for Change,* 248–250
in changing context, 236
and Cold War, 4–5
and commercial best practices, 285
and commercial products, 246
consequences of, 273, 287
and contractor's process, 269
and defense contracting, 243
DSB (1986) study, 193
dysfunctions arising from, 307
experts in, 313
and factors of production, 302
failures associated with, 267
and foreign dependency issue, 197
implementation of, 256, 288–289, 320
and industry data sharing, 294–295
and JAN S stockpiling, 280
key elements dismantled by, 273
and method of enforcement, 265
and military standards, 292–293
minimalist approach in, 291
paperwork reduction in, 256–257
and performance based contracts, 258
Sec. Perry's DoD Specifications and
     Standards Reform in, 255
problems associated with, 268, 270
recovery from, 313
revolution in, 206
and role of DSB, 239–240
and satellite failure, 4–5
and SBIRS, 265–266
SCDs in, 297
SECDEF's, 200
Gen. Slay's approach to, 146
and space qualified parts, 244
support for, 235–236
widespread changes of, 258

*Implosion: Lessons from National Security, High Reliability Spacecraft, Electronics, and the Forces
Which Changed Them,* First Edition. L. Parker Temple III.
© 2013 the Institute of Electrical and Electronics Engineers. Published 2013 by John Wiley & Sons, Inc.

Acquisition Reform. A Mandate for Change,
  247
Acquisition streamlining, 207, 208–209,
  226
  FASA, 246–247
  and JAN S stockpiling, 280
Acquisition Task Force, 188
Ad Hoc Study Group on Space Launch
  Vehicles, 129, 132, 134. *See also*
  Launch vehicles
Ad Hoc Study on Electronic Parts
  Specification Management for
  Reliability, 38, 48, 49–50, 52
Advanced Extremely High Frequency
  (AEHF) communications satellites,
  272
Advisory Group on Reliability of Electronic
  Equipment (AGREE), 32–33, 38
  strengths and weaknesses of, 37
  task groups of, 33–34
Aerospace, military investment in, 165
Aerospace Corp., 46, 220, 259, 311
  Satellite Systems Division of, 136
  Space Parts Working Group of, 121
Agnew, Vice Pres. Spiro T., 89
Aircraft
  acquisitiions, 161
  compared with spacecraft, xiii
  near-real-time reconnaissance provided
    by, 123
  stealth programs, 160–161
Air Force, U.S., 46; *see also* Rome Air
  Development Center; Space and
  Missile Systems Organization
  Ad Hoc Study Group on Space Launch
    Vehicles, 129, 132, 134; *see also*
    Launch vehicles
  Air Force Plant Representative Office
    (AFPRO) surveillance, 148
  Air Force Scientific Advisory Board
    (AFSAB) review, 270
  Air Force Space Command (AFSPACE),
    167, 169, 209, 306
  Air Force Systems Command (AFSC),
    55, 287
    acquisition records of, 97
    laboratories of, 79
    Manufacturing Technology Quality
      Improvement Program, 165–166

national defense space systems owned
  and operated by, 122
Program Offices, 105
and Program Review Board, 105
375 series manuals of, 94
Air Research and Development
  Command (ARDC), 19
Assistant Secretary for Space, 225
Ballistic Missile Division (AFBMD), 29,
  46
ballistic missile program of, 27
Contract Management Division
  (AFCMD), 103
Electronic Systems Division (ESD), 191
and IDEP, 52
laboratory capacity of, 226, 260
leadership of, 153, 209
and maintainable electronic systems,
  11–12
Manufacturing Technology Laboratory,
  59
Materials Laboratory of, 41
military satellites developed by, xiv
Minimum In-plant Surveillance program
  (MIP)
  missile responsibility of, 35
Molecular Electronics of, 41
Satellite Communications System of, 88
Space Division (SD), 213, **213,** 213n
space expenditures of, 169
Strategic Air Command (SAC), 123
Strategic Missiles Evaluation Group of, 18
and *Towards New Horizons II,* 125
Western Development Division (WDD),
  19
Wright Air Development Center
  Manufacturing Technology
  Laboratory of, 42
Air Force Materiel Command, 287
Air Force Space Plan, 209–210
Air Launched Miniature Vehicle (ALMV)
  anti-satellite, 133–134, 163, 167,
  184, 185, 210. *See also* Anti-satellite
  systems
Air Research and Development Command
  (ARDC), WDD of, 19
Ajax anti-aircraft missile, 25, 26
Alberts, R. D., 42
Aldridge, Edward C. (Pete), 208, 225, 270

ALERT system
  IDEP, 80
  NASA's, 74
Allen, Brig. Gen. Lew, 105
Allison 250-C3OR Helicopter Engine, 240
Anti-satellite (ASAT) systems, 125;
      see also Air Launched Miniature
      Vehicle antisatellites; Zeus anti-
      ballistic missile
  Bold Orion, 125
  High Virgo, 125
  nuclear, 125
  Project 437, 124
  Project 505, 124
  SAINT, 53, 54, 88
  Soviet, 124, 125, 133
Anti-trust laws, 156
Apollo Guidance Computer, 59, 60, 62
Apollo program, xi, 89, 99
  adequate quality supply for, 99
  devastating loss of, 94
  winding down of, 93
Application specific integrated circuits
      (ASICs), 140–141, 142
  cost of, 171
  development of, 154
  increasing complexity of, 298
  and lack of standardization, 215
  and Mead-Carver revolution, 149
  process certification for, 200, 201
  QPL approach for, 302
Armed Services Electro-Standards Agency,
      49
Armour Research Foundation, of IITRI, 70
Army, U.S.
  and C41 capability, 263
  Diamond Ordnance Fuze Labs, 40
  and IDEP, 52
  missile responsibility of, 35
  TENCAP program of, 91
Army Signal Corps
  Auto-Sembly process developed by, 41
  Micro-Module program of, 40–41
  and solid state technology, 11
  and transistor technology, 16, 63
ARPANET, 142
ASM-135, 133. *See also* Air Launched
      Miniature Vehicle
A Specification, 97, 98, 99

Aspin, Les, 246
*Assembly-Through-Flight Test and
      Evaluation Guide* (SAMSO), 94, *96,*
      96–97, 99, 259
Aston, William J., 137
Atlas missile, 19, 25, 44–45
Audits
  DESC, 223
  in quality assurance, 103
Augustine, Norman, 187
Automobile industry, specialized ICs
      produced by, 246
Autonetics Division, of North American
      Aviation, 31, 60, 61, 99
Auto-Sembly process, 41

B-2, Northrop stealth bomber, 161
B-29, Boeing Superfortress, 11, 18
B-36, Convair Peacemaker, 13
B-47, Boeing Stratojet, 124–125
B-58, Convair Hustler, 125
"Balkanization," 295–296
Bardeen, John, 10, 27
Base Realignment and Closure Commission
      (BRAC), 260
  Air Force chief scientists' letter to, 298
  hearings, 285
  recommendations of, 263
Basic research, loss of sponsorship for,
      286–287. *See also* Research
Bateman, Herbert H., 235
"Bathtub curve," *108,* 108–109, 115, 135,
      304
Battelle Memorial Institute analysis, of
      parts test data, 51
Battle of the Bulge, 9
Bell Telephone Laboratories (Bell Labs),
      10, 62, 72
  data sharing contract with, 12
  semiconductor technology of, 15
  and transistor technology, 11, 14
Berlin Wall, fall of, 231
Best commercial practices, 245, 292, 294
*Beyond Horizons: A History of Air Force in
      Space3, 1947–2007,* 272
Bloch, Felix, 7
Blueprint for Change (PAT), 248–250
Blue Ribbon Commission on Defense
      Management, President's, 188;

see also Acquisition streamlining;
    Packard Commission
and commercial operating environments,
    190–191
intention of, 189–190
Blue Ribbon Commission Report
on commercial electronics, 191
recommendations of, 189, 191, 192
release of, 191–192
Boardman, Allan, 163, 164
Boatright, James, 260
Bockelman, Col. David C., 174
Boehlert, Rep. Sherwood, 264
Boeing B-29 Superfortress, 11, 18
Bold Orion, 125
Bombs; *see also* Proximity fuzes
    atomic, 11, 17
    fission devices, 13
    fusion devices, 16–17
    hydrogen, 13
Boost Surveillance and Tracking System
    (BSTS), 169
Bown, Ralph, 10, 11
Bradburn, Maj. Gen. David D., 129, 130,
    131, 132, 158, 267. *See also* Ad Hoc
    Study Group on Space Launch
    Vehicles
Brattain, Walter, 10, 27
Brauer, Joseph B., 70
Bridges, James M., 38, 41
Broad Area Review (BAR), of launch
    vehicles, 267. *See also* Launch
    vehicles
Brown, Harold, 278
B Spec, 97, 98, 99, 102
Buckley, Oliver E., 10
Budget cuts, 298. *See also* Acquisition
    reform
Burnett, James R., 193, 194, 196, 197, 200,
    288. *See also* DSB study
Bush, Pres. George H. W., 224
Bush, Vannevar, 8–9

Calculators, hand-held, 120
Capacitors
    development of, 9
    tantalum in, 219
Carhart, R. R., 33
Carter, Pres. Jimmy, 133, 142, 146

Casey, Lt. Gen. Aloysius G., 202–203, 208,
    211, 212, 213, 220
Cell phones, 304
Centralab, Globe-Union, Inc.'s, 9
Central Intelligence Agency (CIA), 14
Challenger, 224, 284
    1986 loss of, 208, 210
    recovery, 225
Charyk, Joseph V., 260
Cheney, Vice Pres. Richard, 224, 225
    Defense Management Review of, 226,
        236, 246, 269
    reforms of, 289
Chip technology, 140, 169. *See also* Very
        High Speed Integrated Circuits
Class A parts program, 80, 99, 102
    in DSB study, 197
    and Minuteman experience, 149
Class B parts program, 80–81, 102, 116n
Class C parts program, 81, 102
Class S parts program, 121, 126, 175, 304
    classification of, 213–214
    costs of, 164, 175
    effects of high reliability, 154
    evolution of, 134
    and hybrids and microcircuits, 116n
    initial specifications for, 118
    JAN Class S Operating Stock Program
        stockpile
    long lead times of, 278
    and loss of 25th Titan IIIC, 130
    and market forces, 310
    and mission duration, 137
    for more complex satellites, 155
    on QPL, 135–136
    requirements for, 116, **117–118,** 118, 135
    and satellite technology, 128
    SD Survey of need projections for, 181,
        182
    shrinking of marketplace for, 183
    standards, specifications and guidelines,
        132
    stockpiling of, 280, 281
    success of, 174–175, 280
    suppliers, 259
    wafers, 157
Clean room practices, 172
Clinton administration, defense drawdown
    of, 234

Cold War, 1, 55
  and acquisition reform, 4–5
  adversary's preparations during, 21
  ending of, 231, 232, 282, 297–298
  and history of high reliability parts, xiv
  penetration flights during, 22
  preventing nuclear Pearl Harbor during, 30
  and Reagan administration, 178
  and Soviet Union, 212
  technological superiority during, 202,
    250, 251
  and U-2 overflights, 48–49
Command, Control, Communications,
    Computers, and Intelligence (C4I),
    261, 263
Command destruct systems, high reliability,
    107
Commercial contracts, 145, 147. *See also*
    Contractors
Commercial industry, and military
    standards, 159–160
Commercial item descriptions (CIDs), 249
Commercial marketplace
  development of, 158
  and DoD, 237
  functional specifications in, 237–238
Commercial-Off-The-Shelf (COTS)
    equipment, 240
Commercial parts
  definition of, 244
  in military systems, 245–246
Commercial products, SIA/GPC definition
    of, 244
Communications satellites, 123–124;
    see also Anti-satellites; Satellites
  AEHF, 272
  clones for, 319
  FLTSATCOM, 215
  IDCSP, 88, 123
  market for, 284
Communist Party of Soviet Union (CPSU),
    231
Complementary MOS (CMOS) technology,
    76
Complexity
  dealing with, 298–299
  handling rising, 321
Compliant Hybrid microcircuit, 223.
    See also Hybrid microcircuits

Component Quality Assurance Program
    (CQAP), 65. *See also* Minuteman
    program
Component Reliability History Survey
    Program (CRHS), of U.S. Navy, 74
Components
  and crossover concept, *96,* 96–97, 99
  defined, 94–95
  and mission success, 95
Computers; *see also* Personal computers
  general purpose *vs.* task-specific,
    153–154
  technological developments in, 17–18
  VHSIC spaceborne, 216
Computer software, 98
Computer technology, and acquisition
    reform, 312
Conant, James B., 23
Concurrency approach, 19, 45
Conservatism, defined, 314
Consumer electronics, 156, 290
  explosive rise of, 283
  market for, 282–284; *see also* Electronics
    industry
Contractors
  and acquisition policies, 286
  in acquisition reform, 288
  lack of insight into activities of, 295
  and SCDs, 296–297
Contracts
  commercial, 145, 147
  cost-plus, 146
  performance-based, 258
Convair B-36 Peacemaker, 13
Conway, Lynn, 141–143
Cooper, Thomas C., 207–208
Corona reconnaissance satellite, 3, 24, 25,
    44, 45, 49, 53, 307. *See also*
    Discoverer satellite
Cost consciousness, 178
Cost containment
  and cost-based contracting, 238
  and quality assurance, 146
Cost control, 168. *See also* Process controls
Costello, Robert B., 205
Cost growth
  for defense programs, 242–243
  and military standards, 148
Cost-plus contracts, 146, 148

Costs
and DoD reforms, 239–240
infrastructure, 285–286
storage, 155
Counterfeiting, in parts industry, 257–258,
257n
Cromer, Lt. Gen. Donald L., 220
Crossover concept, *96,* 96–97, 99, 232
C Specs, 97, 102
Cuban Missile Crisis, 59. *See also* Cold War
Culbertson, Brig. Gen. A. T., 73
Cyclical improvement path, need for, 316

D'Amato, Sen. Alfonse, 264
Darnell, Paul S., 38, 48
Darnell Report, 49–51, 101
Data item descriptions (DIDs), 253
Data sharing, of Bell Lab, 12. *See also*
Information; Information sharing
David, Edward, 158
Davis, Harry, 71–72
Defective Parts and Components Control
Program, 120
Defense, Dept. of (DoD); *see also* Revolt of
the Admirals
and acquisition reform, 246
and Ad Hoc Study on Electronic Parts
Specification Management for
Reliability, 38, 48, 49–50
and best commercial practices, 239–240,
249
IDEP's success within, 74
industrial base of, 249
inter-service rivalry and, 35, 36–37, 52
intransigence of, 242
nongovernment standards adopted by, 254
Office of Defense Mobilization's
Scientific Advisory Committee
(ODM/SAC), 23
research and development priorities of,
35–36
Section 800 Panel (*see* Acquisition Law
Advisory Panel)
5000 series of directives and instructions,
97, 226–227
VHSIC program of, 143
Defense Acquisition System, 97
Defense Advanced Research Projects
Agency (DARPA), and VLSI, 141

Defense Contract Administration Services
(DCAS), 103
Defense Electronic Supply Center (DESC),
114, 175–176, 265
and acquisition reform, 259
certification process of, 222
Class S manufacturer certification, 135–136
and class S parts, 135
and high reliability space qualified parts
efforts, 132
stockpiling of, 176, 177, 180, 204–205
Defense Logistics Agency (DLA), 205, 220;
see also Mill Run Products Quality
Assurance Program
Defense Electronics Supply Center, 114
Defense Supply Center, Columbus
(DSCC)
Ogden, Utah, facility of, 183, 205
Defense Management Review, Vice Pres.
Cheney's, 224, 226, 236, 246, 269
Defense Meteorological Satellite Program
(DMSP), 88, 123
Defense procurement
cost of, 250
mistaken criticisms of, 185–186
Defense Reorganization Act (1958), 36
Defense Reorganization Act (1986), 192
Defense Science Board (DSB), 187–188,
285, 289
on Class S issues, 197
on foreign dependency issue, 197
and military standards, 217–218
and order of precedence, 199
on plastic encapsulated parts, 196, 241–242
politics and, 200
proposals of, 194
on quality of commercial parts, 197–198
reactions to, 195
task force of, 238
Defense Science Board (DSB) studies
goals of, 193
results of, 201
Semiconductor Dependency study,
201–202
summer study on defense acquisition
reform, 246
Use of Commercial Components in
Military Equipment summer study,
201

Defense Semiconductor Dependency, Task
Force on, 201
Defense Standards Improvement Council,
255
Defense Supply Agency (DSA), 81
military parts and materials
responsibilities of, 101–102
and Program Review Board, 105
Defense Supply Center Columbus (DSCC),
265
Defense Support Program (DSP), 88, 123,
265
Defense systems, U.S., technological
superiority of, 201–202
DeLauer, Richard D., 160, 161, 162, 166, 187
Delta, 266
Deming, W. Edwards, 172
Denfield, Adm. Louis E., 13
Department Of Defense Authorization Act
(1982), 162–163, 162n
De-rating, in parts industry, 159
Desert Storm, Operation, 4, 231, 250, 282
Deutch, John M., 236
Digital Equipment Corp., 240
Discoverer satellite, 24, 44
Discovery (space shuttle), 172
Discrete device, defining, 39n
Documentation
importance of, 94
necessity of, 180
DoD Activity Code (DODAC) number, 211
Dodson, G. A., 72
Dryden, Hugh, 23
DuBridge, Lee A., 23
Dummer, Geoffrey William Arnold, 15
Dyna-Soar, 53, 68, 88, 278, 306

Early Warning (EW), 21, 23, 90
Eckelberger, Rear Adm. James E., 205
Education policy, and space race, 31–32
Efficiency, and DoD reforms, 239–240
Egan, James, 134
Eisenhower, Pres. Dwight David, 21
and arms race, 22
dependence on science advice of, 22–23
DoD reorganization of, 36–37
and inter-Service rivalry, 35, 36–37, 52,
192
and U-2 intelligence, 30

Electromagnetic pulse, and solid state
devices, 75
Electronic devices; *see also* Integrated
circuits; transistors
capacitors, 9
component level testing of, 52
hand-held calculators, 120–121
hybrids, 132–133, 170–171
microcircuits, 101
procurement programs for, 106–107
resistors, 9
standardization for, 48
Electronic parts; *see also* Parts
costs for commercial, 201
high reliability space-qualified, 92,
173–174
Electronic parts industry, 257–258, 257n;
see also Parts industry
high reliability space qualified, 190
and military standards, 159
moving overseas, 155
Electronics; *see also* Solid state electronics
high reliability, 229; *see also* High
reliability electronics
military investment in, 165
Electronics industry
acceleration of, 220
and acquisition reform, 203
advances in reliability of, 1
and Cold War, 202
complexity in, 85
expansion of, 77
globalization of, 244
hand-held calculators in, 120–121
high technology
extreme performance parts in, 28
laboratory support for, 287
impact of Minuteman program on, 31,
58–59, 61, 65
innovations in, 216
microelectronics manufacturers in, 40–41
military in, 166
miniaturation in, 42–43, *43*
plastic parts in, 241–242
standards proliferation in, 139
statistics for, 166
and VHSIC, 216
Electronics market, high reliability *vs.*
commercial consumer, 238

Electronics miniaturization, 12. *See also*
  Miniaturization
Electronic Systems Division (ESD), Air
  Force, 191
Electron tubes, in World War II, 11–12.
  See also Vacuum tube technologies
Encapsulation, plastics used for, 241–242
Engineering, soft ware systems, and
  acquisition reform, 312
Engineers
  development of, 32
  lost to acquisition reform, 290–292, **291**
  space program, 94
Enthoven, Alain C., 57
Evans, Gen. William J., 129
Everhart, Thomas, 141
Expendable Launch Vehicles (ELVs), 272
  commercial, 158
  eventual replacement of, 208
  performance improvements for, 155
Experimental Stealth Testbed, 161
Explorer I satellite, 3, 6

F-15, McDonnell Douglas, 134
F-22 electronics, 240
F-111, General Dynamics, 55
F-117, Lockheed stealth fighter, 161
Factors of production, 309
Failure; *see also* Physics of failure
  and acquisition reform, 309
  analysis, 64
  costs of, 110
  Discovery space shuttle, 172
  end-of-life, 304
  longer times between, 177
  vs. notable anomalies, 266
  orbital, 119
  space system, 4
  studies of, 302
  tolerance for, 168
Failure modes and effects research, 253, 259
Failure Rate Data (FARADA) Program,
  Navy's, 107–108, 112
Failure rates; *see also* Physics of failure
  "bathtub" curve, *108,* 108–109, 115
  and Bradburn review, 138
  and built-in quality, 174
  vs. calculated rates, 109, **109**
  collecting data, 69

vs. operational experience, 110, **110**
  predicting, 70
  shortcuts and, 300
  sources for, 70–71, *71*
Failure-rate testing, 51
Failure Review Board, 275
Fairchild Semiconductor Corp., 62, 99
  Micrologic product line
  transistors produced by, 66
False shortcuts, 281–282
"Faster-Better-Cheaper," NASA's, 269–270,
  272
Feature size issues, 219, 304
Federal Acquisition Regulations (FARs), 97,
  160
  needing deviation to, 211
  and stockpiling, 202–203
Federal Acquisition Streamlining Act
  (FASA), 246, 256. *See also*
  Acquisition Reform
Federal Contract Research Center (FCRC)
  changing role of, 46–47, **47**
  STL as, 46–47, **47**
Federally Funded Research and
  Development Centers (FFRDCs),
  112, 294
  and acquisition reform, 312
  SAMSO as, 81
Feedback
  in on-orbit experience, 126
  and Program Review Board, 104
"Field return data," 81
Film-return capsules, 308
Flax, Alexander, 225
"Fog of war," 91
Force application, 209, 210
Force enhancement, 209
Forrestal, James, 13
Fort Monmouth, 261, 263
Fowler, Charles A., 194, 195, 288
Franchised, misuse of term, 257n
Front End Terminals, 127, 128
Fuhrman, Robert A., 187, 193, 236
Future Imagery Architecture (FIA) program,
  272

Gallium arsenide, VHSIC's use of, 169n
Gansler, Jacques, 193, 236
Gantt approach, to systems acquisition, 98

General Accounting Office (GAO), on
     streamlined acquisitiion, 226, 227
General Electric
  military investment in, 16
  specialized computation circuits
     developed by, 9
  transistor production at, 20
Germanium, 39
Germany
  reunification of, 231
  tactical ballistic missiles of, 12
Getting, Ivan, 105
Gilpatrick, Roswell L., 46
"Glue logic," 215
Goldwater, Sen. Barry M., 192
Goldwater-Nichols reforms, 224, 226, 286
Goldwater-Nichols Reorganization. *See*
     Defense Reorganization Act (1986)
Gorbachev, Mikhail, 178, 203, 232
Government-Industry Data Exchange
     Program (GIDEP), 111, 253,
     256–257, 268, 294, 296
  and acquisition reform, 312
  creation of, 103
  important function of, 120
  mangement of, 104
Guidance systems
  in Minuteman production, 60–61
  missile, 28

Hagerty, Patrick E., 77
Hall, Col. Edward N., 29–30, 29n, 31
Hall, Theodore Alvin, 29n
Hammer, the $400, 185
Hanscom Air Force Base, 79, 260–261, 263
Hartinger, Gen. James V., 167
Have Blue, 161
Helmer, Dick, 264
Henderson, Brig. Gen. Don, 160, 163
Henry, Lt. Gen. Richard C., 137, 144, 156,
     159, 162, 163, 166, 167, 176, 187
Hermann, Robert J., 236, 238
Hicks, Donald A., 193
High reliability
  cost of, 110–111, **111**
  process controls for, 198–199
  quality expressed as, 293
  success in achieving, 168
  technology trajectory for, 303

High reliability electronics, xi, xiii, 89,
     210
  active *vs.* passive, xi–xii
  evolution of, xiv
  feedback process in, 93
  importance of, xi
  limits of, 2
  manufacture of, xiv
  military investment in, 166
  technological innovation in, 74
High reliability electronics industry
  compared with commecial electronics,
     238
  cost growth in, 243
  and military standards, 248
High reliability parts industry, military and,
     69
High reliability spacecraft
  basis for, 314
  factors affecting production of, 277–278
     complexities of marketplace, 287–290
     cost of infrastructure, 285–286
     demand for greater reliability, 280–281
     end of Cold War, 297–298
     false shortcuts, 281–282
     increasing device complexity, 298–299
     lack of pragmatic standards approach
        across device types, 296–297
     long lead time, 278–279
     loss of engineers, 290–292, **291**
     loss of industry data sharing, 294–295
     loss of in-factory government plant
        representatives, 296
     loss of insight to contractor activities,
        295–296
     loss of sponsorship for basic research,
        286–287
     loss of standards, 292–294
     minimum-buy requirements, 279–280
     push for advanced technology,
        299–300
     rise of commercial space, 284–285
     rising consumer electronics market,
        282–284
  perfect storm in, 273
  qualification for, 219
High reliability space programs
  cost *vs.* benefit in, 316
  cyclical improvement path for, 316

maintainability approach to, 317
smallest possible feature size electronics
for, 318
staying course for, 314–317, 321
High reliability space qualified parts industry
changes to, 302
standardization in, 251–252
High reliability systems
equipment failure rates in, 79
plastic encapsulated devices in, 196
High Virgo, 125
Historical jackets, 259
Hopping, Col. Cecil R., 220
House Armed Services Committee, Gen.
Slay's testimony before, 155
Howard, B. T., 72
Hubble Space Telescope, 317
Huckaby, Brig. Gen. Richard H., 205
Hybrid devices, 132–133
classification as, 223n
individual part costs for, 221–222
military standards for, 172
problems with, 170–171
and QPL rules, 171–172
qualification of, 163
Hybrid industry
certification process in, 223
and element evaluation, 173
Hybrid microcircuits
military specification for, 221n, 226
QPL approach for, 302
requirements for, 163
Hybrids
cost of, 171, 177
increasing complexity of, 298
qualification of, 204
testing of, 173
Hydrogen bomb, 13

IBM, 127
IDEA Corp., 19
Illinois Institute of Technology Research
Institute (IITRI), 70
Indications and Warning (IW), 21, 23
Industry, and scientific research, 8–9.
See also Electronics industry
Inertial Upper Stage (IUS), 209. See also
Launch vehicles
"Infant mortality," 79, 108

Information; see also Intelligence
and acquisition reform, 312
near-real-time, 91–92
Information sharing
and acquisition reform, 294–295
forums, 256–257
process, 51
Information technology (IT)
and high reliability spacecraft, 283
workforce for, 291
Infrastructure costs, in production of high
reliability spacecraft, 285–286
Initial Defense Communication Satellite
Program (IDCSP), 88, 123
Insight, defined, 256
Inspections, and high reliability, 165
Institute of Electrical and Electronics
Engineers (IEEE)
Electron Devices Group, 76
Reliability Group, 76
Institute of Radar Engineers (IRE), annual
Electronic Components Conference
of (1952), 15
Institute of Radio Engineers, 42
Institution, defined, 305–306
Integrated circuit devices
timing of production of, 156–157
wafers, 157
Integrated circuit engineering, 181
Integrated Circuit Engineering consultants,
183
Integrated circuits (ICs); see also Application
specific integrated circuits
advantage of, 52
and Apollo program, 62
commercial, 241, 241n, 245
custom, 154, 215
definition for, 39n
five basic approaches to, 40
government purchases of, 63–64, **64**
highly reliable (Hi Rel), 107
hybrid, 132, 132n, 133, 153
increased use of, 66
level of integration for, 43, 43n
military and space applications of, 59
military investment in, 42
in Minuteman program, 61
molecular electronics' approach to, 41
monolithic, 52–53, 133

Integrated circuits (ICs) (*cont'd*)
  move to multiple core, 304
  and "planar process," 42
  plastic encapsulated, 195–196
  proliferation of, 170
  task-specific, 140
Integrated circuits (ICs) industry, expansion
    of, 77
Intel Corp., 78n, 140, 170
Intelligence
  CIA, 14
  EW, 21, 23, 90
  IW, 21
  near-real-time, 88
Intelligence community, space systems of,
    122
Intercontinental ballistic missiles (ICBMs)
  analog autopilot technology for, 60
  development of, 18, 19
  timelines for, 18n
Intermediate Nuclear Force Treaties, 178
Intermediate range ballistic missiles
    (IRBMs), 18, 59
International Geophysical Year, 30
International Traffic in Arms Regulations
    (ITAR), 155, 311
Interservice Data Exchange Program
    (IDEP), 52
  combined with CRHS, 74
  and quality assurance programs, 103

JAN Class B parts, 174n, 194
JAN Class S industry, 164
JAN Class S Operating Stock program, 202
JAN Class S parts, 178, 194
  in Blue Ribbon Commission's Report, 189
  DESC stockpile of, 176
  increasing volume of, 214
  mandating use of, 211–212
  minimum-buy requirements for, 279–280
  Operating Stock Program for, 221
  prices of, 212, 212n
  procurement problems with, 181
  on ready inventory, 181
  and SCDs, 297
  space application components, 196
  specifications, 217
  value of, 282
JAN Class S program, 174, 174n

JAN-1 electronic standard; *see also* Military
    standards
  development of, 7
  purpose of, 293
"JAN" mark, 101
JAN Microcircuits, Certification
    Requirement for, 139
JAN parts
  compared with commercial parts, 198
  compared with MIL parts, 199
Japan, in American defense market, 164–165
JEDEC JC13.5 Task Group, 221
JEDEC Specification 1, 163
Jet Propulsion Laboratory (JPL), 212
Johnson, Clarence "Kelly," 44
Johnson, Louis, 13
Johnson, Pres. Lyndon Baines, 78, 124
Joint Army-Navy. *See* JAN-1
Joint Chiefs of Staff, 37
Joint Electron Device Engineering Councils
    (JEDEC), 81
  history of, 48n
  standardization work of, 48
Joint Logistics Commanders, 103, 104
Jones, Col. Bobbie L., 176, 180
Jupiter-C rocket, 36
Jupiter missile, 25, 59

Kahn, Robert, 141
Kaminski, Paul G., 237, 256
Kane, Francis X., 158
Kennedy, Pres. John F., 55
Kennedy Administration, reform approach
    of, 55
Kennedy Space Center, 127
Keyhole series of reconnaissance satellites,
    307
Kilby, Jack S., 39, 41, 42
Killian, James R., Jr., 23
Kravchuk, Leonid, 232
Kulpa, Maj. Gen. John E., Jr., 129
Kutyna, Gen. Donald J., 231

Laboratories; *see also specific laboratories*
  Air Force, 226, 260
  superlaboratories, 260
Laboratory Joint Cross-Service Group
    analysis, 260
Laird, Melvin, 88

Lamar, William, 278
Land, Edwin, 23, 88
Large Scale Integration (LSI), 140, 170, 214
Launch services, commercial, 284
Launch vehicles, 74, 128; *see also* Missiles
   Ad Hoc Study Group on Space Launch Vehicles, 129, 132, 134
   Atlas F sustainer engine accident, 131
   Broad Area Review (BAR) of, 267
   Centaur upper stage, 132
   costs of failures for, 267
   Delta, 266
   ELVs, 155, 158, 208, 272
   failures, 128, 128n, 131
   growth of, 306
   high reliability approaches to, 284–285
   IUS, 209
   Jupiter-C, 36
   production lines, 208
   reliability of upper stages, 131
   reusable, 87
   and satellite weight, 75
   Titan III, 128, 129, 130
   Titan IIIC, 130, 131
   Titan IV, 4
   Vanguard, 36, 44
Launch Vehicles Directorate, 131
LeMay, Gen. Curtis E., 46n
Leverton, William F., 136, 311
LGM-30 A and B, 58. *See also* Minuteman program
Licensed, misuse of term, 257n
Life testing, 64, 65, 72, 90
Little, John B., 14
Lockheed Martin F-117 stealth fighter, 161
Logan AFB, Utah, 205
"Long Lead Funding Applications" (Henry), 156–157
Long lead time, in high reliability spacecraft production, 278–279
Lusser, Robert, 37

MacKenzie, Donald, 303, 305
Magellan program, 212
Malthus, Thomas, 219, 220
Manhattan Project, 11, 17
Manned Orbiting Laboratory (MOL), 68–69, 88, 93

Manual M204, 50
Manufacturers
   and buyers of defective parts, 257
   domestic, 156
   and elimination of standards, 310
   and military standards, 258
   and minimum buy requirements, 176
   and minimum-buy requirements, 279
   outsourcing of, 244
   QML, 223
Manufacturing lines, "captured," 315
Manufacturing Technology Quality Improvement Program, AFSC's, 165–166
Maresky, Joseph J., 33, 69
Mark, Hans, 225
Marketplace
   and high reliability processes, 93
   for high reliability spacecraft, 287–290
   IC, 245
Marsh, Gen. Robert T., AFSC Commander, 164, 167, 174, 287
Marshall Space Flight Center, 132
Martin Co., 19, 54
Martin Marietta, 128, 131
Martin Titan III booster, 53. *See also* Launch vehicles
Massachusetts Institute of Tehnology (MIT)
   Instrumentation Laboratory, 60, 62
   Lincoln Laboratory, 107
Material review actions, 165
Mathis, Gen. Robert, 264
Mayaguez incident, 123, 125
McCartney, Lt. Gen. Forrest S., 168, 175, 176, 180, 181, 194, 196, 197
McCurdy, Rep. David, 162
McLucas, John, 158
McNamara, Sec. Robert S., 68, 161, 278
McNamara reforms
   administrative costs of, 55–57
   Concept Formulation Phase, 56
Mead, Carver, 141, 142
Mead-Conway VLSI revolution, 143, 149, 154
Medium Scale Integration (MSI), 140, 214
Metal oxide semiconductor field effect transistor (MOSFET), 75
Metal oxide semiconductor (MOS) devices, 75

Metal oxide semiconductor (MOS)
technologies, 75–76, 76n
Metal Oxide Silicon Implementation
Service (MOSIS), 142
Microchips, 190, 191, 214. *See also*
Integrated circuits
Microcircuits, 62
general specification for, 153, 154
JAN, 139
lack of standards in market for, 296
in national security and national defense
communities, 139
Microelectronics, 39
Japanese manufacture of, 201
technology, 133
Microelectronics industry, orders of
precedence in, 194
Micro-Module program, 40
Midas program, 265
Midas satellite, 24, 45, 53
Military drawings, 191. *See also* Standard
Military Drawings
Military handbooks
"guidance only handbooks," 255
MIL-HDBK-339, 217
MIL-HDBK-217, 69
MIL-HDBK-217-A, 107, 109
MIL-HDBK-339, 170
Military Qualified Products List (MQPL), 50
Military reporting chains, 208, 209
Military Standard Requisitioning and Issue
Procedures (MILSTRIP), 211
Military standards and specifications, xi, 85,
251; *see also* Defense Science
Board studies
and acquisition reform, 5, 242–243, 289,
293
Blueprint's Process Action Team on, 248
and Blue Ribbon Commission, 191
bureaucracy of, 93
compliance with, 161–162
and contractor's practices, 161
and cost growth, 148
cultural changes in, 250
DESC-EQE-44, 132
and infrastructure cost, 285
JAN-1, 7
M-0038510, **117–118**
maintenance and update of, 217

MIL-E-1, 48
MIL-H-38534, 221, 221n, 222, 223.224,
243, 244
MIL-I-38535, 243, 244, 258
MIL-M-38510, 100–101, 114, 172, 173,
177, 221, 243
MIL-M-38510A, 114, 116, 153
MIL-M-0038510B(USAF), 116
MIL-M-38510G, 204, 206
MIL-PRF-38535, 258
MIL-S-19500, 48, 54, 72, 138
MIL-STD-454, 194, 200, 201, 217
MIL-STD-750, 73, 80
MIL-STD-883, 80–81, 100, 112, 132,
133, 172, 173, 177, 221, 222
MIL-STD-976, 132, 139
MIL-STD-1540, 255, 255n
MIL-STD-1541, 253
MIL-STD-1543, 137, 253
MIL-STD-1546, 159, 179, 253, 310
MIL-STD-1547, 213, 310
MIL-STD-1556, 120
MIL-STD-1772, 172, 173, 177, 221, 222,
223, 224
MIL-STD-38510, 134
MIL-STD-154773-2C, 158–159
M-0038510(USAF), **117–118**
overreach of, 292–293
performance based, 254–255, 258
problems with, 236
qualification approach in, 219
reducing use of, 193
Revision B, 255, 255n
Revision C, 255, 255n
SAMSO-STD 73-2, 119
SAMSO-STD-77-2, 137
SD-STD-73-2, 157
SD-STD-73-2C, 158, 174
and solid state electronics industry, 82
tailoring, 160
waiver option for, 234–235
Mill Run Products Quality Assurance
Program, 102, 116
MILSPEC, 193
Milstar system, 272
MILSTRIP, 211
Miniaturization, electronics, 12, 42–43, *43*
as dominant focus, 64
limitations of, 141

Minimum-buy requirements, 279–280
Minimum In-Plant Surveillance (MIPS)
    program, 148, 296
"Minuteman Hi Rel," 90
Minuteman I, specification for, 60
Minuteman II Guidance System, 59
Minuteman III, 65–66, 74
Minuteman missile, 31, 45
    adequate quality supply for, 99
    high reliability of, 110–111
    manufacturing of, 58
    reliability of, 59, 60, 67
Minuteman program, 90, 115, 164
    costs of, 61, 64
    and CQAP, 65
    financial backup for, 72
    guidance systems for, 60–61
    and high reliability electronic industry,
        149
    impact on electronics industry of, 58,
        65–66
    integrated circuits for, 63–64
    measuring, predicting and testing
        reliability in, 68
    technical information gathered by, 72
"Minuteman quality," 304
"Minuteman reliability," xiii, 65–66, 66
Missiles
    downtime for, 79
    ICBMs, 18, 18n, 19, 60
    IRBMs, 18, 59
    launch reliability of, 26
    Miniature Homing Vehicle (MHV);
        see Anti-satellite
    Peacekeeper, 134–135
    reliability in development of, 45
    solid fuel, 29
    strategic missile systems, 29
    submarine launched ballistic, 35
Missiles, long range
    accuracy of, 26
    feasibility studies of, 1
    inter-Service rivalry in development of,
        12
    as launch vehicles, 36
    and procurement policies, 13
    solid state electronics for, 28
Missiles, U.S.
    Ajax, 25, 26

Air Launched Miniature Vehicle
    (ALMV), 133–134, 163, 167, 184,
        185, 210
Atlas, 19, 25, 44–45
Jupiter, 25, 59
Minuteman, 31, 45; see also Minuteman
        program
Nike, 12, 25, 26, 27
Polaris, 29
Thor, 24–25, 59, 124
Titan, 4, 19, 45, 51, 68, 266
Zeus, 25, 26, 124
Mission assurance, 168, 312
    for high reliability space programs, 314
    human resources focused on, 316
    processes, 308
    in space acquisition culture, 314
Mission Assurance Conference, April 1978,
        144
Mission Assurance Improvement Task
        Force, 310
Mission success
    vs. cost, 270–271
    defined, 95
Molecular electronics, 41, 42
Monolithic integrated circuit technologies,
        39
Monolithic Microwave Integrated circuits
        (MMICs), 261
Monopoly laws, 156
Moon landings, 290
Moore, Gordon E., 77
Moore's Law, 77–78, 141, 142, 214, 215,
        219, 283, 303–304, 305
Morgan, Brig. Gen. Thomas W., 105, 106,
        129, 134, 144
Morton, J. A., 52
Motorola Corp., 62, 115–116
Moynihan, Sen. Daniel Patrick, 264
Murphy, Col. George, 129
Mutually Assured Destruction (MAD),
        doctrine of, 169

Naka, F. Robert, 260
Nanotechnology, 305
National Aeronautics and Space Agency
        (NASA)
    ALERT system of, 74
    Apollo Guidance Computer, 59, 60, 62

National Aeronautics and Space Agency
  (NASA) (*cont'd*)
  Apollo program, xi, 89, 93, 94, 99
  Challenger, 60, 62
  civil space programs of, xi
  Class A parts program of, 99, 113
  and class S parts, 135
  creation of, 36
  Faster-Better-Cheaper, 269–270, 272
  Jet Propulsion Laboratory, 212
  and joint SAMSO-NASA preferred parts
    list, 139
  and reusable launch vehicles, 87
  Space Transportation System (Space
    Shuttle), 127–128, 167, 172, 210
  Topological Experiment Program, 212
National Bureau of Standards, 9
National Command Authority, space
  systems of, 122
National defense, and advanced technology,
  82. *See also* Defense, Dept. of
National Polar Orbiting Environmental
  Satellite System (NPOESS), 272
National Reconnaissance Office (NRO), 78,
  91, 102; *see also* Satellites
  control of, 126
  Corona, 3, 24, 25, 44, 45, 49, 53
  declassification of, xiii
  military satellites developed by, xiv
  space systems owned and operated by,
    122
National security
  and advanced technology, 82
  electronic component technology in, xii
National Security Council (NSC), NSC 139
  document of, 14
National Semiconductor Corp., 194
National Space Policy, of Reagan
  administration, 157–158
National Stock Numbers (NSNs), 202
Naval Ammunition Depot (NAD), 81
Navy, U.S.
  and C41 capability, 263
  Component Reliability History Survey
    Program, 74
  Failure Rate Data Program of, 107–108,
    112
  Fleet Satellite Communications System
    of, 88

Guided Missile Data Exchange Program,
    74
and IDEP, 52
information sharing of, 51
MIL-HDBK-217 released by, 69
missile responsibility of, 35
Naval Weapons Support Center CRANE,
    132
strategic missile systems of, xiii
thin film technologies of, 40
the *United States,* 13
Nelson, Richard, 80
Nichols, Rep. Bill, 192
Nike defensive missile programs, 25, 26,
    27. *See also* Zeus anti-ballistic
    missile
Nike missiles, 12
Nixon, Pres. Richard M., 89, 127, 203
Nongovernment standards (NGS), 249,
    254
North American Aviation, Autonetics
    Division of, 31, 60, 61, 99
North Atlantic Treaty Organization (NATO),
    satellites for, 88
Northrop B-2 stealth bomber, 161
Noyce, Robert, 42
Nucci, Edward J., 32
Nuclear Effects and Space Radiation
    Conference (NESRC), IEEE's, 76
Nunn, Sen. Sam, 162
Nunn-McCurdy Amendment, 162–163,
    162n, 184, 185, 270, 271

Obsolescence, and technological innovation,
    217
Office of Defense Mobilization, Scientific
    Advisory Committee (ODM/SAC),
    23
Operating stockpile, 211
Operating stock program, 205–206,
    220–221. *See also* Stockpiling
Operational advocacy, shift toward, 226
Oppenheimer, J. Robert, 23
Opportunity cost, in spacecraft production,
    154
Orbits, low earth *vs.* geosynchronous, 315
Order of precedence
  and DSB study, 199
  MIL-STD-454, 217

of SAFSP, 179
  SD order of, 179
Original equipment manufacturers (OEMs),
  257
Overseas production, issue of, 156
Oversight, defined, 256
Oxcart triple-sonic reconnaissance aircraft,
  44, 48. *See also* A-12 Lockheed
  reconnaissance aircraft

Packard, David, 188, 191, 195
Packard Commission, 188, 204, 224, 246;
  see also Blue Ribbon Commission
  on Defense Management;
  Standardization
  implementation of, 207
  report of, 293
  results with, 269, 289
Palo Alto Research Corp. (PARC), Xerox,
  141
Paperwork reduction, unintended
  consequences of, 256
Part numbering schemes, 294
Parts; *see also* Stockpiling
  high reliability manufacturing of, 156
  and problems with acquisition reform,
  312
  qualification of, 159
  sources of, 99
Parts, materials, and processes (PM&P),
  effective management of, 118, 119,
  126
Parts buying, reactive nature of, 181
Parts control programs, 174, 174n
Parts industry, and military standards, 159.
  See also Electronic parts industry
Parts management
  Darnell Report on, 49–50
  early warning for, 119
Parts standards, 148. *See also* Qualified
  Parts List; Standardization
Pataki, Gov. George, 264
Pathfinder aircraft programs, 290–291
Patton, Gen. George S., 9
Peace dividend, 229, 282, 297–298
Peacekeeper missile, 134–135
Pearl Harbor-like surprise, 79. *See also*
  Cold War
Peierls, Rudolf, 7

*Perestroika: New Thinking for Our Country
  and the World* (Gorbachev), 203
Perry, Sec. William J., 103, 187, 196, 197,
  200, 233–234, 234, 235, 247, 254,
  288. *See also* Defense Science
  Board; Defense Science Board
  studies
Perry initiatives, 289
Perry Mandate, 247
Personal computers, 126, 170. *See also*
  Computers
  first successful commercial, 140
  and high reliability spacecraft, 283
  IBM, 127
Phillips, Lt. Gen. Sam C., 102, 104, 111,
  121, 134, 146, 273
Photography, limitations of, 307
Photonics, 305
Physics of failure, 65
  qualification and, 136
  and reliability, 71
  and reliability calculation problems, 108
Physics of Failure in Electronics Symposia,
  149
  first (1962), 70, 72
  second (1963), 73
  third (1964), 73
  fourth (1965), 73
  fifth, 76
  ninth meeting of, 89
Physics of Failure program, 70
Piece parts; *see also* Working Group for
  Electronic Piece Parts
  commonality of, 157
  national importance of, 279
Piece part testing, 97
  evolution of, 95
  and "historical jackets," 95
Planning Programming Budgeting System,
  55
Plastic encapsulated parts issue, 195–196,
  241–242
Polaris missile, submarine-launched
  solid-fuel, 29
Policy implementation, xiii–xiv
Politics
  and defense acquisitions, 151–152
  and DSB study, 200
  and technology change, 229

Powell, Gen. Colin, 231
Power, Gen. Thomas S., 28, 45
Powers, Francis Gary, 48
Presidential Science Advisory Committee
(PSAC), 23. *See also* Office of
Defense Mobilization's Scientific
Advisory Committee
Preston, Colleen, 236, 246, 254, 256
Process Action Teams (PATs); *see also*
Acquisition Reform
Blueprint for Change report of, 248–250
focus teams, 250–253
recommendations of, 253–254
Specifications and Standards focus
groups of, 250–251
and standardization and maintainability,
252
Systems Acquisition Process, 258
Process certification, concept of, 200, 201
Process controls, 64, 65
and commercial best practices, 245
in commercial market, 282–283
high reliability, 198–199
of microelectronic design, 195
Process standards, overly prescriptive, 251
Procurement issues, 177. *See also*
Acquisition
Procurement policies, 189. *See also* Blue
Ribbon Commission on Defense
Management
Procurement specifications, 249
Production function, 302
Program Executive Officers (PEOs), 189
Program Review Board (PRB)
Class A piece parts policy of, 112–113
establishment of, 104
Phase 1 of, 104–105
Phase 2, 105–106
scope of, 104
Proprietary markings, widespread use of, 295
Prototype Miniature Air-Launched Segment,
133
Proximity fuzes, development of, 9
Publications; *see also* Military handbooks
*Acquisitions Reform. A Mandate for
Change,* 247
*Assembly-Through-Flight Test and
Evaluation Guide,* 94, *96,* 96–97,
99, 259

Blueprint for Change, 248–250
*Quality and Reliability Assurance
Procedures for Monolithic
Microcircuits* (RADC), 80
RADC Reliability Notebook, 69
*Reliability Factors for Ground Electronic
Equipment,* 33, 69
The Reliability Handbook, 52n
*Reliability of Military Electronic
Equipment,* 35
"Report on National Space Policy," 187
"Report on National Space Policy
Enhancing National Security and
U.S. Leadership Through Space,"
157–158, 157n
"Specifications and Standards–A New
Way of Doing Business" (Perry),
254
Towards New Horizons II, 125, 127
"Unreliability of Electronics–Cause and
Cure," 37
Public opinion
and moon landings, 290
and reconnaissance programs, 44
Public policy analysts, xiii

Qualification
"box level," 95
end item, 173
for high reliability spacecraft, 219
of manufacturing processes, 218
MPQL procedures for, 50
of parts, 159
and process, 251
for space flight, 216
and technological innovation, 101n
VLSI, 219
Qualified Manufacturers, DoD, 200
Qualified Manufacturers List (QML), 232,
302
and acquisition reform, 259
certificaion steps for, 222
limits of, 293
microcircuits on, 243, 258
and MIL-H-38534, 224
modern synthesis of, 298
move from QPL to, 264
performance-based specifications and, 251
QPL approach to, 244

Qualified Parts List (QPL), 101, 232, 292
    changing process for, 222
    Class S parts on, 135–136
    and costs of qualification, 136
    hybrid devices on, 133
    limits of, 293
    and VLSI, 215
Quality
    costs of, 165
    and reliability, 198–199
    TQM approach to, 172–173, 173n
Quality and Reliability Assurance
    Procedures for Monolithic
    Microcircuits (RADC), 80
Quality Assurance Program, Mill Run
    Products, 116
Quality assurance (QA), 53, 144
    audits in, 103
    and process, 251
Quality control
    and class A parts program, 99
    of RADC, 82
Quality Horizons, 155
    and "commercial practices," 147
    1979 final report of, 145
    and fixed-price contracting, 147
    objectives of, 145
    and product assurance, 148, 149n
    recommendations of, 146, 147–148
    results with, 149
Quantum physics
    application of, 10
    and electronics technology, 1
Quantum theory of electrons, 7
Quarles, Donald A., 11, 14, 24–25

RADC Reliability Notebook, 69
Radiation
    and feature sizes changes, 304
    and solid state devices, 75
Radiation hardness testing, 311
Radiation tolerance, of spacecraft, 75
Radio Corp. of America (RCA)
    military investment in, 16
    transistor production at, 20
Radio Electronics Television Manufacturers
    Association, 16
Radio Electron Tubes, Joint Army-Navy
    Specification JAN-1A for, 8

Ramo, Simon, 158
Ramo Wooldridge Corp., 19
RAND, 46
Randolph, Gen. Bernard P., 174, 207
Rankine, Maj. Gen. Robert, 216, 218
Raytheon
    military investment in, 16
    transistor production at, 20
Reagan, Pres. Ronald, 191, 269
Reagan administration, 157, 167, 178
    and advanced weaponry, 214
    extreme high technology and, 299
    technical workforce during, 290
Rechtin, Eberhardt, 163, 164
Reconnaissance
    national, 90
    near-real-time, 178
    space-based, 178
Reconnaissance aircraft, high-altitude,
    23–24, 24n
Reconnaissance information, instantaneous
    availability of, 92
Reconnaissance satellites, 278. *See also*
    Satellites, Keyhole series
The Regency (transistor radio), 19
Reliability, 53, 304; *see also* High
    reliability; Standardization
    and acquisition reform, 293
    demand for, 25, 149, 280–281
    DoD process for ensuring, 195
    improved electrical component, 66
    and improving part quality, 89
    management attention to, 111
    of Minuteman missile, 59, 60, 65–66, *66,*
        67
    and mission success, 144
    and physics of failure, 71
    quality and, 198–199
    requirements for, 33–35, 64–65
    of spacecraft, 100, 116, 122
    as specialized knowledge area, 32
    and technology trajectory, 306
    testing in, 45, 45n
Reliability calculation problems, 108
Reliability engineering, xiii
Reliability estimates, and maintenance
    requirements, 69
Reliability Factors for Ground Electronic
    Equipment (Maresky), 33, 69

The Reliability Handbook, National
     Semiconductor's, 52n
Reliability Maintainability Data
     Interchange, 112
*Reliability of Military Electronic
     Equipment,* 35
Reliability Physics Symposia, 89, 115
Reporting chains, military, 208, 209
"Report on National Space Policy
     Enhancing National Security and
     U.S. Leadership Through Space,"
     157–158, 157n
Research; *see also* Defense Science Board
     studies
  and industry, 8–9
  loss of sponsorship for, 286–287
Resistors, development of, 9
"Revolt of the Admirals," 13
Reynolds, Col. Jon F., 213, 217, 218, 220
Rice, Donald B., 225
Riley, Paul H., 38
Rocketry, long range, 290. *See also* Launch
     vehicles
Romania, 231
Rome Air Development Center (RADC);
     see also Rome Laboratories
  Applied Research Laboratory, 70
  Design Analysis Branch, 130
  Design and Diagnostics Branch, 130
  establishment of, 14
  facilities and equipment of, 79
  and failure rate calculations, 109, **109**
  hybrid microcircuit standard of, 204
  personnel of, 80
  physics of failure work of, 65
  Radiation Hard 32-Bit Processor, 216
  relationship with industry of, 81
  Reliability and Diagnostics Branch,
     129–130, 200
  Reliability Notebook of, 69
  Reliability Physics Branch, 70, 130
  slash sheets produced by, 101
  Technical Report TR-67-108, 107–108,
     109
Rome Laboratories, 226, 242, 259, 269, 285
  approach to space quality parts of, 265
  BRAC decision, 298
  closure of, 260, 264
  Design Analysis Branch, 262–263

Electronics Reliability Division, 261
  evolution of, 261
  Reliability and Diagnostics Branch,
     261–262
  Reliability Physics Branch, 262
  strengths of, 263
Roosevelt, Pres. Franklin Delano, 8
Root cause determination, 119
Russell, Gordon, 99
Russo, Lt. Gen. Vincent M., 205

SAINT program, 53, 54, 88
SAMOS satellite, 24
Sandia National Laboratory, 112
Satellite images, intelligence information
     from, 92
Satellite Interceptor (SAINT), 53, 54, 88
Satellites, xi, 29; *see also* Anti-satellites;
     Class S parts
  AEHF comunications satellite, 272
  and American way of war, 183
  Boost Surveillance and Tracking System
     (BSTS), 169
  Class S parts in, 174–175
  commercial, 284
  communications, 123–124, 272, 284, 319
  Corona reconnaissance, 3, 24, 25, 44, 45,
     49, 53; *see also* Discoverer
  cost growth for, 270–271, **271**
  cost of repairing, 317
  demonstrated capabilities of, 122
  Discoverer, 24; *see also* Corona
     reconnaissance satellite
  Dyna-Soar, 53, 68, 88, 278, 306
  early study of, 30
  evolution of, 6
  Explorer I, 3, 6
  failures of, 4, 79, 267
  FLTSATCOM, 215
  high reliability, 183, 284–285, 301, 303,
     305
  Hubble, 317
  and ICs, 153
  IDCSP, 88, 123
  Keyhole series, 307
  lifetimes for, 154, 183, 305
  loss of awe about, 225
  Midas, 24, 45, 53
  Milstar, 216

NPOESS, 272
nuclear effects experienced by, 75
reconnaissance, 78, 126; *see also*
    Keyhole series
reliability of, 144
SAMOS, 24
SBIRS, 265, 266, 270, 271
SolarMax, 317
Sputnik, 30–31
SSTS, 169
21st-century failure of, xi
testing of, 268, 285–286
Vela, 24
Viking, 112
Scaff, Jack H., 10
Schriever, Gen. Bernard A., 19, 24, 31, 45,
    55, 158
Schultz, Maj. Gen. Kenneth W., 105, 113
Science, The Endless Frontier (Bush), 8
Scientists, development of, 32. *See also*
    Engineers
SECDEF, acquisition reform initiative of,
    200
Secretary of Air Force Special Projects
    (SAFSP), 128, 160, 179. *See also*
    National Reconnaissance Office
Section 800 Panel, 246
Semi-Automatic Ground Environment
    (SAGE), 33
Semiconductor Dependency study (DSB),
    201–202
Semiconductor devices
    process certification for, 200, 201
    standardization of, 48
Semiconductor electronics, 62
Semiconductor industry
    distribution of sales in, 62–63, **63**
    history of, xii, 62–63, **63**
    military influence on, xii
    quality control in, 82
Semiconductor Industry Association,
    Government Procurement
    Committee (SIA/GPC) of, 244
Semiconductor manufacturing equipment,
    cost of fabrication facilities for, 214n
Semiconductor market, Texas Instruments
    in, 17
Semiconductor materials, behavior of
    electrons in, 7

Semiconductor parts
    classes of, 80–81
    DoD control of, 195
Semiconductors, 10
    Arrhenius model of temperature stress
        for, 72, 90
    early patents for, 15–16, **16**
    and VHSIC, 216
Service Acquisition Executive, 207–208.
    See also Acquisition streamlining
Shockley, William, 10, 15, 27
Shushkevich, Stanislav, 232
Signal processing architectures, prototyping
    of, 130
Silicon, 39
    etching, 43
    transistors, 17
    wafer, 115
Silveira, Milton A., 175–176
Single chip microprocessors, 214
Slash sheet 608, 249
"Slash sheets," 101, 191, 223
Slay, Gen. Alton D., 145, 146, 147, 149,
    155, 156, 202, 207, 238
Small Scale Integration (SSI), 43
Smith, Adam, "invisible hand" of, 303, 305
Software development, and acquisition
    reform, 312
Software Management and Assurance
    Program, NASA's, 232
SolarMax, 317
Solid state devices; *see also* Hybrid devices
    compared with vacuum tubes, 27
    and military standards, 82
Solid state electronics, 1–2, 82, 303–304
    compared to vacuum tubes, 304–305, 306
    and demand for reliability, 26
    density of early, *43*
    determining reliability in, 32
    development of, 7, 12
    military interest and investments in, 11, 48
    in military systems, 100
    and radiation tolerance, 75
    and role of RADC, 80–82
    and temperature ranges, 199
Solid state electronics industry, 76–77
    impact of Minuteman on, 58–60
    military support of, 27
    space programs and, 83

Solid state revolution, beginning of, 10
Sony, 19
Source control drawings (SCDs), 170–171,
177, 293
in acquisition reform aftermath, 317
alleged abusive use of, 177
costs of, 178
criticism of, 296–297
eliminating, 179, 187
and stockpiling effort, 203
Southeast Asia War, reconnaissance
capabilities during, 91. *See also*
Vietnam War
Soviet Union; *see also* Cold War
ABM system of, 78
and Cold War, 212
collapse of, 232
co-orbital anti-satellite program of, 87
and deterrence, 24
fission device detonation of, 13
mass-manufactured satellites of, 318
Sputnik launched by, 30–31
thermonuclear fusion device detonated
by, 16–17
Space, commercial activities in, 158
Space Act (1958), 36
Space and Missile Systems Organization
(SAMSO), 67, 81; *see also* Program
Review Board
acquisition records of, 97
Aerospace Select Committee on Piece
Parts of, 134
*Assembly-Through-Flight Test and
Evaluation Guide* of, 94, *96,* 96–97,
99
auditing for, 103
and class S parts, 135
Commander's Policy Book of, 113–114
Electronic Parts Management Board, 112
Engineering Support Branch, 120
and high reliability space qualified parts
efforts, 132
IC requirements of, 112
and joint SAMSO-NASA preferred parts,
139
quality assurance activities at, 102
Quality Assurance Working Group, 105
and Quality Horizons recommendations,
146, 147–148

Space Parts Working Group of, 121
WGEPP established for, 106, 107, 110–111
Space-Based Infra-Red Satellite (SBIRS),
265, 266, 270, 271
Space Command, Air Force, 209. *See also*
Air Force, U.S.
Space Command, U.S. (USSPACECOM),
169, 192, 231
Space control activities, 209, 210
Spacecraft; *see also* High reliability
spacecraft *and names of individual
satellite programs*
acceptance of, 151
acquisitions, 161
compared with aircraft, xiii
costs of, 138
development of, 154, 164
evolution of, 129
feedback process for, 93
military capabilities provided by, 125–126
piece parts for, 157
reliability of, 122
robotic, 3
Spacecraft industry
and acquisition reform, 260
testing facilities in, 220
Space Division (SD), program parts usage,
213, **213,** 213n
Space flight
human, 54
qualification for, 216
Space launch vehicles, reusable, 85.
See also Launch vehicles
Space missions, 209
Space Parts Working Group (SPWG), 121,
180–181
operating stock program of, 181, 183
on stockpiling, 202
Space programs, 82–83, 85, 169; *see also*
High reliability space programs
acceptance of, 151
advanced technologies for, 300
civil, 88, 89
compartmentalization of, xiii
costs of, 78
defining reliability requirements for, 107
expectations for, 53
military, 82, 167
national defense, 209

national security, 83, 88, 209
national security *vs.* national defense, xiv
transformation of, 154
Space quality parts, 235. *See also* Parts
Space race, and education policy, 31–32
Space Shuttle program, 178, 208, 210, 284
  Discovery, 172
  economics of, 127–128, 305
  and mainframe computers, 127
  and Reagan Administration, 167
  retirement of, 317
Space support, 209
Space Surveillance and Tracking System
  (SSTS), 169
Space systems; *see also* Satellites;
  Spacecraft
  acceptance of, 206
  control of, 127
  early skepticism about, 87
  failure rates in, 3
  and fixed-price contracts, 147
  high reliability of, 6–7
  increased complexity of, 232
  mainstream, 86
  military standards and, 294
  national defense, 85–86
  national security, 179
  near-real-time, 92
  owned by AFSC, 167
  peacetime capabilities of, 114
  and quality assurance, 146
  robotic, 88
  special nature of, 225
  unique requirements of, 240–241
Space Systems Division, 55. *See also*
  Western Development Division
Space Task Group, 1969, 89, 314
Space Technologies Laboratory (STL), 30
  emerging role of, 46
  as FCRC, 46–47, **47**
  task list for, 47, **47**
Space Transportation System (STS), 89,
  127–128, 167, 172, 210
"Space war," Operation Desert Storm as,
  231
Specifications; *see also* Military standards
  and specifications
  need to rejuvenate, 309–311
  use of term, 250

"Specifications and Standards–A New Way
  of Doing Business" (Perry), 254
Spintronics, 305
Spires, David, 272
Sputnik, 30–31, 290
Sputnik II, 31
Standardization; *see also* Military standards
  and specifications
  affecting semiconductor devices, 80
  benefits of, 8
  commercial alternatives to, 194
  compliance to government-unique
    requirements in, 238–239
  Darnell Report on, 49–50
  for high reliability spacecraft, 292
  in high reliability space qualified parts
    industry, 251–252
  hybrid-related, 173
  need for, 7, 48
  process approach to, 243
  role of RADC in, 80–81
  of transistor technology, 16
Standard Microcircuit Drawings (SMDs),
  249
Standard Military Drawings (SMDs), 101,
  191. *See also* Standard Microcircuit
  Drawings
Standards; *see also* Military standards and
  specifications
  commercial *vs.* military, 240–241
  need to rejuvenate, 309–311
  updating, 312
  use of term, 250
"Star Wars," 169
Stealth programs, 160–161
Stever, H. Guyford, 260
STG
  1969, 127
  recommendations of, 94
Stockman, Harvey, 23, 307
Stockpiling; *see also* Operating stock
  program
  and billing and deviation problems, 203
  of Class S parts, 280, 281
  at DESC, 177, 204–205
  drawbacks of, 176
  handling procedures for, 205
  initiative for, 190n
  SPWG effort on, 202

Stockpiling (*cont'd*)
  startup problems with, 206
  viability of, 180
Storage costs, 155
Strategic Air Command (SAC), national
    security space systems supported by,
    123
Strategic Arms Limitation Treaty, 178
Strategic Arms Reduction, 178
Strategic Defense Initiative (SDI), 169
  and Reagan administration, 178
  success of, 210
Strategic Missiles Evaluation Group, Air
    Force, 18
Stressing tests, 65
Submarines, and MIL-HDBK-217, 69
Sutherland, Ivan, 141, 143
Sylvania
  military investment in, 16
  transistor production at, 20
System Program Offices (SPOs),
    management items of, 171, 171n
Systems acquisition, "waterfall" approach
    to, *98,* 98–99
Systems engineering, and acquisition
    reform, 311
Systems thinking, 94
System test, 110. *See also* Tests and
    Testing

Tactical Data System, Naval, 33
Tactical Exploitation of National
    Capabilities (TENCAP), 91
Tantalum, 219
Task Force on Defense Acquisition Reform,
    BSB, 236
Task Force on Defense Semiconductor
    Dependency, 201
Taylor, Rep. Gene, 235
Teal, Gordon, 14
Technological Capabilities Panel (TCP),
    ODM/SAC's, 23, 24
Technological innovation
  and ability to test, 220
  acceleration of, 214–215
  in Air Force, 126
  in high reliability electronics, 74
  information sharing in, 51

and Minuteman program, 64
obsolescence and, 217
and role of Darnell group, 49
Technologies; *see also* Solid state
    technologies; Standardization;
    Transistor technology
  Auto-Sembly process, 41
  complementary MOS (CMOS)
    technology, 76
  metal oxide semiconductor field effect
    transistors (MOSFET), 75
  monolithic integrated circuits, 39
  monolithic integrated semiconductor
    circuitry, 42
  moving overseas, 155–156
  photoreconnaissance, 44
  silicon etching, 43
  thin film, 40, 53
Technology bubble, 283–284
Technology change
  and acquisition management, 227
  and military standards, 248–250
  politics and, 229
  potential anarchy of, 303
  rate of, 301
  and technology trajectory, 303
  and VLSI acceleration, 218
  and world politics, 212
Technology trajectory, 306
Telemetry systems, high reliability, 107
Temperature ranges, and solid state
    electronics, 199
Teradyne tester, 261–262
Terhune, Charles H., 158
Testing
  and acquisition reform, 311–312
  contractor, 286
  piece parts, 95, 97
  of satellites, 268, 285–286
  and technological innovation, 101n
Test optimizations, and acquisition reform,
    265
Tests, and high reliability, 165
Texas Instruments (TI), 19, 27
  military specification format used by,
    115–116
  silicon transistors developed by, 17
TFX, 55. *See also* F-111

Theuerer, Henry C., 10
Thin film technologies, 40, 53
Thor Intermediate Range Ballistic Missile, 24, 25, 59, 124
Throw-away culture, 59
"Tinkertoy" assembly, 40
Titan launch vehicle, 306
Titan missile, 4, 19, 45, 51, 68, 266
Toilet seat, the $600, 185–186
Topological Experiment Program (TOPEX), NASA's, 212
Total Package Procurement Concept (TPPC), 55, 161
  problems with, 56
  success of, 57
Total Quality Management (TQM), 172–173, 173n
Total System Performance Responsibility (TSPR), 57, 161, 161n, 266, 269, 270
  in national defense space programs, 272
  philosophy, 270
  for space systems, 271
Towards New Horizons II, 125, 127
Transistorized Digital Computer (TRADIC), 17
Transistor radios, 283
  first commercially available, 19
  military applications of, 20
Transistors
  field effect, 75
  increased production of, 27
  military interest in, 16
  MOS, 75
Transistor technology
  of Bell Labs, 14
  introduction of, 11
  "point-contact," 14
  silicon in, 17
  standardization of, 16
  symposium (1952) on fundamentals of, 15
  symposium (1951) on military applications of, 14–15
Truman, Pres. Harry S., 8, 14
TRW, Inc., 30, 46
TSPR approach, 290–291
"Tyranny of numbers," 52

The *United States,* 13
United States, technological superiority of, 299. *See also* Cold War
"Unreliability of Electronics–Cause and Cure" (Lusser), 37
U-2 overflight program, 44, 48
U-2 overflights, 307
U-2 reconnaissance aircraft, 23–24, 24n

Vaccaro, Joseph, 70, 72
Vacuum-tube technologies, 26
  ICs compared with, 52
  in World War II, 7
Valve behavior, one-way, 10
Vandenberg AFB, California, 127
Vanguard launch vehicle, 36, 44
V diagram, 232–233, *233,* 267, 286
Vela satellite, 24
Very High Speed Integrated Circuit (VHSIC) technologies, 143, 226, 283
  Phase 1 of, 169
  VHSIC Hardware Description Language, 215
Very Large Scale Integration (VLSI), 214
  and feature size issues, 219
  qualification of, 216
  Qualification Working Group, 218, 220
  revolution, 143, 153, 215
  university courses in, 142
VHSIC Spaceborne Computer, 216
Vietnam War, 122, 287. *See also* Southeast Asia war
Viking space exploration program, 112
Vito, Carmine, 44, 307
Von Neumann, John, 17, 18
V-2 type rockets, 12

Wafer lot, 135
Wafer processing, and Class A parts program, 99
Wafers, banking of, 157
War; *see also* Southeast Asia War; Vietnam War
  American way of, 183, 210
  and military space program, 225
  role of space systems in, 4
"Waterfall" approach, *98,* 98–99, 232

Waterman, Alan, 23
Weapon system; *see also* Minuteman
    missile
  concept of, 94
  Weapon System 119, 124
  Weapon System Q, 30, 31
Weinberger, Casper, 210
Western Development Division (WDD), of
    Air Research and Development
    Command, 19
Western Electric, Laureldale facility of, 16
Westinghouse, 41, 42
Westinghouse Modular Avionics Radar, 240
Wheelon, Albert (Bud), 158
Wilson, Alan, 7
Wilson, Charles E., 35
Wooldridge, Ramo, 30
Working Group for Electronic Piece Parts
    (WGEPP), 106, 107, 110–111
  Class A piece parts policy of, 112–113
  companion to MIL-M-38510B(USAF) of,
    118

  and quality improvement, 145
  recommendations of, 112
  standard for highest reliability parts of,
    116
  success of, 111–112
Wright Air Development Center
    Manufacturing Technology
    Laboratory, 42
Wright Patterson AFB, 41

X-20 Dyna-Soar space plane, 53, 68, 88,
    278, 306
Xerox Palo Alto Research Corp. (PARC),
    141

Yarymovych, Michael I., 125, 126, 158,
    260
Yeltsin, Boris, 232
Young, A. Thomas, 270
Young report, 271, 272

Zeus anti-ballistic missile, 25, 26, 124

Printed and bound by CPI Group (UK) Ltd, Croydon, CR0 4YY

27/10/2024

14580337-0004